Planning Twentieth Century Capital Cities

Planning, History and Environment Series

Editor:
Professor Dennis Hardy, Middlesex University, UK

Editorial Board:
Professor Arturo Almandoz, Universidad Simón Bolívar, Caracas, Venezuela
Professor Nezar AlSayyad, University of California, Berkeley, USA
Professor Eugenie L. Birch, University of Pennsylvania, Philadelphia, USA
Professor Robert Bruegmann, University of Illinois at Chicago, USA
Professor Jeffrey W. Cody, Getty Conservation Institute, Los Angeles, USA and Chinese University of Hong Kong, Hong Kong
Professor Robert Freestone, University of New South Wales, Sydney, Australia
Professor David Gordon, Queen's University, Kingston, Ontario, Canada
Professor Sir Peter Hall, University College London, UK
Professor Peter Larkham, University of Central England, Birmingham, UK
Professor Anthony Sutcliffe, Nottingham, UK

Technical Editor
Ann Rudkin, Alexandrine Press, Marcham, Oxon, UK

Published titles

The Rise of Modern Urban Planning, 1800–1914 edited by Anthony Sutcliffe
Shaping an Urban World: Planning in the twentieth century edited by Gordon E. Cherry
Planning for Conservation: An international perspective edited by Roger Kain
Metropolis 1980–1940 edited by Anthony Sutcliffe
Arcadia for All: The legacy of a makeshift landscape by Dennis Hardy and Colin Ward
Planning and Urban Growth in Southern Europe edited by Martin Ward
Thomas Adams and the Modern Planning Movement: Britain, Canada and the United States by Michael Simpson
Holford: A study in architecture, planning and civic design by Gordon E. Cherry and Leith Penny
Goodnight Campers! The history of the British holiday camp by Colin Ward and Dennis Hardy
Model Housing: From the Great Exhibition to the Festival of Britain by S. Martin Gaskell
Two Centuries of American Planning edited by Daniel Schaffer
Planning and Urban Growth in the Nordic Countries edited by Thomas Hall
From Garden Cities to New Towns: Campaigning for town and country planning, 1899–1946 by Dennis Hardy
From New Towns to Green Politics: Campaigning for town and country planning 1946–1990 by Dennis Hardy
The Garden City: Past, present and future edited by Stephen V. Ward
The Place of Home: English domestic environments by Alison Ravetz with Richard Turkington
Prefabs: A history of the UK temporary housing programme by Brenda Vale
Planning the Great Metropolis: The 1929 Regional Plan of New York and Its Environs by David A. Johnson
Rural Change and Planning: England and Wales in the twentieth century by Gordon E. Cherry and Alan Rogers
Of Planting and Planning: The making of British colonial cities by Robert Home
Planning Europe's Capital Cities: Aspects of nineteenth-century urban development by Thomas Hall

Politics and Preservation: A policy history of the built heritage, 1882–1996 by John Delafons
Selling Places: The marketing and promotion of towns and cities, 1850–2000 by Stephen V. Ward
Changing Suburbs: Foundation, form and function edited by Richard Harris and Peter Larkham
The Australian Metropolis: A planning history edited by Stephen Hamnett and Robert Freestone
Utopian England: Community experiments 1900–1945 by Dennis Hardy
Urban Planning in a Changing World: The twentieth experience edited by Robert Freestone
Twentieth-Century Suburbs: A morphological approach by J.W.R. Whitehand and C.M.H. Carr
Council Housing and Culture: The history of a social experiment by Alison Ravetz
Planning Latin America's Capital Cities, 1850–1950 edited by Arturo Almandoz
Exporting American Architecture, 1870–2000 by Jeffrey W. Cody
Planning by Consent: The origins and nature of British development control by Philip Booth
The Making and Selling of Post-Mao Beijing by Anne-Marie Broudehoux
Planning Middle Eastern Cities: An urban kaleidoscope in a globalizing world edited by Yasser Elsheshtawy

Titles published 2005

Globalizing Taipei: The political economy of spatial development edited by Reginald Yin-Wang Kwok
New Urbanism and American Planning: The conflict of cultures by Emily Talen

Titles published 2006

Remaking Chinese Urban Form: Modernity, scarcity and space. 1949–2005 by Duanfang Lu
Planning Twentieth Century Capital Cities edited by David L.A. Gordon

Planning Twentieth Century Capital Cities

edited by
David L.A. Gordon

First published by Routledge, 270 Madison Ave, New York, NY 10016

Simultaneously published in the UK
by Routledge
2 Park Square, Milton Park, Abingdon,
Oxfordshire OX14 4RN

Routledge is an imprint of the Taylor & Francis Group, an informa business

© 2006 Selection and editorial material David L.A. Gordon; individual chapters: the contributors

Typeset in Palatino and Humanist by PNR Design, Didcot, Oxfordshire
Printed and bound in Great Britain by Bell & Bain Ltd, Glasgow

This book was commissioned and edited by Alexandrine Press, Marcham, Oxfordshire

All rights reserved. No part of this book may be reprinted or reproduced or utilised in any form or by any electronic, mechanical or other means, now known or hereafter invented, including photocopying and recording, or in any information storage or retrieval system, without permission in writing from the publishers.

The publisher makes no representation, express or implied, with regard to the accuracy of the information contained in this book and cannot accept any legal responsibility or liability for any errors or omissions that may be made.

British Library Cataloguing in Publication Data

A catalogue record of this book is available from the British Library

Library of Congress Cataloging in Publication Data

Planning twentieth century capital cities / edited by David L.A. Gordon.
    p. cm. — (Planning, history, and the environment series)
  Includes bibliographical references and index.
  ISBN 0-415-28061-3 (hb)
    1. City planning—Cross-cultural studies. I. Gordon, David L. A. II. Series

HT166.P5433 2004
307.1'216'—dc22

ISBN10: 0-415-28061-3 (hbk)
ISBN10: 0-203-48156-9 (ebk)
ISBN13: 978-0-415-28061-7 (hbk)
ISBN10: 978-0-203-48156-1 (ebk)

Contents

Foreword *Anthony Sutcliffe*		vii
Acknowledgements		xi
Illustration Sources and Credits		xiii
The Contributors		xv
1	Capital Cities in the Twentieth Century *David L.A. Gordon*	1
2	Seven Types of Capital City *Peter Hall*	8
3	The Urban Design of Twentieth Century Capitals *Lawrence J. Vale*	15
4	Paris: From the Legacy of Haussmann to the Pursuit of Cultural Supremacy *Paul White*	38
5	Moscow and St Petersburg: A Tale of Two Capitals *Michael H. Lang*	58
6	Helsinki: From Provincial to National Centre *Laura Kolbe*	73
7	London: The Contradictory Capital *Dennis Hardy*	87
8	Tokyo: Forged by Market Forces and Not the Power of Planning *Shun-ichi J. Watanabe*	101
9	Washington: The DC's History of Unresolved Planning Conflicts *Isabelle Gournay*	115
10	Canberra: Where Landscape is Pre-eminent *Christopher Vernon*	130

11 Ottawa-Hull: Lumber Town to National Capital 150
David L.A. Gordon

12 Brasília: A Capital in the Hinterland 164
Geraldo Nogueira Batista, Sylvia Ficher, Francisco Leitão and Dionísio Alves de França

13 New Delhi: Imperial Capital to Capital of the World's Largest Democracy 182
Souro D. Joardar

14 Berlin: Capital under Changing Political Regimes 196
Wolfgang Sonne

15 Rome: Where Great Events not Regular Planning Bring Development 213
Giorgio Piccinato

16 Chandigarh: India's Modernist Experiment 226
Nihal Perera

17 Brussels: Capital of Belgium and 'Capital of Europe' 237
Carola Hein

18 New York City: Super-Capital – Not by Government Alone 253
Eugenie L. Birch

19 What is the Future of Capital Cities? 270
Peter Hall

Bibliography 275

Subject Index 295

Index of Towns and Cities 299

Index of Persons 301

Foreword

Anthony Sutcliffe

There is always something special about a capital city. I always claim to have been brought up in one, even though I lived fifty metres outside the Greater London Council boundary. My mother was proud of living in Essex, which sounded rural, or 'countrified' as she would put it, but my golden bough was the newly modernized Central Line which took me with many a bump, rattle and flush of air into the smoky heart of the Empire at an average speed of twelve miles an hour. In winter the searing fogs rasped at my throat as soon as I stepped off the escalator. By the evening the collar of my best white school shirt was flecked with smoky grime. Yet never for a moment did I doubt that London was the only place where I, or anyone else, could ever want to live.

Other capitals soon had me in their thrall. I tarried in Oxford for a while but went on to Paris. I was never a true Oxonian but easily became a true Parisian, despising the uncouth provinces and all foreign climes. As a tourist I came to prefer plebeian Rome to all the treasures of Florence and Venice. I have worked hard to know and value the world's greatest uncrowned capital, New York, ever praising that great mayor who has enabled me to study it on foot, and without constantly looking over my shoulder. For a while I worked in Ottawa, not the most statuesque of capitals but one that still calls me.

For the last forty years, however, I have been embraced by English provincial life, only a hundred or so miles from London but worlds away. Worse, I feel increasingly ill at ease on my visits there. Changes to which a Londoner might have gradually adjusted seem abrupt and disturbing – traffic jams on a Sunday, the suffocating motor fumes which, unlike the old fogs, build up on the sunniest summer days, breakdowns on the Underground, the NatWest Tower and its band of faceless waifs and strays lowering over St. Paul's, the neo-Dickensian fairyland of cobbles, restoration and pastiche on the South Bank, the comic book environment of the Docklands, arguments with taxi drivers, and the constant Babel-babble of the streets. Today, I am completing a history of London architecture on a sylvan David Wilson estate not far from Sherwood Forest, rarely going to London, losing my way in places that I once knew well, not keeping up with the flow, and often working from a distorted memory of the past. But perhaps that is what history is all about.

Whatever their personal experience, historians of urban planning are easily attracted to the capital city. Its size, complex structure, multiple external links and aura of wealth and power may be daunting as well as fascinating, but they can usually draw on a wealth of published studies, media contributions and interviews with participants. They will normally be aware of links with the provinces, and they will be tempted to detect a process of competition and diffusion involving other capital cities. Such interactions can be attractive themes for the historian, linking cities to economic, social and strategic change on a world scale. More than just symbols of, say, capitalism, immorality, pride, communism, religious belief, anarchism, suffering, gluttony, sybaritism and revolution, the capitals are easily viewed as forces of change in their own right. Their architecture and design seem to represent their values, while their culture is better known and respected than that of their provincial satellites. To cap it all, books and articles on capital cities will attract publication more readily than those on provincial places, and the capital city expert – including the historian – will appear on national radio and television more often than his provincial counterparts.

These features can attract a variety of scholars, including urban historians, historical geographers, and a multitude of cultural studies people, area studies specialists, and even the obscure band of critical theorists. What draws the planning historian to the capital is the interaction of forces of uncontrolled development determined mainly by economic and demographic forces, and deliberate direction of the city's growth by public authority in accordance with conscious goals and coherent theories. As an open, political process, planning brings features and trends in the city's growth more clearly into view. Planning, when broadly defined, can be detected throughout the six thousand year history of urbanisation. In this broader form it is part of the organic process of urban development rather than a brutal imposition by authority. At times it can become an expression of narrow political power, as in Hitler's Berlin or Nero's Rome, but in such cases the conversion of authority and ambition into forms and spaces has its own interest for the historian. Meanwhile, the response of the physical nature of the city to plans and planning decisions adds a dimension of understanding which is more readily available to the planning historian than to other city specialists.

Planning historians are often drawn into comparative work but inter-capital studies are normally too demanding to be undertaken readily by a single author. Peter Hall's astounding grasp of world urbanization, stretching far beyond his original disciplinary base in geography, brought the 'world city' concept from its venerable German origins into today's world of major urban problems and potential solutions through planning. Bestriding our narrow world like a Colossus, Hall has inspired more than one generation of comparative planning historians without generating a true emulator. Although Thomas Hall's *Planning Europe's Capital Cities* [1997] in this series is a unique account of the nineteenth-century European city planning and Steen Eiler Rasmussen and Donald Olsen draw the planning of two or three capitals together in memorable books, the collaborative volume under the direction of an editor or editorial team is more common. Three such comparative volumes (Elsheshtawy: *Planning Middle Eastern Cities* [2004], Almandoz: *Planning Latin America's Capital Cities, 1850–1950* [2002], and Sutcliffe: *Metropolis 1890–1940* [1984]), have already been published in this series, and there are many more studies in this comparative mode, dating back at least to the 1930s and including conference proceedings, such as the forerunner of the current volume, *Capital Cities – Les Capitales*, edited by John Taylor, Jean Lengellé and Caroline Andrew in 1993.

The rise of the capital city as a subject for the planning historian has been accompanied by the progress of the Routledge series in which this book appears. Planning, History and Environment published its first book on planning history in 1980, three years after the First International Conference on the History of Urban Planning which, meeting in London, had launched the world-wide study of planning history in its current form. With thirty-six titles now published, the series attracted David Gordon at an early stage of his enterprise. The author of this Foreword, then editor of the series, gave immediate support. It is appropriate that this unique volume should appear in the Routledge series where it will complement so much earlier work.

The seven-part typology of capital cities provides the prime structure for the book. Derived from Peter Hall's evolving thinking over the last forty years, it is convincing and it allows a choice of case studies which will not be too controversial. It allows the inclusion of giant cities and some small ones, stressing the importance of function rather than mere size. The subject, however, is not the cities themselves but their planning, mainly since the emergence of modern town planning concepts and powers since the end of the nineteenth century.

Capital cities exist not by virtue of their own size or economic importance, but because of their relationship to a national State. The national state, in its current form, has emerged slowly since the later Middle Ages as the most common and effective solution to the government of the more advanced areas of the world. Even federal states, such as the United States and Canada, have normally had a national capital. The variety of these states and their origins is such that their capitals have no universal features, including physical features. Perhaps their only common feature is perceptual, in that each capital is usually seen as representative of the state which it serves. Even the short-lived capital of the German Federal Republic, Bonn, a medium-sized provincial city selected mainly because it had no serious associations with the Hitlerian past, was often seen as reflecting the anti-militarist, hard-working character of post-war Germany.

Nevertheless, the city or national authorities often harbour a general belief that the face of the capital can be physically altered to produce an image appropriate to national or civic mythology. They also seek greater efficiency in order to allow the capitals to carry out their evolving functions. The result is often larger buildings and streets, but open spaces and artistic embellishments often play their part. The essays in this volume deal with these efforts, but they are also set in a context of power, wealth and conflict which involves the population at large and the distribution of capital in the city and in the nation as a whole. The capital city thus becomes a participant in, and an expression of, broader historical forces. These are the most difficult issues for an author, but they link the capital city to history as a whole. This book will contribute to the flow of historical debate as well as to current concerns in planning history.

Editor's Acknowledgements

My first debt is to the contributors for making this book a reality, while thanks are also due to the many people who helped and advised along the way – Anthony Sutcliffe, Stephen Ward, Peter Hall, Lawrence Vale, Michael Hebbert, Denis Hardy and Ann Rudkin have all given help and advice.

The Canadian Centre for the Study of Capitals provided a venue for the discussion of the book's ideas, with Caroline Andrew and John Taylor as hosts. We appreciated the support of Canada's National Capital Commission for this symposium.

An enthusiastic and talented team of research assistants helped in preparing the manuscript. I offer my heartfelt thanks to Laura Evangelista, Wesley Hayward, John Kozuskanich, Kelly McNicol, Jeffrey O'Neill and Jo-Anne Rudachuk and to Angie L'Abbe who provided administrative support for the project at the Queen's University School of Urban and Regional Planning.

Two Canadian Social Sciences and Humanities Research Council grants supported the book and my Ottawa research. A Fulbright Senior Fellowship supported a sabbatical year at the University of Pennsylvania, where most of the project was organized. Gary Hack and Eugenie Birch were gracious hosts at Penn's Graduate School of Fine Arts.

My greatest debts are to Katherine Rudder and Sarah Gordon, who cheerfully allowed me to abandon them for extensive periods of research, travel and writing. I owe them more than I can repay.

David L.A. Gordon

Kingston, Ontario (Canada's first capital…)

March 2006

Illustration Credits and Sources

The editor, contributors and publisher would like to thank all those who have granted permission to reproduce illustrations. We have made every effort to contact and acknowledge copyright holders, but if any errors have been made we would be very happy to correct them at a later printing.

Chapter 3
- 3.1 *Source*: Lawrence Vale, personal collection
- 3.2 © Lawrence Vale
- 3.3 *Source*: US Commission of Fine Arts Collection
- 3.4 *Source*: National Capital Planning Commission, Washington DC
- 3.5 *Source*: National Capital Commission, Ottawa
- 3.6 *Source*: National Archives of Australia
- 3.7 *Source*: Bierut (1951)
- 3.8 *Source*: Bierut (1951)
- 3.9 *Source*: Department of City and Regional Planning, Middle East Technical University, Ankara
- 3.10 *Source*: Constantinos Doxiadis Archives
- 3.11 © Lawrence Vale
- 3.12 *Source*: Kenzo Tange Associates; *photo*: Osamu Murai
- 3.13 *Source*: James Rossant, Conkin Rossant Architects
- 3.14 *Source*: Sri Lanka Urban Authority
- 3.15 © Lawrence Vale
- 3.16 © Lawrence Vale
- 3.17 © Wesley Hayward

Chapter 4
- 4.1 © Paul White
- 4.2 *Source*: Paul White redrawn from Rouleau (1988)
- 4.3 *Source*: Evenson (1979)
- 4.4 *Source*: Institut Paul Delouvrier
- 4.5 *Source*: Atelier Parisien d'Urbanisme
- 4.6 © Anthony Sutcliffe
- 4.7 © Paul White
- 4.8 © Anthony Sutcliffe
- 4.9 © Anthony Sutcliffe
- 4.10 *Source*: Noin and White (1977)

Chapter 5
- 5.1 © Michael H. Lang
- 5.2 © Michael H. Lang
- 5.3 © Michael H. Lang
- 5.4 *Source*: Punin (1921)
- 5.5 *Source*: *Gorodskoye khozyaistvo Moskvy* (1949) no. 12, following p. 6
- 5.6 © Tretyakovskaya Gallery, Moscow
- 5.7 *Source*: *Arkhitektura SSSR* (1935) nos. 10/11, pp. 26–27
- 5.8 © Michael H. Lang
- 5.9 © Michael H. Lang
- 5.10 *Source*: Kaganovich (1931), cover photograph.
- 5.11 *Source*: Astaf'eva-Dlugach (1979), p. 39
- 5.12 *Source*: Tretyakovskaya Gallery, Moscow

Chapter 6
- 6.1 *Source*: National Board of Antiquities, Helsinki
- 6.2 *Source*: Helsinki City Museum
- 6.3 *Source*: Helsinki City Museum
- 6.4 *Source*: Helsinki City Archives
- 6.5 *Source*: Museum of Finnish Architecture, Helsinki
- 6.6 *Source*: Helsinki City Museum
- 6.7 *Source*: Helsinki City Museum
- 6.8 *Source*: Lehtikuva Oy
- 6.9 *Source*: Helsinki City Museum

Chapter 7
- 7.1 *Source*: Museum of London
- 7.2 *Source*: Museum of London
- 7.3 © Jane Woolfenden
- 7.4 *Source*: Museum of London
- 7.5 *Source*: Purdom (1945)
- 7.6 *Source*: Royal Institute of British Architects (above) and Greater London Authority (below)
- 7.7 © Jane Woolfenden

Chapter 8
- 8.1 *Source*: Fujimori (1982), figure 50
- 8.2 *Source*: Collection of Shuni-ichi J. Watanabe
- 8.3 *Source*: Fukuda (1919), figure 48
- 8.4 © Kawasumi Photograph Office, *photo*: Akio Kawasumi
- 8.5 *Source*: Collection of Shuniichi J. Watanabe

Chapter 9
- 9.1 *Source*: Moore (1902), p. 35
- 9.2 *Source*: Allen (2001), p. 456
- 9.3 *Source*: US National Archives and Records Administration

9.4   *Source*: Moore (1902), frontispiece
9.5   *Source*: National Capital Planning Commission (1961), p. 47
9.6   *Source*: President's Council on Pennsylvania Avenue (1964), p. 55
9.7   *Source*: National Capital Planning Commission (1997), p. 7

Chapter 10
10.1   *Source*: National Capital Authority, Canberra
10.2   *Source*: National Library of Australia
10.3   *Source*: National Archives of Australia
10.4   *Source*: National Archives of Australia
10.5   *Source*: National Archives of Australia
10.6   *Source*: National Archives of Australia
10.7   *Source*: National Capital Authority, Canberra
10.8   *Source*: National Capital Authority, Canberra
10.9   *Source*: National Capital Authority, Canberra
10.10 *Source*: National Capital Authority, Canberra
10.11 *Source*: National Capital Authority, Canberra

Chapter 11
11.1   © Jeffrey O'Neill
11.2   *Source*: Federal Plan Commission for Ottawa and Hull (1916), figure 6
11.3   *Source*: Gréber (1950), diagram 143
11.4   *Source*: Library and Archives Canada, PA-201981143
11.5   *Source*: Gréber (1950), diagram 128
11.6   *Source*: National Capital Commission, negative 172-5

Chapter 12
12.1   *Source*: Correio Braziliense – CD Brasilía 40 anos, 2000
12.2   *Source*: Governo do Distrito Federal/SEDUH
12.3   *Source*: Correio Braziliense – CD Brasilía 40 anos, 2000
12.4   *Source*: Governo do Distrito Federal/SEDUH
12.5   *Source*: Governo do Distrito Federal/SEDUH
12.6   *Source*: Correio Braziliense – CD Brasilia 40 anos, 2000

Chapter 13
13.1   *Source*: Irving (1981), p. 54
13.2   *Source*: Delhi Development Authority (1990) redrawn by Souro D. Joardar
13.3   *Source*: Souro D. Joardar
13.4   *Source*: Irving (1981), p. 77
13.5   *Source*: National Capital Region Planning Board (1988) redrawn by Souro D. Joardar

13.6   *Source*: Delhi Development Authority (2001)

Chapter 14
14.1   *Source*: Lehwess (1911)
14.2   *Source*: Häring and Wagner (1929)
14.3   *Source*: Schäche (1991)
14.4   *Source*: Bundesminister für Wohnungsbau Bonn und Senator für Bau- und Wohnungswesen Berlin (1960)
14.5   *Source*: Verner (1960)
14.6   *Source*: Zwoch (1993)

Chapter 15
15.1   *Source*: Archivio Capitolino, Rome
15.2   *Source*: Archivio Capitolino, Rome
15.3   *Source*: Archivio Capitolino, Rome
15.4   *Source*: Archivio Cenrale dello Stato, Rome
15.5   *Source*: Archivio Cenrale dello Stato, Rome
15.6   *Source*: Comune di Roma

Chapter 16
16.1   *Source*: Sarin (1982)
16.2   *Source*: Nihal Perera adapted from Kalia (2002)
16.3   *Source*: Nihal Perera adapted from Kalia (2002)
16.4   © Nihal Perera
16.5   © Nihal Perera
16.6   © Nihal Perera

Chapter 17
17.1   *Source*: Gouvernment Belge (1958)
17.2   *Source*: *Journal Le Soir*, Brussels
17.3   *Source*: European Parliament
17.4   *Source*: SCAB (1979).
17.5   *Source*: ARAU (1984)
17.6   © Carola Hein
17.7   *Source*: European Parliament, Brussels

Chapter 18
18.1(a) *Source*: Rockefeller Center Archives
18.1(b) *Source*: Rockefeller Center Archives
18.1(c) *Source*: Rockefeller Center Archives
18.2   *Source*: Dudley (1994)
18.3   *Source*: Newhouse (1989)
18.4   © Downtown Lower Manhattan Association
18.5   © Downtown Lower Manhattan Association
18.6   © Downtown Lower Manhattan Association
18.7   *Source*: Newhouse (1989)
18.8   *Source*: Newhouse (1989)

Contributors

Geraldo Nogueira Batista is Professor in the Department of History and Theory, School of Architecture and Urban Planning, University of Brasília. He is author of *Guiarquitetura de Brasília*.

Eugénie L. Birch is Professor and Chair of the Department of City and Regional Planning, and Co-director of the Penn Institute for Urban Research, University of Pennsylvania. She co-edited the *Journal of the American Planning Association*, was President of the Society of American City and Regional Planning History and the Association of Collegiate Schools of Planning and currently serves on the editorial boards of several professional journals. Her present research focuses on downtown living in American cities. A member of the New York City Planning Commission in the 1990s, in 2002 she served as a member of the jury to select the designers for the World Trade Center.

Dionísio Alves de França is an Architect and is currently enrolled in the architecture course at the Graduate School of Engineering, Institute of Technology of Nagoya, Japan.

Sylvia Ficher is Professor in the Department of History and Theory, School of Architecture Urban Planning, University of Brasília. Her books include *Arquitetura moderna brasileira* and *Preservação do patrimônio arquitetônico*.

David L.A. Gordon is Associate Professor in the School of Urban and Regional Planning, Queen's University, Canada. He is the author of *Battery Park City: Politics and Planning on the New York Waterfront* and numerous articles on plan implementation and Ottawa planning history. As a practitioner, in 1991 and 1992 he shared the Canadian Institute of Planners National Award of Distinction.

Isabelle Gournay is Associate Professor in the School of Architecture, University of Maryland. Her publications include *The New Trocadéro* and *Montréal Métropole* and articles in architecture, art history and preservation journals in the US, Canada, France and Italy. Her current research focuses on cross-currents between Western Europe and America in the field of architecture and urbanism.

Dennis Hardy is Emeritus Professor of Urban Planning at Middlesex University and works freelance as a writer and consultant. He is the Editor of the Routledge series, *Planning, History and Environment* for which he has authored three books: *From Garden Citites to New Towns*, *From New Towns to Green Politics* and *Utopian England: Community Experiments, 1900–1945*. He is President of the International Communal Studies Association. His latest book is *Poundbury: The Town that Charles Built*.

Peter Hall is Bartlett Professor of Planning and Regeneration, University College London. He is author or editor of nearly forty books on urban and regional planning including *Cities of Tomorrow*, *Cities in Civilization* and (with Colin Ward) *Sociable Cities*. He was awarded the 2005 International Balzan Foundation Prize for the *Social and Cultural History of Cities since the beginning of the 16th century*.

Carola Hein is Associate Professor at Bryn Mawr College (Pennsylvania) in the Growth and Structure of Cities Program. She has published widely on contemporary and historical architectural and urban planning – notably in Europe and Japan. For twenty years she has worked on an examination of visions for a capital of Europe, the reality of the three headquarters cities, and the issue of the emerging polycentric capital. Her book *The Capital of Europe. Conflicts of Architecture, Urban Planning and European* was published in 2004.

Souro D. Joardar is a Professor in the School of Planning and Architecture, New Delhi. Previously he was a Professor at the National Institute of Urban Affairs, New Delhi. He has also held academic posts at the Indian Institute of Technology, Kharagpur and at King Saud University, Riyadh and has worked in government and private planning offices.

Laura Kolbe is Professor at the Department of History, University of Helsinki. She is the author of *Helsinki, the*

Daughter of the Baltic Sea, editor of *Finnish Cultural History I–V* and co-editor of the series *History of Metropolitan Development in Helsinki – Post 1945*. Her research is in Finnish and European History, urban and university history. Her latest research deals with urban governance and policy making in Helsinki during the twenty-first century. Professor Kolbe is founder and chair of the Finnish Society for Urban Studies and member of the IPHS Council.

Michael Holavko Lang is Professor and Chair of the Department of Public Policy and Administration, Graduate School, Rutgers University-Camden. He is the author of *Designing Utopia: John Ruskin's Urban Vision for Britain and America* and articles on Moscow planning history. He teaches courses on planning and planning history as well as courses on the city planning of St. Petersburg and Moscow.

Francisco Leitão is an architect and urbanist. He teaches architectural history in the Technology Department of the Centro Universitário de Brasília. From 1991 he has been engaged in urban planning and historical preservation activities for the Government of the Federal District.

Giorgio Piccinato is Professor of Urban and Regional Planning and Director of the Department of Urban Studies, University of Rome Three. President of the Association of European Schools of Planning 1992–1994, consultant to the United Nations and the European Union for programmes dealing with urban and regional planning, urban conservation, professional education. His publications include *La costruzione dell'urbanistica. Germania 1870–1914*, *Città, territorio e politiche di piano in America Latina*, *Alla ricerca del centro storico* and *Un mondo di città*.

Nihal Perera is Associate Professor of Urban Planning and Director of the CapAsia programme at Ball State University, Indiana. His publications include *Decolonizing Ceylon: Colonialism, Nationalism, and the Politics of Space in Sri Lanka* and *Society and Space: Colonialism, Nationalism, and Postcolonial Identity in Sri Lanka*. His main research interest is the politics of space.

Wolfgang Sonne is Lecturer in History and Theory of Architecture at the Department of Architecture at the University of Strathclyde in Glasgow. He studied art history and archaeology in Munich, Paris and Berlin and holds a PhD from the Eidgenössische Technische Hochschule in Zurich. He has previously taught at the ETH Zurich, at Harvard University and at the Universität in Vienna. His publications include *Representing the State. Capital City Planning in the Early Twentieth Century*.

Anthony J. Sutcliffe is Emeritus Professor of History, University of Leicester. He was founding editor (with the late Gordon Cherry) of the journal *Planning Perspectives* and the book series *Studies in History, Planning and Environment*. He is the author or editor of many books including *The Rise of Modern Urban Planning: Towards the Planned City, Metropolis, 1890–1940, Paris: An Architectural History;* and *London: An Architectural History*.

Lawrence J. Vale is Professor and Head, Department of Urban Studies and Planning, Massachusetts Institute of Technology. He is the author of *From the Puritans to the Projects, Reclaiming Public Housing*, and *Architecture, Power and National Identity* (winner of the Spiro Kostof Award from the Society of Architectural Historians) and co-editor (with Sam Bass Warner) of *Imaging the City*, and co-editor (with Thomas Campanella) of *The Resilient City*.

Christopher Vernon is Senior Lecturer in Landscape Architecture, School of Architecture and Fine Arts, University of Western Australia. Design advisor to the National Capital Planning Authority. He is the author of several articles on Canberra and currently preparing a book on Walter Burley Griffin.

Shun-ichi J. Watanabe is Professor of Urban Planning, Department of Architecture at Tokyo University of Science. Author of many books including *The Birth of City Planning: Japan's Modern Urban Planning in International Perspective*. He has won many awards for his work including The City Planning Institute of Japan Award. Professor Watanabe is a Board member of the International Planning History Society.

Paul White is Pro-Vice Chancellor and Professor of Geography, University of Sheffield. He has a long-standing research interest in Paris, which resulted in the publication of a major book, *Paris* with Daniel Noin (Wiley 1997). Other recent work there has included studies of creative literature depicting social issues in the Paris suburbs.

Chapter 1

Capital Cities in the Twentieth Century

David L.A. Gordon

The twentieth century witnessed an unprecedented increase in the number of capital cities worldwide. In 1900 there were only about forty nation states with capital cities; half of these were in Latin America, created as a result of the break up of the Spanish and Portuguese empires in the late nineteenth century. But things were set to change. World War I and its aftermath sounded the death knell of the Austro-Hungarian, German, and Ottoman empires; the period following World War II saw the gradual disintegration of the French and British empires; and the 1980s and 1990s witnessed the demise of the Soviet Union and fragmentation of the former Yugoslavia. Thus, by 2000 there were more than two hundred capital cities. And this, surely, is reason enough for a book devoted to the planning and development of capital cities in the twentieth century.

However, the focus is not only on recently created capitals. Indeed, the case studies which make up the core of the book show that, while very different, the development of London or Rome presents as great a challenge to planners and politicians as the design and building of Brasília or Chandigarh. Put simply, this book sets out to explore what makes capital cities different from other cities, why their planning is unique, and why there is such variety from one city to another.

To help map this journey we turn to Peter Hall's 'seven types of capital city' – Multi-Function Capitals; Global Capitals; Political Capitals; Former Capitals; Ex-Imperial Capitals; Provincial Capitals; Super Capitals – which he discusses in chapter 2, identifying the functions and characteristics of each, and distinguishing their overlapping roles. Each of the capitals in the book may be classified as one or more of these types, for example, New York is both a Provincial Capital and a Super Capital, Tokyo is a Multi-Function and a Global Capital, and London is a Global and Multi-Function Capital, but also an Ex-Imperial Capital.

Following Peter Hall's classification of cities in terms of their functions and the reasons for their ascendancy, in Chapter 3 Lawrence Vale turns to urban design. As he says, 'the planning and design of national capitals is inseparable from the political, economic, and social forces that sited them and moulded their development'. He analyses twentieth-century urban design policy and action in capital cities against the background of three key developments: the dismemberment of empires, the emergence of new federal systems, and the growing importance of super-national groupings. He concludes that whichever of these affected a capital, all have striven to maintain their image and 'symbolic centrality' and that 'urban design remained a vital part of the public projection and reception of capital cities in the twentieth century'.

Chapters 2 and 3 set the stage for the cities which are our chosen case studies. The first of these is Paris, an archetypal Multi-Function Capital. However, as Paul White explains, the first sixty years of the twentieth century were a period of inaction in the city's planning history. Indeed, even today inner Paris remains much as Haussmann planned it, while the schemes introduced in the 1960s were more to do with the city as a large urban area than as a national capital. Strategic planning over the last forty to fifty years has focused on managing development and ameliorating serious imbalances within the capital region. Against this background and in the face of growing global competition, increasing attention has been paid to the need for Paris to maintain its place as a major world capital. Politicians and planners alike have sought to achieve this by enhancing the city's cultural image not by creating a single area of capital city attractions, but by placing new developments throughout inner Paris.

Michael Lang, in Chapter 5, provides an overview of the planning history of Russia's present and former capitals, both as he says 'indelibly marked by the cruel hand of totalitarian rulers'. St Petersburg, the eighteenth-century creation of Peter the Great, was capital only until 1918. During those early years of the twentieth century, for all the architectural splendour of its centre, the city had the worst housing and services of any capital. Following the Bolshevik Revolution, Lenin moved the capital back to Moscow, where it had been before Tsar Peter set out to build the city which 'would soar as an eagle'. Without doubt Stalin was the master planner of Socialist Moscow. However, many pre-Revolutionary architects and planners remained in Russia to help build the ideal communist city, blending foreign notions with Russian design traditions to meet the needs of the new Socialist society. Competition with the West was a further driving force in the way the city developed. The result was a capital where growth was uncontrolled and whose population was inadequately housed. It is perhaps too early to tell how the long-term development of either Multi-Function Moscow or Former Capital St Petersburg will be affected by the fall of communism and the re-introduction of the private market.

In 1812, Helsinki, the subject of the next chapter, was made capital of Autonomous Grand Duchy of Finland by decree of the Russian Emperor Alexander I. Laura Kolbe describes how urban planning measures were introduced in response to the city's rapid growth. The first master plan drawn up in the 1910s was, however, never confirmed. Finland gained independence in 1917; bloody civil war followed in 1918, but so too did Saarinen and Jung's master plan Pro Helsingfors. This plan was to have considerable influence on Helsinki throughout the twentieth century. The years of war with the Soviet Union were a turning point for the development of the city, but it was not until 1959 that regional planning and master planning became mandatory. Five years later Helsinki's first city planning department was created and Alvar Aalto produced his second plan for the city centre. Although little of this plan was realized, urban policy since then has been characterized by the maintenance of a strong city centre.

No book such as this could omit London. The twentieth century saw the capital transformed from an Imperial to a Global City. Dennis Hardy's account of the city is in three sections – the first a summary of the immense changes that transformed the city in the twentieth century; second a review of the nature and extent of public intervention in relation to its capital status; and third an inquiry into why London's continuing dominance as a capital has not been fully reflected in its architecture and civic design. He argues that the city's success has little to do with political support or positive planning. While Patrick Abercrombie's Greater London Plan of 1944 may have provided the benchmark for postwar planning, there was no overall structure to make it happen, while the 1969 Greater London Development Plan was fraught with political conflict and opposition, and it is probably too early to judge the outcome of The London Plan published in draft form in 2002. Nor can one guess at the effects of recent terrorist activity or, indeed, the award of the 2012 Olympics to the city.

Tokyo, the focus of Chapter 8, like London is a Global Capital. At the beginning of the twentieth century, as Shun-ichi Watanabe explains, the urban form of the old castle town of Edo had almost disappeared. The 1919 City Planning Act had some influence on urban structure, but weak land-use controls hindered improvement of urban spaces. However, disaster struck the city when, in 1923, a massive earthquake destroyed much of the city. The following seven years of reconstruction emphasized modernization and protection from future earthquakes, but the Second World War saw destruction of a different kind and in 1945 planners, faced with reconstruction once again, provided the foundation that enabled the subsequent rapid growth of Tokyo. That growth saw large-scale development schemes, but also concern for the over-concentration of the nation's political, economic, and cultural activities in the city. The 1990s witnessed the end of the economic boom and the city's future development is likely to be more modest and on a more human scale.

Unquestionably, Washington is a Political Capital. It is perhaps this which is at the root of what Isabelle Gournay identifies as the city's 'unresolved conflicts and endemic tensions'. In Chapter 9, she suggests that three factors give rise to these: 'the notion that Washington belongs to all US citizens, rather than to its inhabitants, is fixed in the national psyche' which resulted in attention being given to ceremonial symbols and not neighbourhood improvement; 'the imbalance between the city's demographic, economic and cultural significance and its political stature', particularly given the ethnic diversity of the city; and, finally, 'taxation without representation' – although the city does now have a non-voting delegate in Congress, whose role is to lobby on behalf of the city's inhabitants, the preparation and implementation of plans continue to depend upon 'feudal' annual congressional appropriation. Gournay provides an illuminating survey of Washington's planned development from the McMillan Plan of 1902 via the work of the National Capital Park and Planning Commission and its successor, the National Capital Planning Commission.

Canberra, too, is a Political Capital and, as Christopher Vernon says, it is 'Australia's greatest achievement in landscape architecture and town planning'. Following the search for a site for the new capital within a larger federal territory, in 1912 the international competition for the city's design was won by Walter Burley Griffin. The design, the work of Griffin and his wife, was a sensitive response to the site's natural features. The grandeur of the site was to be the surrogate for the cultural and monumental artefacts found in cities of the Old World, but lacking in the new nation. Griffin's replacement in 1921 by a succession of advisory bodies resulted in numerous departures from the original plan and increasing antipathy to the new capital; for more than three decades there was little development. However, in the 1950s the city found a champion in the then Prime Minister Robert Menzies, who invited William Holford to make design proposals for the city's development. Implementation of Holford's scheme began in 1958, overseen by the National Capital Development Commission established in the same year. In 1988 the Commission was replaced by the National Capital Planning Authority. Today, as Vernon says, 'the picturesque reigns triumphant at Canberra' and Canberrans hold dear the presence of 'nature' within their city.

Unlike Canberra, Ottawa was not a greenfield site nor, as I explain in Chapter 11, was there a master plan for the capital of the United Canadas. Moreover, the small lumber town was not a place where politicians or civil servants wished to live. Following an initial period of neglect, the Ottawa Improvement Commission (1899–1913) appointed landscape architect Frederick Todd to design the city's park and parkway system. Criticism of the Commission's work led to the creation of a different body, the Federal Plan Commission under whose auspices Edward H. Bennett prepared a plan for the capital. The First World War, lack of funding and political support resulted in a period of inertia, but the interwar years saw some limited progress in the capital's development. After World War II, Prime Minister Mackenzie King established the National Capital Planning Committee and his chosen architect, Jacques Gréber, became head of the National Capital Planning Service; his National Capital Plan, which was to be a landmark in Canadian planning history, was published in 1950. With the establishment of the National Capital Commission in 1959, development moved apace and the erstwhile lumber town was transformed into the green and spacious capital city of today.

Moving the capital of Brazil was first mooted some three hundred years before it became a reality. Geraldo Nogueira Batista and his colleagues describe how the establishment of a new capital on the central plain was a precept of the 1891 Constitution of the Republic, but despite selection of a site little real progress was made for a considerable time. Indeed it was not until 1956, after further reconnaissance and technical reporting, and with President Kubitschek's support, that authorization was given for the relocation of the capital from Rio de Janeiro to Brasília and the establishment of the Federal District. The same year saw the creation of the Company for Urbanization of the New Capital, the appointment of Oscar Niemeyer to direct architectural design, and the call for a competition for the design of the city's Pilot Plan. Lúcio Costa's winning design introduced the superblock, the most distinctive and inspired physical-spatial element of Brasília. The President set 16 April 1960 for the city's inauguration – a date met albeit with much construction unfinished. From then on the population of the Pilot Plan and the Federal District grew rapidly, inducing urban dispersion over the entire territory, the development of satellite towns and slum settlements, despite attempts to control development. 'The expectation that a planned core would induce an orderly occupation of the territory – an essential utopia of Modernism – did not come to pass', yet Brasília today is a remarkable achievement – a Political Capital that offers opportunities to rich and poor alike.

In Chapter 13, Souro Joardar describes the planning and development of New Delhi, which spanned the first three decades of the twentieth century. Thereafter, he argues, 'it became more and more, physically and administratively, an integral part of the exploding and impersonal metropolitan Delhi and its region, especially after India's Independence'. The 1911 announcement of the proposed move of the capital of Imperial India from Calcutta was supported by Charles Hardinge, Viceroy of India, who had considerable say in the site selection, but less in the make up of the Delhi Town Planning Committee and the appointment of Edward Lutyens as chief planner. It was accepted that a key concept of the planning and design of the new capital was connectivity between the major capital elements and the landmarks of historic Delhi. Lutyens's plan with its sweeping vistas and vast open spaces and landscaping was in marked contrast to the crowded environment of Old Delhi. Furthermore it took no account of the people of the old city or their livelihoods, nor had any allowance been made for the growth of the capital city of a large and populous country. But by the time New Delhi was inaugurated in 1931, the end of the Imperial era was in sight. Growth pressures both before and after Independence resulted in new administrative bodies and new planning measures. Today, Lutyens's New Delhi represents only 3 per cent of the land area and 3 per cent of the population of the Delhi National Capital Territory, but is subject to debate between those who support its preservation and those who believe it should be developed at higher density.

As Wolfgang Sonne says, Berlin has had 'a chequered planning history [which] offers numerous insights into the factors that lead to the success or failure of capital city planning'. The beginning of the twentieth century saw the city as capital of the German Empire with little need for planning intervention, since the important institutions were already housed in 'monumental splendour', and little inclination for the development of a master plan, given the political tensions between the imperial house and the social democratic city. The First World War brought drastic change; Germany and its capital entered the era of the Weimar Republic. During this period there were design proposals for a democratic government district in the capital, but these were thwarted by the country's failing economy. The National Socialists seized power in 1933. Hitler appointed Albert Speer to realize his ambition to develop Berlin 'into a real and true capital city of the German Reich', but by 1945 what remained of this grandiose vision was little more than a pile of rubble. Sonne suggests that it was 'not until a certain continuity manifested itself during the second half of the century – the existence of the GDR for forty years and the stability of the Federal Republic of Germany since 1949 – that successful capital city planning came within reach' of Berlin – albeit a city divided for much of that time. German reunification in 1990 heralded a fresh chapter in the city's history.

Rome, once capital of the mighty Roman Empire and later of the Vatican, before becoming capital of Italy in 1861, has seen change and development through 'great events' rather than by means of regular planning. So argues Giorgio Piccinato in Chapter 15. He cites, for example, the Great National Exhibition of 1911, celebrating fifty years of Italian unity, as bringing about spatial transformation independently of any plans. The Fascist years brought rapid change to Rome as Mussolini sought to ensure that the city reflected the greatness of Fascism. The 1931 plan, the work of Marcello Piacentini, was soon negated by preparations for the *Esposizione Universale 1942* (or EUR) with its monumental marble public buildings and wide streets. War intervened and

1945 found Rome with an influx of refugees from all over the country, a dire housing problem, and poor transport and services. In 1960 another 'great event' in the shape of the Olympics came to Rome and boosted development, while in 1962 the City Council adopted a new plan. Some forty years later this was replaced by another master plan, but it is the EUR, Piccinato suggests, which is the real success story of post-war urban planning in Rome.

Chandigarh is not a national capital, but serves as the capital of two states, Punjab and Haryana; belonging to neither, it is classified as a Union Territory and is today administered by the federal government. As Nihal Perera explains in Chapter 16, need for the new city arose when Punjab was divided between India and Pakistan and the traditional capital, Lahore, fell within Pakistan's borders. Of course, it is impossible to discuss Chandigarh without mentioning Le Corbusier's role in the city's planning. However, he was not the first choice of the Punjab officials. In 1950, the American firm of Mayer and Whittlesey was appointed, and Albert Mayer and Matthew Nowicki prepared the first master plan. However, following Nowicki's death, Le Corbusier was brought in to execute the plan – instead he revised it radically, turning a plan based on garden city principles into a Modernist vision. Further, while Mayer and Nowicki had some understanding of Indian culture and society, Le Corbusier did not; his plan had scant regard for the traditional way of life and his interest lay in creating a city according to the principles of CIAM. In comparing the two plans, Perera explains why the Punjabi officials and even Nehru were convinced by Le Corbusier's design. Today a city of more than 900,000, Chandigarh has undergone and continues to undergo a process of 'urbanization, familiarization, and Indianization'.

Brussels became capital of Belgium in 1830 and, as Carola Hein explains, when Léopold II came to the throne in 1865 he 'introduced a complete plan for beautifying the city, introducing major parks and green spaces, broad avenues and a uniform design for private buildings'. Since then there have been no such attempts at beautification. In contrast, during the 1960s in particular, 'new office buildings rose quickly and "bruxellization" became a term for urban destruction', while disputes amongst the country's two main language groups led to the creation of separate regional and community organizations, all but one of which chose Brussels as its capital. However the impact of these organizations on the city is but little compared to the impacts resulting from the city's role as 'capital of Europe'. Hein describes how, in the late 1950s, the Belgian government used the presence of the European headquarters to boost Brussels's urban development and traces the tangled processes by which the Quartier Léopold became Brussels's European district, and home to the European Commission (the Berlaymont Building), the Council of Ministers (the Justus Lipsius Building) and the European Parliament.

The final case study city, New York, is without doubt a Super Capital. Eugenie Birch argues that the workings of a tri-partite governmental (city, state and federal) structure, where each level has sharply defined powers, and a system of implementation in which the public and private sectors work together to produce creative funding and administrative structures resulted in 'a "chemistry" of design, politics and finance that catalyzed New York's emergence as a Super Capital'. To reveal this process more clearly, Birch discusses four large-scale developments – the United Nations, the Rockefeller Center, the World Trade Center, and the Lincoln Center for the Performing Arts. By the 1940s as a result of huge population growth, economic dominance, and leadership in culture, communications and style, the city had become the 'capital of capitalism', and by the 1970s it was a Super Capital, but this did not stem from comprehensive planning. That meteoric rise was due to the efforts of small groups of public and private leaders enabling major projects to go forward and so shape the city as it is today. It is a fascinating tale with which to end the case studies.

The case studies reveal much about capital cities in the twentieth century, their planning and design, and the roles of the different actors involved in their development. But what of the capital city in the twenty-first century? As Peter Hall points out in his concluding chapter 'It all depends on the city'. However, two key trends – *globalization* and *informationalization* – together will result in the increasing importance of world cities. Within this global framework there will be dynamic shifts with cities such as Beijing rising to the top range of world cities while others slip backwards. He concludes that the creation of new capitals on anything like the scale witnessed in the twentieth century is unlikely in the twenty-first, though history can always bring surprises.

Chapter 2

Seven Types of Capital City

Peter Hall

Not all capital cities are alike. Some owe that role solely to the fact of being the seat of government; at least one (Amsterdam) is a capital though it is not the seat of government. Capitals in federal systems may have less well developed governmental functions than those in centralized systems. Though most seats of government attract to themselves other national functions (commerce, finance, the media, higher education), not all do so in equal degree. We can usefully distinguish the following cases:

1. *Multi-Function Capitals*: combining all or most of the highest national-level functions (London, Paris, Madrid, Stockholm, Moscow, Tokyo).

2. *Global Capitals*: a special case of (1), representing cities that also perform super-national roles in politics, commercial life or both (London, Tokyo).

3. *Political Capitals*: created as seats of government, and often lacking other functions which remain in older-established commercial cities (The Hague, Bonn, Washington, Ottawa, Canberra, Brasília).

4. *Former Capitals*: Often the converse of (2); cities that have lost their role as seat of government but that retain other historic functions (Berlin from 1945 to 1994, St Petersburg, Philadelphia, Rio de Janeiro).[1]

5. *Ex-Imperial Capitals*: A special case of (3), representing former imperial cities which have lost their empires though they may function as national capitals, and may also perform important commercial and cultural roles for the former imperial territories (London, Madrid, Lisbon, Vienna).

6. *Provincial Capitals*: A special case in federal nations, overlapping with (3); cities which once functioned as *de facto* capitals, sometimes on a shared basis, but have now lost that role, retaining however functions for their surrounding territories. (Milan, Turin, Stuttgart, Munich, Montréal, Toronto, Sydney, Melbourne).

New York is a very special case here, almost *sui generis*, of a global provincial capital.

7. *Super Capitals* functioning as centres for international organizations; these may or may not be national capitals (Brussels, Strasbourg, Geneva, Rome, New York).

Some might argue that not all these cases deserve to be treated as capitals. But all perform roles that are capital-like, and are performed by capital cities elsewhere. In any case, as I shall try to argue, it is important to try to distinguish these overlapping roles, because they are changing in different ways and even in different directions.

The Political Role

The twentieth century saw three important political changes which have profoundly affected the roles of capitals as seats of government. The first is the dismemberment of empires, both land-based (Germany, Austria and now Russia) and sea-based (Britain, France, Portugal). The second is the development of new federal systems (Australia, South Africa, Germany, Spain, and the Soviet Union) and the development of more decentralized systems within a centralized framework (France). The third is the development of new super-national groupings (the League of Nations, the United Nations and its agencies, the Council of Europe, the European Communities). All three trends had precursors in the eighteenth and nineteenth centuries (the dissolution of the Spanish Empire; the creation of the United States and the Dominion of Canada; the Congress of Vienna); but all three exhibited sharp acceleration in the twentieth.

The effects on certain cities have been profound. Vienna lost its role as capital of a land-based empire, and with it much of its political and economic role; ever since 1918, its public buildings have been anomalously too large and too grand. The same occurred to Berlin after 1945. In both cases, the effects were exacerbated by the division of Europe into two rival blocks, with the concomitant loss of trading relations and trading functions. The leading provincial cities of Germany had a new lease of life after 1945 as the effective power-sharing capitals of the Federal Republic; Munich, in particular, regained much of the role it had lost to Berlin in 1871. Thanks to the Treaty of Rome, Brussels acquired an importance and a dynamism that would otherwise have been denied to it.

In all these cases, change occurred suddenly and drastically in the aftermath of war. Elsewhere, the changes were more gradual, even unnoticeable. London and Paris have not self-evidently suffered from the loss of empire; if their economies have experienced partial contraction, deindustrialization, not loss of empire, was the cause – and London's job base is growing again. The major Australian cities have not notably lost importance since the belated rise of Canberra; nor has the autonomy of Barcelona, Bilbao or Seville threatened the primacy or vigour of Madrid. The United Nations is still no more than marginal to the whole New York City economy.

These historical examples point to a number of lessons, all important for the future of capital cities. It requires a rather drastic political change – the sudden and total dismemberment of an empire, the division of a country – to bring about a major shift in the role and the fortunes of a capital city. Otherwise change tends to be marginal, and existing urban economies tend to retain a great deal of resilience. Major global cities may lose political empires but may retain much of the associated economic and cultural hegemony over their former territories. Very large cities are not greatly affected by additions to, or subtractions from, their overall role.

The Economic Role

The categorization of capital cities shows very clearly that there is no rule that a political capital automatically attracts concomitant economic functions. Rather, the capitals that developed such functions did so because of historic contingency. In particular, the great European capitals grew on the basis of centralized regal power in the period between the sixteenth and the eighteenth centuries, which also happened to be the period when great trading empires developed. The two forces interacted and assisted one another; the political dominion and the economic one grew in parallel. On the basis of the trading function developed financial ones. Central power and a trading function demanded legal codification and legal enforcement, engendering a set of specialized functions – courts, lawyers and ancillary functions. Further, because these cities were centres of culture and of conspicuous consumption, local demand gave rise to activities such as universities, theatres, art and architecture, concert halls, newspaper and book publishing, and their twentieth-century media offshoots. These functions tended to assist each other, demand from one being met by supply from another. And with the progressive growth of the service economy, most of these functions have tended to expand in scale and importance.

However, these functions do not necessarily belong together. In states which from the beginning had specialized political capitals, we find that typically many or most of the other functions remain elsewhere, either in the former capital or in the most important existing commercial centres. In the United States, for instance, New York dominates the commercial, financial and entertainment worlds, and has a very important role in law, education and publishing. Washington has developed some independent cultural life in the last quarter-century, but is still a shadow of its near neighbour. In Canada, these functions are distributed among the provincial capitals but are disproportionately clustered in Montreal and Toronto; they are notably underdeveloped in Ottawa. In Australia the situation is precisely the same, with Sydney and Melbourne dominating the others; Canberra has acquired a cultural status through deliberate government action (Australian National University, the National Art Gallery of Australia), but still cannot compete with the older-established centres. Notably, in every one of these cases the political capital was a relatively late arrival, when the initial urban hierarchy was already well developed.

Even in Europe, the continent where the all-powerful multi-function capital is best developed, it is not absolutely universal. States that from the start were federal or confederal may share economic and cultural roles among several centres, as Switzerland shows. In Italy, where commercial life was already well advanced from Roman times onward in the Northern Plain, Milan and Venice retained their commercial and cultural roles after unification; Milan in particular has remained the dominant high-level service city of Italy, only slightly behind London and Paris.[2] In Germany, the federal structure after 1949 only underlined a long tradition of urban autonomy going back to the Middle Ages, in which Hamburg, Frankfurt and Munich retained the functions and the prestige they had partly lost in 1871. In the Netherlands, Amsterdam has always been the primary commercial, financial and cultural centre (and, by reason of the presence of the Royal Palace, the capital) though the government has been located in The Hague. Notably, The Hague has attracted some headquarters functions such as Royal Dutch Shell, but it still remains fundamentally a mono-functional city. In all these cases, accidents of historical evolution explain the separation of functions; but these are not rare anomalies.

Forces for Change

We can distinguish a number of possible forces for change in the next two decades: political, technological, economic.

1. Political

For the next decade, as in the last, the most momentous political change seems certain to be the effective dismemberment of the Russian empire both within its 1917 boundaries, and outside them. Nationalism has become a major political force once again, just at the point when in Europe it seemed to be surrendering to super-nationalism. Within Eastern Europe, this seems to spell a return to the political geography of 1918–1939, with strong national capitals. But the unknown factor is the impact of German reunification on that country's urban hierarchy. Berlin has again become the political capital, with residual functions left in Bonn. What is still unclear is whether this will lead to a reconcentration of other aspects of national life in the capital, including finance, commerce, culture and the media. With the emergence of Frankfurt as an economic super-capital housing the European Federal Bank, this seems less certain. The related question is whether the May 2004 enlargement of the European Union will allow Berlin and Vienna to recapture part of their pre-1914 roles as Imperial capitals. Given the strength of the nation-states and national capitals that replaced these land empires, it seems doubtful.

2. Technological

Two virtually certain developments, already in progress, seem likely to affect the relationships between capital cities and the other centres in their national urban systems. These are the informational revolution, and the development of new systems of high-speed ground transportation.

Information. A good deal of recent research on information-based services seems to agree that higher-level producer services, dependent on face-to-face information exchange, remain concentrated in the cores of the most highly-developed central metropolitan areas of the most highly-developed national economies (London, Paris, New York, Tokyo). However, specialist activities, such as research laboratories and routine producer services, may decentralize either to sub-centres within easy travelling distance of the major metropolitan centre, or to provincial cities offering lower rents and availability of the right kind of medium-skill labour.[3] An open question is whether certain types of 'head office' activity are also decentralizing to 'edge city' locations, leading to the development of polycentric metropolitan areas, as observed both in the San Francisco Bay Area and South East England.[4] But in this process the major metropolitan regions as a whole continue to expand – in particular the 'command and control' global cities, which increasingly relate more to each other than to the rest of the world: London, New York, Tokyo.[5] Second-order provincial capitals, and smaller national capitals in Europe, have also performed strongly, however.[6] Thus the cores of the great metropolitan regions may be shedding lower-level functions to other centres, including both sub-centres within their own spheres and provincial capitals, while they continue to dominate the most information-rich activities.

High-Speed Ground Transportation. An equally important development is the spread of high-speed train systems. It seems virtually certain that by 2010 Europe will have a network linking the national capitals and leading provincial cities, and taking much of the present air traffic up to a critical limit of about 800 kilometres, as has already been observed in Japan and France. Observations in these countries suggest that the new systems aid their terminal cities (Tokyo/Osaka, Paris/Lyon) while weakening intermediate cities (Nagoya). A crucial role will be played by a relatively few interconnection points between rail and intercontinental air services, as at Paris Charles de Gaulle and Amsterdam Schiphol.

The Overall Impact of Technology. Technological change is therefore likely to fortify rather than weaken the roles of the major cities, including the national capitals. But the effect will not be uniform, because the high-speed trains will find their optimal locus in the range from about 300 to about 600 kilometres. High-order cities, including capital cities, bunched within these limits may enjoy some advantage over the rest. The effect will be most noticeable in Europe, where the new trains should give a real comparative advantage to the 'Golden Triangle' bounded by London, Paris and Frankfurt as against more peripheral centres such as Madrid, Berlin, Copenhagen and even Milan. But much will depend on the operational characteristics, in particular the average speed, of the new system.

4. Economic Change

The most important economic changes are the shift to the informational economy, and the globalization of the corporation. Both will favour the highest-order global cities, but perhaps increase the pressure within them for local deconcentration. These of course are not necessarily capital cities, as New York illustrates. Though there are important linkages – between media empires and governments, for instance – these will necessarily take place in each national capital. But the complexities of control of such vast conglomerates seem likely to concentrate the headquarters operations in one place or at most two. Critical here will be the quality of international information linkages. The largest cities tend to have the richest and highest-quality information technology networks as well as the richest facilities for personal movement: international airports, high-speed train connections. These advantages tend to be cumulative, though they might be weakened by congestion of airspace around major airports and by the progressive build-up of connections at their second-rung competitors.

The Impact of Policy

During the 1950s and 1960s, governments in Europe made vigorous efforts to promote decentralization from their national capital regions. But in the 1980s and 1990s these policies lost force, because they were based on moving manufacturing industry, which has catastrophically declined – not least in these cities. Instead, government policies targeted relatively small zones – typically inner-city areas adjacent to central business districts – for intensive redevelopment through mega-developments carried through by public-private partnerships. This policy shift is now almost complete. However, it has left wide areas of the big cities – east London, the north side of the Paris region – struggling to find a new economic role to replace the factory and goods-handling jobs that have disappeared.

The Ultimate Solution: Moving the Capital

During the last nearly fifty years, a dozen countries in South America and Africa have planed the relocation of their capital city, or have actually established a completely new capital, Washington or Canberra style, in a greenfield location or on the basis of a small existing city.[7] The reasons are varied and invariably involve political motives, but congestion and resulting inefficiency in the old capital are usually cited. Some were logical and even necessary, in that new nation states were being created; many were over-ambitious in terms of available financial and organizational resources, and have proved failures. There are far fewer cases of recent deliberate relocation in the most advanced industrial countries; Bonn's establishment in 1949 as Federal German capital was a reflection of the division of the country a year before. Since 1960 the Japanese have twice seriously considered moving the seat of government from Tokyo; the 1990s explosion of land values triggered the debate for a third time, with Sendai north of Tokyo and Nagoya among the favoured alternatives.[8]

In the United Kingdom, alternative capitals have been discussed from time to time, but the idea has never received serious official consideration.

The likelihood is that governments will draw back both from the direct financial costs and indirect disruption that would inevitably be entailed. The German government has faced

Table 2.1. New capital cities in Africa, Asia and Latin America since 1960.

Year	Country	New Capital	Old Capital
1956	Brazil	Brasília	Rio de Janeiro
1957	Mauritania	Nouakchott	Saint Louis (Senegal)
1959	Pakistan	Islamabad	Karachi
1961	Botswana	Gaborone	Mafeking
1963	Libya*	Beida	Tripoli/Benghazi
1965	Malawi	Lilongwe	Zomba
1970	Belize	Belmopan	Belize City
1973	Tanzania	Dodoma	Dar es Salaam
1975	Nigeria	Abuja	Lagos
1982	Liberia*	? TBA	Monrovia
1983	Ivory Coast	Yamoussoukro	Abidjan
1987	Argentina*	Viedma/Carmen de Patagones	Buenos Aires

* None of these plans came to fruition.
Source: Gilbert, 1989, Table 1.

huge costs in relocating the capital to Berlin while paying for the modernization of the East German economy – not least, the fact that Berlin seems to have been less of a magnet for inward investment than was imagined. Other countries, with no such major political change in prospect, are even less likely to take the plunge.

Apart from the cost and disruption, there are two other reasons why they should be cautious. The first, pointed out by Jean Gottmann, is that capital cities often act as hinges between different regions of the country;[9] it would be very difficult to move them without engendering huge regional rivalries which would express themselves politically. The other is that cities, above all major global cities, now increasingly compete with each other to attract top-level global activities, transnational capital and elite populations.[10] Because of this fact national governments are less likely to countenance a move that could compromise the position of their leading city and, by implication, their country. Therefore, the likelihood is that they will seek to decentralize more routine governmental functions to provincial cities, leaving the capital as an ever-more specialized command and control centre for government and, by implication, their nations' economic and political life.

NOTES

1. Gottmann (1983a).
2. Brunet (1989).
3. Baran (1985); Nelson (1986); Mills (1987).
4. Beers (1987); Buck, Gordon, and Young (1986), p. 97.
5. Castells (1989), pp. 151, 169.
6. Hall (1987); Gillespie and Green (1987).
7. Gilbert (1989).
8. Miyakawa (1983); Anon (1988).
9. Gottmann (1983b).
10. Gastellars (1988); Lambooy (1988).

Chapter 3

The Urban Design of Twentieth Century Capitals

Lawrence J. Vale

Introduction: Urban Design, Capitals, and Twentieth-Century Political History

The planning and design of national capitals is inseparable from the political, economic, and social forces that sited them and moulded their development.[1] Peter Hall's useful typology of capitals set out in Chapter 2 distinguishes such cities chiefly according to the functions they perform in the national and global economy and in terms of the period and reasons for their ascendancy. In terms of economic influence, Hall distinguishes among 'Political Capitals' those which were created chiefly to serve as a seat of government, and those which serve more broadly as 'Multi-Function Capitals'. He also separates out 'Global Capitals' for their super-national role in the global economy, and 'Super Capitals' – cities that house international organizations, but are not necessarily national capitals. Then, acknowledging that circumstances change, Hall distinguishes among various kinds of 'Former Capitals', depending on whether they were 'Ex-Imperial Capitals' or merely important cities, and on whether they retain important status as 'Provincial Capitals' in federal nation-states. This chapter attempts a global comparison of the role of twentieth-century urban design in making capital cities appear as a distinctive type of place.

The urban design and planning of twentieth-century capitals is, above all, inseparable from the broader pattern of political change. The century began with the last gasp of imperial expansion, and was torn by repeated wars that, cumulatively, opened the possibility for well over a hundred new nation-states to emerge, each with its own capital. If one looked at a list of world capitals in 1900 and did so again in 2000, there would be little overlap. More than three-quarters of the cities that served as capitals when the century closed were not the capitals of independent states when the century began. For every London, Paris, and Lisbon that retained its position and centrality (even as it lost its imperial reach) there was a Belmopan (Belize), an Ankara and a Nouakchott (Mauritania) conceived anew.

Moreover, even in some cases where the same city was nominally a national capital in both 1900 and 2000, the intervening years brought vastly transformative interregnums.

Moscow began the century as the *de facto* co-capital of the Russian empire and ended the century as the capital of something still called Russia, but three-quarters of the century saw Moscow as the seat of the Soviet Union, and it was the socialist policies of that state that drove most of the visible design transformation of the capital. While early Soviet theorists explored Garden City ideals, Stalinists actually transformed the city, razing churches, carving out wide boulevards, and erecting retrograde signature skyscrapers, while retaining the Kremlin as a refuge and backdrop for military and ideological display. Similarly, Berlin's monumentally disrupted evolution – ranging from megalomaniacal urban design schemes for Germania to wartime devastation to post-war division to post-wall reunification – has yielded a city where urban design always kept pace with ideological preferences and conflicts. Beijing's tumultuous century began with an imperial 'Forbidden City', but absorbed dramatic changes after the Communist victory in 1949. The urban designers of the People's Republic of China symbolically and literally transformed a key part of the city's famed north-south axis by re-configuring Tiananmen Square: once a narrow T-shaped palace approach, it became an expansive rally ground, intended to accommodate a million Party faithful (see figure 3.1).[2]

To assess the role of design in the planning and development of twentieth-century capital cities one must acknowledge that design takes many forms and capitals perform many functions. Some capital cities began their lives during the twentieth century as deliberately planned acts; for many more capitals, the twentieth century was merely an arbitrarily circumscribed period

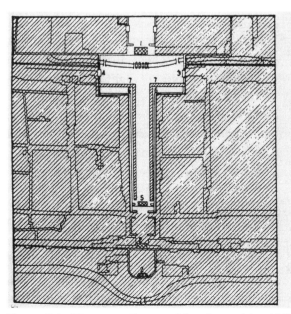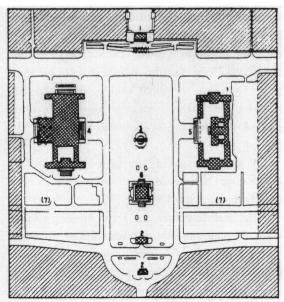

Figure 3.1. Tiananmen Square, old and new. After 1949, Chairman Mao ordered Tiananmen Square widened to create a massive parade ground and setting for new national monuments.

of ongoing development. Even newly designed capitals have arisen in different contexts. In some cases, such as Brasília, Canberra, Abuja (Nigeria) and Dodoma (Tanzania), cities have been willed into existence in predominantly rural areas, distant from previous seats of power. In other cases, such as Imperial Delhi (soon less imperially referred to as New Delhi) or Islamabad, new cities have been constructed adjacent to older ones. Assessing the impact of design always entails assessing the place of the capital in its country's overall pattern of urban development. Moreover, national development patterns are also inflected by more global events, such as the Great Depression and two World Wars, which severely constrained – or at least delayed – spending on urban design. At base, however, the urban design of capital cities is special because so much of it is state-sponsored and so self-consciously imbued with the need to convey national aspirations. In what follows, I call particular attention to three major developments of the twentieth century – the dismemberment of empires, the emergence of new federal systems, and the growing importance of super-national groupings – and analyse the effect of these developments on urban design.

Imperial Dismemberment and Capital City Urban Design

The Urbanistic Vestiges of Colonialism

As the twentieth century began, capital cities experienced the full flowering of the Beaux-Arts inspired efforts in urban design. Wide boulevards, large neo-classical structures and monuments, and vast axial symmetry seemed perfectly suited to conveying the grandeur and centrality of those in positions of power – even in places far removed from the inspiration of Paris.

In its colonialist guise, this is best expressed by the work of Edwin Lutyens and Herbert Baker for New Delhi, known formally as 'Imperial Delhi' until 1926. Lutyens drew heavily on his familiarity with Paris, Versailles and Rome (both its ancient Capitol and the axial boulevards orchestrated by Pope Sixtus V), and both he and Baker praised the vision of Pierre Charles L'Enfant for Washington. For Lutyens, it mattered greatly that his design for the Viceroy's Palace marked the culmination and centrepiece of the plan; he famously broke ranks with his friend Baker over a hillside-grading decision that failed to give the Viceroy his proper visual prominence in the tableaux (see figure 3.2). Although Lutyens did reserve a site for a Council House – looking ahead towards a time when the local population could take increased responsibility for its own rule – this location held a clearly secondary position.[3]

Other twentieth-century capitals also expressed Beaux Arts aspirations but attached them to quite different political regimes. Soon after the Americans gained control of the Philippines, Daniel Burnham planned the expansion of Manila as a city of long tree-lined boulevards, parks and monumental buildings. Burnham's artful marriage of urbanism and boosterism had been famously displayed at Chicago's World's Columbian Exposition in 1893, and he soon became a key member of the Senate Parks Commission that produced the McMillan Plan for Washington, DC in 1901–1902 – the epitome of the American City Beautiful movement.[4] That plan, although it took much of the twentieth century to implement, set the tone for the monumental core of the United States capital, a cross-axial swath of tree-lined greensward, lined with museums and government offices, and terminated by neo-classical monuments to national leaders (see figure 3.3). Although always ambivalent or resistant to charges of imperial ambition, the

Figure 3.2. Compromised approach to power, New Delhi. A hillside grading error caused Edwin Lutyens's Viceroy's Palace to sink slowly in prominence as it was approached, giving greater visual weight to the flanking Secretariat buildings designed by Herbert Baker.

Figure 3.3. McMillan Plan for Washington, 1901–1902. The McMillan Commission produced models of the Washington Mall, showing existing conditions (left) and proposed construction (right). After more than a hundred years of the city's development, the Mall had become a picturesque landscape, interrupted by a rail line. The new plan introduced the axial arrangement of museums, monuments, and office buildings that now characterize the city's monumental core.

Americans nonetheless designed a capital city to express such aims, just in case.

Unlike the New Delhi of Lutyens, however, the Washington of L'Enfant and the Senate Park Commission used grand urban design in service of democracy. Although the geometry and scale adopted the rhetoric of past monarchies and empires, Washington's socio-spatial diagram centred itself on the workplaces of democratically elected leaders, not kings or their designated imperial minions. Unlike New Delhi, the work of Burnham, Charles McKim, and Frederick Law Olmsted, Jr in Washington was centrally dedicated to highlighting the institutions of democratic rule and national culture, centred on the Capitol.[5] As the twentieth century closed, Washington's National Capital Planning Commission released its *Extending the Legacy* plan. This plan – advisory and lacking solid means of implementation – proposed to re-centre Washington even more formally on the Capitol, recognizing that vastly disproportionate investment had gone into the Monumental Core to the west of the Capitol and to the city's Northwest District and Potomac riverfront. Instead, the planners now suggested, federal investment should help develop neglected areas to the north, south, and east of the Capitol, home to the District's least advantaged communities (see figure 3.4).

Design always encodes a politics but both design and politics are mutable. It may be that Washington's politics are insufficiently mutable to embrace the 1997 NCPC plan, but it is worth noting how new political systems can sometimes embrace the places built to serve the old. Post-revolutionary USSR returned the capital to Moscow's Kremlin, seat of the Tsars and the Orthodox Church. And, merely sixteen years after the inauguration of New Delhi, the government

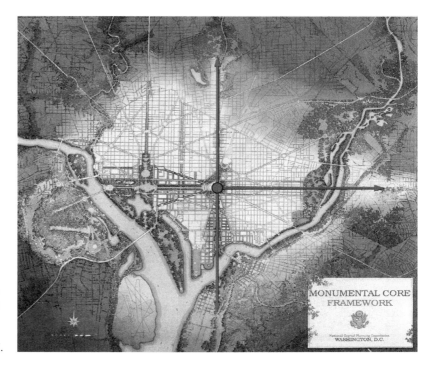

Figure 3.4. A vision for twenty-first-century Washington. The National Capital Planning Commission's 'Extending the Legacy' plan (1997) 'recentres Washington on the Capitol and extends development to the four quadrants of the city'.

of a newly independent India productively appropriated the landscape once intended to showcase British rule. The president of India (a largely ceremonial post) took up residence in the palace of the erstwhile Viceroy, and the axial 'Central Vista' soon embraced parades celebrating India's Republic Day (replete with vestigial bagpipers), a seemingly effortless substitution for the military hoopla of the colonial regime.

Edward H. Bennett (Burnham's co-author on the famous 1909 Plan of Chicago) produced an elaborate City Beautiful scheme for Canada's Federal Plan Commission in 1915. Ottawa, designated Canada's capital in the mid-nineteenth century, most memorably featured a neo-gothic parliamentary complex on a high bluff on the Ontario side of the river, but Bennett's plan reached out across the river to encompass Hull, in Quebec. However politically astute and necessary, the plan was not implemented, though urban design efforts to link Anglophone and Francophone Canada continued for decades. In contrast to the strident axiality of most designed capitals, Ottawa held tight to its more picturesque roots; even the 1950 Jacques Gréber plan still allowed the neo-gothic parliamentary spires to be approached best diagonally, enabling these imposing buildings to convey maximum surprise and visual appeal.[6] As the century closed, a new art gallery, museum and departmental offices clung to the inside of the loop of Confederation Boulevard joining Ottawa and Hull in a wishful composition centred on the Ottawa River itself. Here, as in other multi-cultural nation-states, planners have used urban design to convey a microcosm of the intended society (see figure 3.5).

Walter Burley Griffin's prize-winning design for Canberra – nearly concurrent with the Lutyens's plan for Delhi, adopted both the bi-

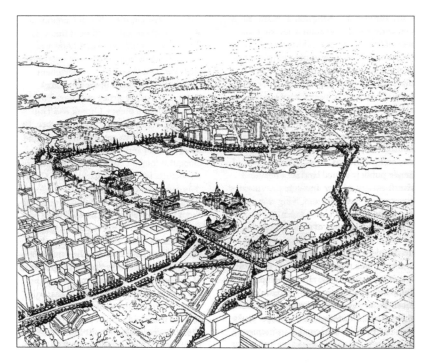

Figure 3.5. Canada's Confederation Boulevard, which uses urban design to promote linkage between Ottawa and Gatineau and, by extension, Ontario and Québec

axial symmetry of the City Beautiful movement and the democracy-oriented symbolism of Washington, DC. At the same time, paralleling the contemporaneous plan for New Delhi, Griffin also employed some unusual hexagonal geometry that had little relation to Beaux Arts practice.[7] Griffin's plan placed the Australian parliament building at the terminus of his 'land axis', though the diagonals of the roadway plan converged on 'Capital Hill', located just above and beyond it (see figure 3.6). Griffin intended the hill to be crowned by a ziggurat-capitol and public park but – three-quarters of a century later – the Australians eventually conflated the two aspects of Griffin's intent, by essentially replacing the profile of 'Capital Hill' with a grass-covered capitol building designed by Aldo Giurgola. Here, the designer intended, democracy could literally be inscribed into the landscape.[8] At its core, Canberra's urban design took full advantage of its physical setting, aligning its axes with mountains, and taming its geometric rigour with a more informal embrace of the undulating geomorphology and a carefully orchestrated deployment of native flora.

War and Reconstruction

Despite the best intents of planners, much planning is necessarily reactive. Cities must respond to unexpected traumatic events such as wars and natural disasters. This kind of destruction often provided significant opportunity to address pre-trauma shortcomings in urban form, but – most famously following the London fire of 1666 – often entrenched property interests produced more inertia than innovation. Still, given the intense destruction brought by two World Wars and countless regional or civil conflicts, the combined forces of physical destruction and regime change significantly directed the nature and scale of urban design intervention in the twentieth century.

Some capitals, notably Tokyo, suffered the double ravages of war and natural disaster. From the devastation of the Kanto earthquake of 1923 to the fire bombings of 1945, Tokyo

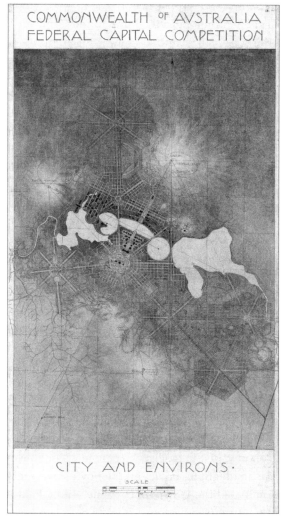

Figure 3.6. Griffin Plan for Canberra. Walter Burley Griffin exported aspects of City Beautiful ideals to the Australian bush, while coupling these with an effort to highlight the institutions of democratic governance, and emphasizing the qualities of the landscape.

faced a repeated need to rebuild. Despite the apparent opportunity occasioned by such horrific destruction, twentieth-century Tokyo is not distinguished by major urban design projects, notwithstanding Kenzo Tange's wild proposal for a new city of 10 million in Tokyo Bay. Instead, the underlying urban pattern of the city remained remarkably undisturbed by the cycles of trauma and reconstruction, and major changes depended on a perceived link to modernization efforts (which included significant amounts of landfill). In this regard, Tokyo came to resemble other Japanese cities, even those (such as Kyoto) that did not experience wartime destruction. Tokyo's planners and designers seemed much less preoccupied with 'image' than those of other capitals, though this market-driven city did tout its 333-metre-high Tokyo Tower (completed in 1958), and gained renewed international visibility as the host of the 1964 Olympics.[9]

Other capital cities destroyed by war undertook more dramatic urban design initiatives. The Germans destroyed 80 per cent of Warsaw during World War II, and the city lost 800,000 of its 1.3 million residents, but planners and designers assiduously erected a selectively edited replica of the Old City soon after the war, improving upon its constricted traffic flow by introducing an underground highway. Many other parts of the reconstructed city took the form of large Soviet-inspired housing estates, intended to demonstrate the centrality of investment in those workers charged with powering a new industry-driven economy. Just in case such economic symbolism was inadequate, however, the Soviet overseers dominated the Warsovian skyline with a Palace of Culture (see figures 3.7 and 3.8).[10] The post-Soviet period also triggered considerable interest in urban design, both in Soviet-dominated parts of Eastern Europe and in the newly-independent constituent republics of the former USSR. As elsewhere, historic preservation entailed reconstruction of a pre-colonial past, coupling the draw of tourism with a highly-visible assertion of nationalist aspirations.

Berlin faced perhaps the world's most complicated series of urban design challenges during the twentieth century, given the variety of regimes that separately – or simultaneously – attempted to express their designs on the city. Whether the urban designer was charged with building 'the word in stone' to support the megalomania of Hitler and Speer's Germania, or whether urban design dramatized the virtues of socialism along East Germany's Stalinallee; whether the goal was to extol the post-war capitalist triumphs in the West in the Modernist Hansa Quarter, or to reassure a still-wary world that a resurgent unified Germany would not threaten the global political order, twentieth-century urban design in Berlin consistently encoded politics. The power of Speer's largely unexecuted north-south axis for the city was so great that, even unbuilt, its ghost exercised a demonic hold over German planners and designers. In the 1990s, competition organizers of the International Competition on Ideas for Urban Design at the Spreebogen (intended as the site for the German Federal Parliament and Federal Chancellery on the spot once designated for Hitler's gigantic Great Hall) warned entrants about its phantom prehistory as the termination of the Nazi north-south axis. It was no surprise that the winning scheme emphasized an east-west-oriented ribbon of buildings.[11]

Other European cities emerged from World War II scarred by the physical and psychic damage, but less permanently politicized than Berlin. The face of twentieth-century London, for instance, though markedly altered by the emergence of high-rise offices in the City, was arguably less affected by the Blitz than by the

Figures 3.7 and 3.8. Destruction and reconstruction of Warsaw. World War II unleashed an unprecedented scale destruction on many cities in Europe. In terms of devastation and loss of life, however, Warsaw stands out dramatically. Its Old City was reduced to rubble (figure 3.7), but carefully (and selectively) reconstructed (figure 3.8). Other destroyed areas of the city were radically rebuilt as workers' housing soon after the war.

more gradual dissolution of the maritime empire, which yielded the corresponding opportunity to redevelop the vast Docklands for new commercial and residential purposes. Urbanistically, London began the twentieth century by expressing its last imperial gasp – the grand neo-classical sweep of the Kingsway – yet closed the century as again a 'world city' in both an economic and cultural sense. In the interim, planners and designers vainly sought to curtail its growth and centrality through devolution into orderly satellite garden cities or new towns, or by construction of a greenbelt. Instead, London continued its slow growth, gaining a double skyline of corporate identity in the City and in Canary Wharf, while dramatically improving the Thames riverfront as a place of culture and residential appeal. London's appeal persisted despite (or perhaps because of) a failure to coordinate its physical planning, while doggedly pursuing its status as a global financial centre.[12]

Paris suffered relatively little destruction during World War II, and was not subjected to the even more brutal possibility of Le Corbusier's unexecuted Voisin plan from 1925. In fact, it experienced relatively little major urban design intervention in the twentieth century until after about 1960. In contrast to most other capitals, major French government functions remained hidden in side streets; only the Louvre – the museumifed home of the former monarchy – remained an urbanistically prominent reminder of governance. Paris retained its economic and political centrality not by constructing ways to highlight the presence of government, but by commissioning a variety of *grands travaux* that emphasized the reinvigoration of cultural institutions in diverse areas of the city. The projects are most durably associated with the presidency of François Mitterrand, but had antecedents all the way to de Gaulle's decision, taken in 1960, to tear down Les Halles. Major projects continued, championed by the presidents throughout the remainder of the century. From the predictable emphasis on the Louvre and the extension of the 'Grand Axe' westward through La Défense to Spreckelsen's dramatically-scaled corporate/ministerial arch, to the less predictable decisions to invest cultural capital in neglected parts of the city such as the Cité des Sciences and Parc de la Villette and the Opera de la Bastille, late twentieth-century Parisian urban design regained global prominence and attention.[13]

In the full scheme of late twentieth-century capitals, however, places such as London and Paris stand out as exceptions. Most of the globe's other capitals have faced frequent challenges to their role on the world's stage. Only the most secure governments seem able to escape the need to demonstrate their command and control through dramatic exercises in architecture and urban design.

Urban Design and the Search for Post-Colonial Identity

Capitals like London or Paris, boasting a long history of carefully secured centrality, predated the twentieth-century predilection to give government functions their own separate district right from the start. To be sure, much of the world has long segregated the ruling institutions in separate quarters, usually a palace or military cantonment (if not both). Yet the twentieth century – as a century marked by a dramatic increase in the proclamation of democratic governance (however unevenly realized) – marks an increased effort to plan and design for the conjoined urban presence of government bureaucracies, not just the symbolic ruling structures. This growth in the mechanics of government, expressed through vast new ministries charged with overseeing all manner of public investment, in turn provided opportunities for designers to propose schemes for whole new government-centred districts, or even whole new cities.

Many such places were encouraged by the success of independence movements, inspiring a fledgling new regime to take flight in a new location. Ankara, strategically placed in Anatolia, seemed an opportune break from the eccentric position of Istanbul after the collapse of the Ottoman Empire, and the resultant design highlighted the presence of a new parliamentary system and government district for the Turkish republic, while also according prime position for a monument to Atatürk (see figure 3.9).[14] Helsinki gained its status as capital of the Republic of Finland in late 1917 following the Russian Revolution. Soon afterwards, its leaders embarked on design efforts to clarify its status as the seat of an independent country, after seven centuries as either a royal Swedish town or an imperial Russian city. Although progress remained slow, the Finns commissioned a new Parliament House and used landfill to extend the central business district and establish an alternative national centre for the capital.[15]

Other new capitals had less direct links to the collapse of empires. Brazil, though formally decoupled from Portuguese colonial rule since the nineteenth century, nonetheless clung to its

Figure 3.9. Diagrammatic plan of Ankara. Hermann Jansen's master plan for Ankara developed a capitol complex near the centre of the city, linked to a district of government ministries. A long boulevard leads from this complex back towards the heights of the original citadel, but the night-lit Atatürk Mausoleum dominates the skyline.

coastal port cities, until the regime of Juscelino Kubitschek pledged to build the inland capital of Brasília during the course of a single five-year term in the late 1950s. Lúcio Costa's competition-winning plan asserted clear possession over the vast landscape, and was structured according to a simple cross-axis – one axis of *superquadras* for residential purposes, and the other dedicated to government – lined with identical ministry buildings and culminating in the 'plaza of the three powers' flanked by iconic Modernist structures housing these principal branches of government. On its fourth side, the plaza was left open to the landscape and cloudscape beyond. Brasília – at once a bold formal gesture, a huge economic gamble and a valiant attempt to cleanse Brazilian governance of the perceived corruption of political life in Rio – remains the quintessential experiment in high Modernist capital city design. In its first half-century, it has gradually attained a better reputation within Brazil. However, despite the radical intent of its principal designers, its *superquadras* failed to house the full range of the Brazilian bureaucracy all together in classless harmony, and the functionless openness of the government axis and plaza seems more likely to attract architectural photographers than sociable citizenry. Brasília succeeded in giving Brazil wide international acclaim for its modernity, but its design intentions did little to serve either the high-end housing aspirations of leaders who preferred to live in lakeside villas, or help the impoverished who could not afford to live within the 'Pilot Plan' at all, yet constitute about 85 per cent of the Federal District's current population.[16]

Decolonization affected the urban design of capitals much more directly by inspiring about a dozen entirely new cities after 1960, as well as the construction of many smaller capitol complexes in existing capital cities. In many cases, rulers of newly independent nation-states viewed urban design as a mechanism to shore up their rule. Following the partition of India, Pakistan commissioned Constantine Doxiadis to design Islamabad, adjacent to the existing city of Rawalpindi (see figure 3.10). Here, too, there is a reliance on axial planning inflected by Modernist urbanism, this time with a long Capitol Avenue leading to the Presidential Palace. As in Brasília, New Delhi and elsewhere, capital city design isolated the capitol complex from the rest of the city. Once again, twentieth-century urban design for new capitals emphasized a separate district for government.

Modernist urbanism exhibited both continuity with its Beaux Arts antecedents as well as

Figure 3.10. Capitol Avenue, Islamabad. Islamabad, planned between 1959 and 1963 by Doxiadis Associates, who envisioned a long avenue leading to an administrative centre, with a linear civic centre shown at left in this model.

notable deviation. In the classic Modernist designed capitals of Chandigarh and Brasília, axes and long-views continued to predominate, yet such places break from Beaux Arts precedent in the composition and importance afforded to culminating plazas. Rather than axes that terminate in a single building, Modernist urbanism emphasizes a complex balanced asymmetry of juxtaposed buildings and framed landscape views. The space between – however dysfunctional in its barren realization – matters as much as the surrounding buildings (see figure 3.11). Modernist urbanism also embraced the automobile to an extent unimaginable at the dawn of the twentieth century. This new emphasis on speed altered the rules of proximity and permitted a vastness to urban conception that far exceeded the carriage-and-pedestrian domains of earlier eras. Although pre-Modernist designed capitals, such as Washington, New Delhi, and Canberra also embraced large-scale conceptions – suggesting that the scale of composition largely depends on the designer's freedom that accompanies any city-sized scheme for a previously exurban site – the additional infusion of Modernist urbanism resulted in an unprecedented open-ness, almost a figure-ground reversal of landscape and built form. At the same time, modern architecture – at least in its earliest pre-corporatist decades – clearly stood out as a key symbol of progress, often associated with left-wing regimes, or at least with conveying an intended break from the neo-classicism associated with European colonial rule.

Many newly-independent countries embarked on new capital city designs after 1960, ranging

Figure 3.11. Capitol Complex, Chandigarh. Le Corbusier's design for the capitol complex at Chandigarh is a complex system of plazas and framed views, but the result is a pedestrian-resistant expanse of paving. The sense of isolation is further enhanced by the artificial mound (on the left), constructed to block the view of the rest of the city.

from Belize to Botswana, and many other more-established countries – such as Argentina, Japan, and South Korea – also seriously discussed moves during the 1980s and 1990s. The two most ambitious moves of the late twentieth century – Nigeria's Abuja and Tanzania's Dodoma – remained substantially unfinished at the century's close. The Nigerian capital, now largely functional, was designed during a brief period of democratic rule in the 1970s when the Nigerians adopted an American-style constitution and wanted a Washington to go with it. The result, launched by North American planners and Japan-based designers from Kenzo Tange's firm, envisioned a monumental city with a vast Washington Mall-like central axis, and aligned with a distant hilltop, just as in Canberra. Abuja's central axis leads to the 'Three Arms Zone', where the three branches of government could demonstrate their separation from each other; unfortunately, the design also served to separate off all three powers from the rest of the emerging city.[17] Construction proceeded slowly due to economic woes and protracted political instability, but the high-security promise of Abuja's protected government precinct carried ongoing appeal to both military and civilian rulers (see figure 3.12).

In urban design terms, though, it is Dodoma that bucks the century's trend towards monumental axiality. Here, in a city centre designed by the same firm (Conklin Rossant) that planned Reston, Virginia, the design intent is radically modest. Attempting to design a city for a low-income country committed to village-based socialism, the designers emphasized residential areas and public transportation, and kept the city centre low-rise and pedestrian-oriented.

Figure 3.12. Envisioning Abuja, Nigeria. At the heart of Abuja, Kenzo Tange's firm designed a wide mall. This served as an axis that would lead to the parliamentary complex, backed by the massif of Aso Hill. The city, currently still under construction, eventually took quite different architectural form.

Instead of an isolated district for government, the designers proposed a mixed-use area whose largest proposed building was a sports stadium (see figure 3.13). Much of the modesty was undercut, however, by a separate proposal by Chinese consultants to erect a Party headquarters and interim parliamentary complex separate and above the cultivated understatement of the main town. Whatever the contrast from typical capitals in design intent, however, Dodoma's many setbacks have placed it firmly alongside Washington and Canberra in the pantheon of long-delayed visions.[18]

Given the high cost and significant disruption of constructing a new capital city, it is not surprising that the majority of countries chose to undertake a more modest investment in urban design. Many places, including such diverse locales as Sri Lanka, Kuwait, Bangladesh, Papua New Guinea and Malaysia, stopped short of building entirely new capitals and instead invested lavishly in new capitol complexes. These capitol complexes (not unlike the capitol complex at what Le Corbusier termed '*la tête*' of Chandigarh) serve as the seat of legislative governance (at least in theory) and also combine other government or 'national' functions into adjacent areas. Given the high premium placed on security in such places, it is hardly surprising that many such complexes function more as

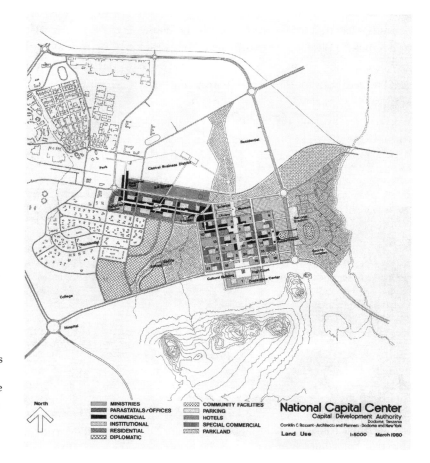

Figure 3.13. Dodoma, Tanzania: an anti-monumental designed capital? In contrast to the usual heavy reliance on axiality and a separate zone for government functions in twentieth-century designed capitals, the plan for Tanzania's Dodoma (designed in the late 1970s) envisioned a much more modest, mixed-use, city centre, and devoted considerable attention to the residential districts of the city.

separate islands for government. In some cases, most literally true in Sri Lanka, the parliament building structure is sited on an actual manmade island (see figure 3.14). Although architecturally quite inclusive in its hybrid references, the urban design of the parliamentary district highlighted facilities designed to serve the Sinhalese Buddhist majority. Further, this complex, constructed during a time of growing Tamil unrest, was provocatively sited adjacent to the fifteenth-century palace/fortress from which the pre-colonial Sinhalese last effectively controlled the whole of the island.[19] Louis Kahn's 'citadel of assembly' for what became independent Bangladesh (the complex was originally designed as East Pakistan's parallel to West Pakistan's Islamabad) is also dramatically set off by water, as well as isolated from the main traffic of Dhaka (see figure 3.15).[20] In every case – whether capital city or capitol complex – urban design has been used as a tool to set off – and thereby highlight – those key parts of the city that the regime wants to regard as representative of national aspirations.

It is too simplistic to see the design politics

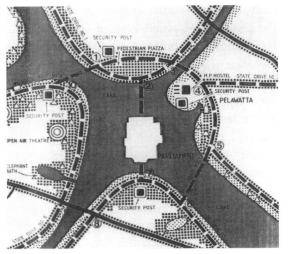

Figure 3.14. Securing the capitol: Sri Lanka. Sri Lanka's capitol complex, constructed in the 1980s, placed the parliamentary structure on a man-made island.

of national capitals as purely an expression of nationalism, however. Or, more precisely, it is essential to see how the concept of nationalism, if carefully unpacked, exists on many scales above and below the 'national'. In enhancing the draw of capital cities as places of cultural centrality through the choreography of ritual space for

Figure 3.15. Isolating the capitol: Dhaka, Bangladesh. The capitol complex for Bangladesh occupies its own district, set well apart from the bustle of the city.

mass display and movement, the attempt in capital cities to constitute 'the national' is also an effort to seek out international recognition, and a means to highlight the presence of dominant groups in a plural society. At the same time, at least in some cases – such as Brasilia, Dhaka, or India's Chandigarh, where a handful of architect/planners have been given free creative rein – the pursuit of 'national identity' can but thinly veil a narrower expression of personal identity, either the design agenda of the architect, or the political agenda of the client, or both. In short, regimes design capital cities and capitol complexes chiefly to serve personal, subnational, and supranational impulses rather than to advance 'national identity'.

Federalism, Capitals and Urban Design

It is tempting to search for distinct urban design trends in those capitals that have been created as a direct outgrowth of the creation of new federal systems. Yet however much such places may be asked to mediate between the national scale and the strong powers accorded to smaller governance units such as states or provinces, these federal capitals face considerable pressure to demonstrate visibly the need for 'the national', not just demonstrate its presence. Domestic visitors from distant jurisdictions need to be reassured by a visit to the national capital, able to see something of their own more locally attached identity expressed through the place designated as the seat of national governance. In urban design terms, federalism encodes a paradox: it increases pressures to build aspects of the 'national' outside the capital city, but also underscores the need to shore up the *image* of the capital. In many cases, the national capital retains the key symbolic presence for a particular government bureau, while back office employment can be decanted into more geographically dispersed locales. Similarly, the presence of other major cities outside the national capital often permits distribution of some national facilities (such as a public university or even a major branch of government) to a provincial location.

At the same time, federal systems also encourage – or at least permit – constituent states or provinces to demonstrate significant aspects of their subnational identity, sometimes driven by the personal agenda of a single charismatic politician. In the United States, this was most famously realized by New York governor Nelson Rockefeller's bombastic 1960s modernism on the Albany Mall, though Louisiana governor Huey P. Long's skyscraper capitol for Baton Rouge, completed in 1932 as the tallest building in the American South, rivalled even Rockefeller for sheer audacity.[21] Elsewhere, from Canada to India, the construction or augmentation of subnational capitals offered opportunities to highlight the dominant presence of some provincially-dominant group. In Ville de Québec, for instance, the provincial government invested heavily in efforts to restore the Lower Town to its condition before the British conquest, presumably anticipating a more prominent future role in a francophone post-Canada system of governance.

Polycentric Capitals and the Cultivation of Urban Image

Both well-established capitals and newly promoted ones share a common desire to cultivate an image of national importance. In many older capitals, urban design techniques have been used to limit development in the historic core, while steering new construction – seen as vital to expanding the economy of the city but less central to its symbolic role – to some new outlying district where it would appear less disruptive to the city's image. Paris retained its height limitations, but channelled high-rise development just to its west, to La Défense. Similarly, Rome's planners developed the EUR district on the city's outskirts, initially intended as a site of a major exposition during the Mussolini era. Ottawa held to height limits for a half-century, permitting Parliament Hill to dominate, but succumbed to development pressures in the 1960s. Washington, DC held out longer on this issue, and continued to limit development height in the Capital District, permitting the principal nineteenth-century structures to retain their centrality. Washington's planners enhanced the monumental core with new memorials (some quite controversial), but could only watch as significant investment in high-rise office clusters flocked to Virginia communities immediately across the Potomac, investment that did nothing to help the shaky finances of the District. This polycentric growth pattern of enhanced midtowns surrounded by an acne of 'edge cities' is hardly unique to capitals, but the pattern is inflected by the desire of national governments to reserve significant central parts of capitals for sites and amenities deemed to be of national importance.

Capitals of federal governments also often use urban design to find ways to acknowledge the existence of component states or provinces. In Canberra, for instance, streets named after the various state capitals radiate outward from Capital Hill. Similarly, many of Washington's principal avenues are named after American states. The problem is more complicated when a federal system masks an uneasy co-existence of contending groups. In such cases, urban designers face a tougher challenge of representation, especially if component states/provinces have markedly different dominant architectural traditions. At base, the key question is: Who deserves to be represented as part of 'the national'? Frequently, invocations of 'the national' have had to overcome stiff opposition from marginalized portions of the population.

Super-National Challenges to Urban Design

Finally, the rise of supranational organizations (such as the League of Nations, the United Nations, and the EU) accelerated the profusion of large campus-type projects in capital cities such as New York and Brussels. The ill-fated League of Nations (whose star-crossed urban design competition took so long to resolve that it almost outlasted the institution it was designed to house) promised a new district for inter-war Geneva. After World War II, plans for the new United Nations started with grand proposals for a 40 square mile 'Capital of the World' to be located amidst the New York City suburbs, but quickly succumbed to the peremptory offer of 17 slum-clearance acres in midtown Manhattan. In short order, New York gained not a satellite world capital, but a more modest new district of and for diplomats. As before, the default option for urban design remained a separate zone for government. Much the same thing occurred at the end of the century in Brussels, as the European Union took hold of a glassy enclave, in the city but somehow not quite *of* it (see figure 3.16).

Conclusion: Imaging the Capital

Capital cities are sites of display, conveying desired elements of national culture to visitors and locals alike. Capitals are places of touristic

Figure 3.16. Brussels: super-national government and local neighbourhoods. The glittering buildings of the European Union stand implacably apart from the city's Quartier Léopold.

pilgrimage; often they are also the economic magnets for those electrified by the prospects of employment. Capitals are host to diplomatic quarters, as well as the site of often undiplomatic efforts to oust or relocate the desperate poor who seek refuge. Urban designers have famously drawn official plans for capitals, but they have also been called in to help design satellite towns for some of the people who did not fit (or were not wanted) so close to the centre of power.

Twentieth-century urban design for capitals has certainly not been immune from the century's most dramatic contribution to city planning theory – the concept of land-use separation.

In many ways, capitals – and especially those designed from scratch – have taken the concept of zoning to the extreme. Although some older capitals have continued to scatter important public buildings widely across the various parts of the city, this has not been the trend. Rather, the most visible legacy of the twentieth century for urban design in capital cities has been the concentration and isolation of government functions.

At the same time, however, capitals have dramatically oriented themselves to attract tourists – both domestic tourists who come to visit national institutions, and foreign

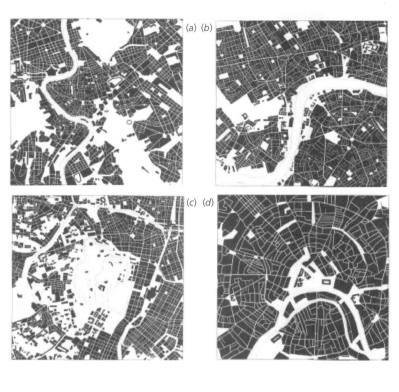

(a) Rome Figure Ground
(b) London Figure Ground
(c) Tokyo Figure Ground
(d) Moscow Figure Ground

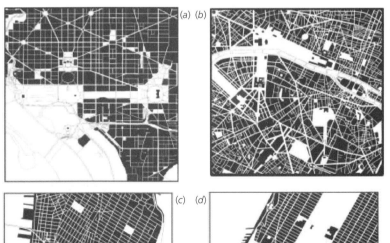

(a) Washington Figure Ground
(b) Paris Figure Ground
(c) New York Figure Ground
(d) Mid-Town Manhattan Figure Ground

Figure 3.17. Capital city urban design compared at a common scale (5 kilometre square). These figure-ground renderings of the fifteen case study capital cities hint at many patterns. Four of them (Canberra, New Delhi, Chandigarh, and Brasília) resulted entirely from twentieth-century design efforts, while the rest represent centuries or even millennia of urban accretion. Not surprisingly, the complexity of small block forms is most obvious in older cities such as London, Paris, and Rome though the latter two clearly demonstrate the axial overlays imposed on the fabric in the nineteenth century (Haussmann's Parisian boulevards) or twentieth

(a) Helsinki Figure Ground
(b) Ottawa Figure Ground
(c) Berlin Figure Ground
(d) Brussels Figure Ground

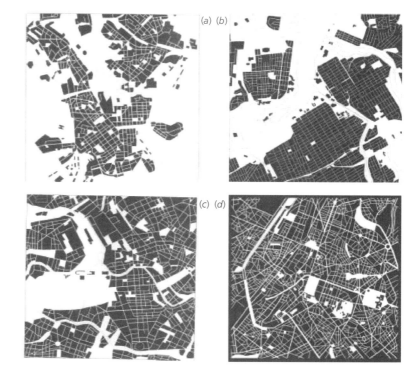

(a) Canberra Figure Ground
(b) New Delhi Figure Ground
(c) Chandigarh Figure Ground
(d) Brasillia Figure Ground

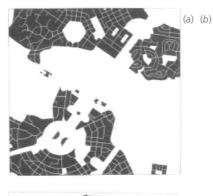
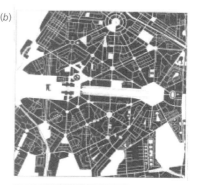
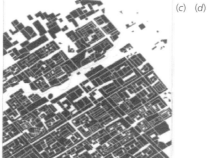
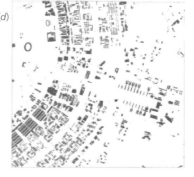

century (Mussolini's version of these for Rome). Similarly, Moscow reveals the clear trace of planned concentric growth, while midtown Manhattan is organized by the plaid of its early-nineteenth-century grid interrupted by the occasional superblock incursion of twentieth-century modernism as in the 17-acre site occupied by the UN complex. By contrast, most of the capitals created in the twentieth century have been designed at a broader scale, epitomized by the vast superblock streetlessness of Brasília and the sector planning of Chandigarh, but also apparent in the openness of Canberra and New Delhi.

visitors, some of whom are being courted as potential investors and need to be cultivated and encouraged by the modern appearance of the capital. Urban design has moved a long way from the axial bombast of the Beaux Arts inspired era that opened the century, yet when it comes to capital cities, the degree of change is sometimes rather less apparent. Few capitals have gained new public spaces that mimic the multi-use pedestrian-scaled streets and plazas of older cities. Capitals, as self-consciously 'national' cities, often lack the intimacy of older civic centres, and instead – at least the newly-designed ones – are characterized by automobile-oriented boulevards leading to privileged hilltop precincts. In an age of increasing security concerns, designers who work in capitals face enhanced need and pressure to pursue protection through separation. Capital city urban design is the last refuge of grand axial planning, which has survived even the ignominy of Speer's proposal for Berlin. And, in some places, even the once archetypal reliance on moated castles has not wholly disappeared from the repertoire of preferred techniques.[22] Even in a century marked by the emergence of modernism in all fields, many aspects of capital city urban design continued to embrace a premodern sensibility, rooted in a drive for hierarchy, rank and clarity of expression.

In assessing this seemingly retrograde trend, it is important to distinguish between designs and regimes. Surely, axiality is far more palatable in service of post-revolutionary Washington than it is when marshalled to celebrate Hitler's Germania. In the context of democratic rule, clarity of urban design expression may serve quite useful purposes, enabling visitors and locals alike to apprehend the key monuments and major socio-political relationships of their capital. Because capitals necessarily have a didactic role, perhaps they may be properly expected to demonstrate hierarchy by showing who and what matters most to the State. Still, the continued grandiosity of most recent capital city urban design can only be appreciated in relation to the degree of actualized democracy each particular nation-state has achieved. Grandiosity in service of tyranny offends, but grandeur in recognition of respectful democratic partnership may legitimately inspire.

In all this, it is striking how much investment has occurred in each capital city to manipulate and sustain its image and symbolic centrality. From the imperial neo-classicism of capital city urban design that greeted the twentieth century to the high Modernist variants that gave bombast a different skin, urban design remained a vital part of the public projection and reception of capital cities in the twentieth century.

NOTES

1. Vale (1992).
2. Hung (1991), pp. 84–117.
3. See Irving (1981), pp. 82–83, 142–154, 311–312.
4. See Wilson (1989), pp. 53–95.
5. See Gillette (1995), pp. 88–108; and Gutheim (1977), pp. 118–136.
6. Taylor (1986).
7. Such hexagonal urban geometry was frequently employed during the first third of the 20th century, and seems to have its orgins in Christopher Wren's plan for post-1666 London, as re-popularized and reinterpreted in the work of Barry Parker and Raymond Unwin. See Ben-Joseph and Gordon (2000), pp. 238–243.
8. Vale (1992), pp. 73–88.
9. Hein (2005*a*).
10. Goldman (2005).
11. Helmer (1985), pp. 27–48; Ladd (1997), pp 127–235; Wise (1998), pp. 57–80, 121–134.
12. Hebbert (1998).
13. Mission Interministérielle de Coordination des Grandes Opérations d'Architecture et d'Urbanisme (1988); Curtis (1990), pp. 76–82.

14. Vale (1992), pp. 97–104.

15. The Finnish struggle also underscores the frequent problem of capitals that find themselves promoted from provincial to national status. Accommodating the full repertoire of national institutions (from ministries to courts to diplomatic quarters) is often difficult in those places where a smaller government once dominated a geographically constrained place (as with Ottawa's Parliament Hill).

16. Vale (1992), pp. 115–127.

17. *Ibid.*, pp. 134–147.

18. *Ibid.*, pp. 147–160.

19. *Ibid.*, pp. 190–208.

20. *Ibid.*, pp. 236–271.

21. Bleecker (1981); Goodsell (2001).

22. Vale (1992), p. 293.

Chapter 4

Paris: From the Legacy of Haussmann to the Pursuit of Cultural Supremacy

Paul White

Paris as a Planned Capital

Of the cities considered in this book, Paris claims to have the longest history as a capital. In 486 AD, the Frankish king Clovis, following his defeat of the Romans, chose the site as his administrative centre. Then in 987 the founding of the Capetian dynasty established Paris as the centre from which the concepts of France and a French polity were gradually extended.[1] The growth of Paris has therefore been intimately linked to its functions within France, and beyond, for over a millennium.

This growth of Paris has, however, often been perceived in the rest of France as involving unwarranted domination. Over the two hundred years since the overthrow of the absolutist *ancien régime*, France has remained marked by intense centralization, with only the years since 1981 producing any apparent shift. A recurrent discourse has seen Paris as a 'hypercapital', drawing lifeblood from the rest of France.[2] A feature of political life has been rivalry between the interests of Paris and of France as a whole for the attention of government. This is clearly exemplified in attitudes to the planning of Paris, both as a city in its own right and as France's capital.

Peter Hall's classification in Chapter 2 places Paris in the first rank as a Multi-Function Capital, and in this respect other French cities offer no competition – whether in terms of business, education, or culture. However, claims can also be made for Paris as a 'Super-Capital' whose influence stretches far beyond the general internal control and external gate-keeping roles of capital cities. In the later years of the nineteenth century, Paris was seen as the capital of the *Belle Époque*, with a reputation and prestige that were world-wide. Urban design associated with the Second Empire (1852–1870) was of considerable significance in making Paris the exemplar of what became known as the 'City Beautiful'.[3] In a number of ways, Paris has also served as a 'model' capital for the rest of the world.[4] As will be seen later, the end of the twentieth century saw arguments within France to reassert Paris's role on a wider stage, particularly in terms of the opportunity for the city to take its place as the capital of Europe – if not in political or economic terms then in relation to culture and prestige.[5] Allied to this ambition has been the claim, largely unnoticed by Anglo-Saxon commentators, of Paris as the capital of the French-speaking world (*la Francophonie*).[6]

One might imagine that these claims about Paris's major capital city status would result in constant strategic thinking about the planning of the city, and its embellishment to reflect the glories of France. Paradoxically, for much of the twentieth century the reality was very different. Although there was considerable discussion about the city for decades, it was not until the 1960s that strategic designs for Paris were finally accepted and implemented. Even then such schemes related more to the city as a large urban space than as a capital of a world power. By 1960 nearly a hundred years had passed since the construction of the projects that had made Paris the epitome of modern capital city organization and created its image as the reflection of the French Second Empire. Sutcliffe, writing about the inner city of Paris, has fairly characterized the century after the fall of the Empire in 1870 as the 'defeat of town planning'.[7]

The impact of Haussmann's activities during the 1850s and 1860s in redesigning and redeveloping Paris have justly been emphasized in planning and architectural histories.[8] The legacy of the Second Empire was a modernized urban context in which the basis for a functioning great metropolis had been established. It is, however, important not to overstress what Haussmann achieved: in the spaces between the major new axes and *boulevards* a remarkable degree of old Parisian life remained unchanged.[9] Nevertheless, by 1870 Paris had a ground plan and a set of functional spaces that have changed remarkably little over the following one hundred and thirty years. The enhanced activities of a twentieth-century capital city could be inserted with little major impact on the urban fabric, or need for redevelopment.

One paradox of planned intervention in Paris over the last century is that whilst it has been the laboratory in which French planning theories, designs and strategies have been created and tested, such developments have only infrequently encompassed what might be termed 'capital city' objectives within them. The status of Paris as the capital of France has been taken for granted, with little perceived need for further legitimation through planning action. The planning of Paris specifically as a 'capital' city is thus predominantly a legacy of nineteenth- rather than twentieth-century activities. To provide a consideration of wider urban planning endeavours in the Paris agglomeration through the twentieth century is far beyond the aims of this chapter: instead the focus here is specifically on planning thought and intervention as it has related to the representation of the city as the capital of France.[10] Although Paris may have been in part a model for the design of many more recent capital cities, the capital city functions of Paris itself were not at the centre of French planning action for much of the twentieth century.

As a capital, inner Paris is a monumental city and not simply a city of monuments, and the spaces associated with monumentalism (such as the axis from the Tuileries Gardens to and beyond the Arc de Triomphe, or the Esplanade des Invalides) were already in existence by 1900 or even earlier. Rapoport's comment that 'political meaning is increasingly communicated by single elements, fixed and semifixed, rather than the city or even parts of it'[11] applies strongly in Paris: it remained the case there into the period of the *grands travaux* (major projects) of the late twentieth century that have enhanced the city's international image. Recent additions to the Paris landscape have been fitted into the existing ground plan with little accompanying creation of vistas or prestige axes (unlike many of the changes instituted over two centuries up to Haussmann): the only means of drawing attention to their presence has sometimes been through their height.[12]

There has been similarly little attempt to create specific zones associated with capital city functions, although most of these are actually located within the wealthier western half of the inner city as a result of centuries-old organic developments. The palaces housing the offices of the President of the Republic (the Elysée) and the Prime Minister (the Matignon) are located on side streets. Major Ministries are scattered throughout the dense urban landscape. Until the construction of La Défense started in the 1950s, the city's central business district had not generated any planning or architecture to demonstrate its wider prestige. Cultural and educational facilities (such as the national libraries, art galleries, and academies) can be found almost anywhere. Indeed, for much of the twentieth century little was added to the city's built environment to reflect its capital status: the legacies of earlier centuries were in many ways the most obvious testament to the importance of Paris – and of France.

However, Paris is not France. The tension implied in this simple statement explains much of the inaction in capital city planning through the first sixty years of the twentieth century. The remainder of this chapter on Paris as a twentieth-century capital city deals with two themes: first, the general stagnation of planning activity up to 1960; and second, the new strategic thinking of the last forty years which has made Paris one of the more strongly planned urban areas of the world. The explanation of the evolution from the first to the second period lies in discourses about the relationship between France as a whole and its capital city.

Paris and France

The planning of a capital city has a number of aspects to it, of which three are of significance here. The first concerns the operation of the city itself as a complex metropolitan unit, such complexity being greater than in other cities precisely because of the capital city function. As already indicated, this aspect is not considered in detail here: many highly significant developments relate only indirectly to the capital city role and have resonances within other major agglomerations.

The second aspect concerns the relations of the capital with the rest of country. The third aspect, of increasing importance in a period of enhanced globalization, relates to the international competitiveness of the capital within the network of global cities. In each of these aspects, strategic thinking can play a formidable role in articulating the development agenda, in creating mechanisms to ensure implementation, and in projecting the needs of coming decades. Taken together, these three aspects underlie the aims (or lack of aims) in planning for Paris during the twentieth century. They have fluctuated considerably in their relative significance. The relationship between the capital and France has been particularly important.

Paris was a problematic city for the Third Republic (1871–1940). Paris and its inhabitants had been tainted by what had become seen as the excesses of Napoleon III's regime, and by criticism of the works undertaken during the Second Empire (1852–1870).[13] In Olsen's words,

Paris had 'deliberately been made the expression of the values of the discredited Empire'.[14] Its image had been further damaged by the periods of revolt throughout the nineteenth century that had culminated in the Commune of 1871. Already in that year a commentator observed 'there is less of France in Paris than one thinks. Paris forms a separate nation, and thinks of itself as the capital of the world rather than as the capital of France'.[15]

Haussmann and his Emperor had viewed Paris as belonging to France rather than to its citizens, and argued that if Parisians were consulted about great urban projects nothing would get done.[16] After French local government reform in 1884 and 1887, Paris was the sole *commune* in France that was not allowed to elect its own mayor – something that remained true until 1977 when Jacques Chirac won the right to occupy the *Hôtel de Ville*. Even in the later years of the twentieth century, the state retained the power to approve, by decree, the strategic plans for the Paris region as a whole, but within a France that was more urban in orientation and mentality.[17] The 1975 legislation which created the mayor's office for Paris, along with other local reforms, marked the final (relative) independence of the city from state power. Decentralization measures introduced at the national level by the socialist government in 1981 further strengthened the powers of the Paris region, but without removing all state tutelage.

The political elites of the Third Republic sought to restrain Paris (or not to encourage its accretion of further power), arguing that such restraint would 'benefit' the provinces. Such reasoning was well-established in French thinking, and its influential climax came during the Fourth Republic (1944–1958) in the publication of Jean-François Gravier's 1947 study *Paris et le Désert Français*.[18] This argued that Paris was draining the rest of France through population migration, capital flows, entrepreneurial initiative, and institutional control, and that it had been doing so since at least the first third of the nineteenth century. These arguments underpinned much of the new system of French central planning, from the first national plan of 1947–1953 onwards, attempting a 're-equilibration' of the relationship between Paris and the French provinces. When attention was finally devoted to the Paris Region, the first major regional planning scheme (of 1960) was imbued with the need to contain the growth of Paris.[19]

Only with the creation of the Fifth Republic in 1958 did attitudes to Paris and its region begin to change, these becoming fully apparent in planning thinking from 1965 onwards. Expansionist Gaullist views held that the interests of France were best served by enhancing the international role of Paris, by its transformation into an efficient modern metropolis, and by seeking to couple the international prestige of the country and its capital. The first appearance of a 'world city' theme in actual planning thinking concerning Paris (as opposed to theoretical exercises) occurred in the *Schéma Directeur d'Aménagement et d'Urbanisme* (SDAU) of 1965, which began the emphasis on Paris as a major global player of European, if not wider, significance.

Whatever the relationship of Paris and France, the capital city region continued to grow in population throughout the twentieth century, and planning interventions were called for (but not always forthcoming) to deal with the issues of growth. Figure 4.1 indicates the growth of the resident population of both the City of Paris itself and of the wider agglomeration of which it formed the central part.[20] In 1901, 75 per cent of the population of the agglomeration lived within the city, but over the following century massive suburban growth around Paris reversed the balance completely.[21]

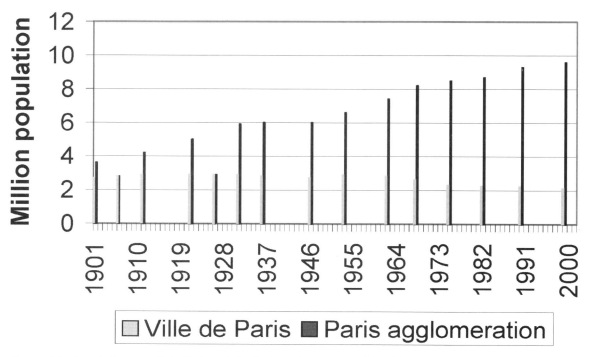

Figure 4.1. Population growth in Paris and the Paris agglomeration, 1901–2001.

Throughout the twentieth century the suburbs and outlying parts of the agglomeration were the sites of greatest development. Planned intervention in the nineteenth century had been confined to the city. Although the boundaries of the *commune* of Paris were extended in 1859, under Haussmann,[22] they have never again been revised, so that the administrative City of Paris (the *Ville de Paris*) covered the same spatial extent in 2000 as in 1870. As Sutcliffe has observed, 'A government which fears the independence of the capital will naturally hesitate to entrust a larger area to it'.[23] Increasingly, the vast unplanned suburban expanses with highly fragmented administrative structures came to dominate the region, and the perceived problems of these areas loomed ever larger. A major driving force behind the creation of real planning policies for the whole Paris Region in the early 1960s lay in concerns about the efficient operation of the Paris suburbs, within a context of projected large-scale future population growth.

The Years of Relative Inaction, 1900–1958

Haussmann's demission from office left a number of projects unfinished, and the following fifty years saw continued activity in some areas (see figure 4.2).[24] During the early years of the twentieth century some thought went into further intervention in the city, notably under the government architect Eugène Hénard between 1903 and 1909, but the outcomes were extremely restricted (as were all developments in Paris up to the Second World War) by a reluctance on the part of government to set up suitable financial measures.[25]

With no overall vision from the state for either the City of Paris or its rapidly growing suburbs, the first half of the planning history of Paris in the twentieth century can be characterized as an era of ideas, projects and pragmatics. Much of the thinking was highly advanced in planning terms, but only a small proportion of the projects ever came to fruition, and then generally only in limited zones of activity. Among the major schemes were Le Corbusier's *Plan Voisin* of 1925 (which was not implemented), followed by his further plans exhibited at the 1937 World Fair (again not carried through), and Henri Sellier's *cités-jardins* (which were actually built).[26] Most thinking in the inter-war years was dominated not by concerns for Paris's capital city roles but by the prevailing physical circumstances of the city itself – a city that had been largely neglected since the inception of the Third Republic and in which slum housing, poverty and overcrowding were the daily reality for substantial sectors of the population, both in the inner city and in the burgeoning suburbs beyond the city boundary.[27] Given the continuation of wartime rent controls, landlords were uninterested in redevelopment.

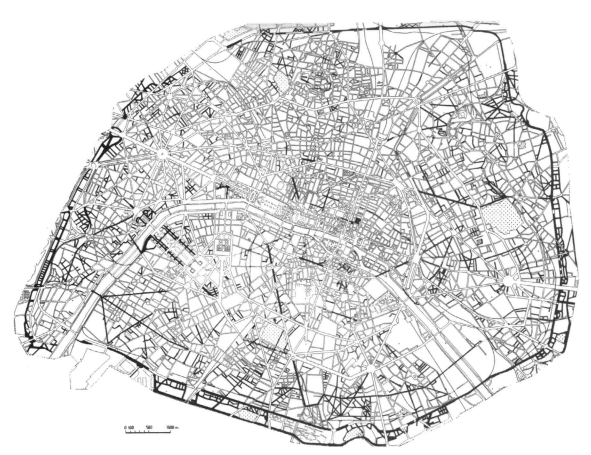

Figure 4.2. Haussmann schemes completed during the twentieth century (indicated in bold on the map).

Further issues concerned finance and control, since the French Senate was unwilling to vote money for projects in Paris, and the State Council was very restrictive in interpreting rights of expropriation for the public good.[28] It was difficult to argue the public interest in the absence of an overall strategy.

Architectural and planning competitions are a distinct Parisian tradition, dating back to the 1870s and still operating today. A major series of these was held in 1919–1920, but little happened to implement the prize-winning entries.[29] The Paris World Fairs of the post-Haussmann period – in 1878, 1889, 1900 and 1937 – all had the potential for enhancing capital city visions. However, each was centred in the same general area of the city along the Seine downstream of the city centre, and resulted in the addition of important individual buildings to the city,[30] rather than radical restructuring or large-scale redevelopment; nor did they act to rebalance urban space or power.[31]

The first effective strategic planning thinking for the Paris Region started with the creation, in 1928, of the *Comité Supérieur d'Aménagement et d'Organisation Générale de la Région Parisienne* (CARP) alongside legislation of the same year concerning suburban land division.[32] This led to the 'Prost' Plan of 1934 (see figure 4.3), approved in 1939 and operationalized in 1941, which remained in force until 1960 and which spawned a series of detailed plans at local and inter-*commune*

Figure 4.3. The Prost Plan of 1934.

level up until 1971. This, like the PADOG (see below) that followed it in 1960, aimed to restrict urban sprawl around Paris by limiting physical expansion.[33] Development in each *commune* was to be limited to what that *commune* could afford in terms of urban facilities, but with an overall regional plan also being drawn up.[34]

Unlike certain other French cities such as Le Havre, Caen or Dunkirk, Paris suffered little damage during the Second World War. Central funds for post-war reconstruction therefore went elsewhere. Arguments favouring the provinces continued, having been accentuated under the Vichy regime. But whilst the leaders of the post-war Fourth Republic favoured the interests of *la France profonde* (rural and provincial France), the people of *la France profonde* were leaving their roots and migrating to the big cities, and particularly to Paris, during precisely this period. Between 1946 and 1975 the share of the total French population living in the expanding Paris agglomeration rose from 11.5 per cent to 16.3 per cent.[35] This brought a massive burst of housing construction, but in a manner that was unrelated to any strategic plan for the managed growth of the capital since there was none – indeed, at the heart of strategic thinking lay the assumption that Paris should not grow in size or power.

The rapid growth of the suburbs created considerable problems over housing standards, infrastructure provision, and economic and social balance within the Paris region. It also provoked political concern, particularly through the accentuation of the long-standing juxtaposition of a relatively wealthy inner city (the City of Paris itself) and an underprivileged Communist 'Red Belt' in the surrounding industrial suburbs.[36] These growing anxieties led to the first integrated planning statement for the Region since the Prost Plan of 1934. The new strategy was approved in 1960 as the *Plan d'Aménagement et d'Organisation Générale de la Région Parisienne* (or PADOG), the fruit of planning legislation that had constituted some of the first measures of the new Fifth Republic that had been created in 1958. The PADOG still bore testimony to earlier thinking, with its desire to turn the tide of population growth in the Paris region. Certainly, it was now recognized that putting a lid on the growth of Paris was implausible but, in a move that was to have important direct and indirect consequences, a growth pole strategy was suggested based on the development of a major new area of high-density land use at La Défense in the western suburbs – in many ways the lasting legacy of the plan. However, within a year of the approval of PADOG came a new turn in national thinking about Paris, related to the political philosophy that has come to be known as 'Gaullisme' after Charles de Gaulle, President of France between 1958 and 1969.

Planning Paris under the Fifth Republic, 1958–

With the inauguration of the Fifth Republic in 1958 the political character of France took on a new and forward-looking direction, with a determination to solve some of the problems that had bedevilled the Third and Fourth Republics (1871–1940 and 1944–1958, respectively). Amongst the aims of the new leadership there was now a palpable desire to consider the future of one of Europe's greatest cities. The systems that were set in motion were to lead to drastic change in the character of many aspects of life and landscape within the Paris agglomeration as a whole, yet had fewer effects on the urban structure of Paris as a capital city than might have been expected. Central government control remained strongly in evidence, and it was not until the early 1980s that a Socialist president introduced what some have seen as half-hearted measures of political decentralization.

Just as with the relationship between Napoleon III and Haussmann, another pairing of political leader and bureaucrat was crucial. De Gaulle in 1961 appointed Paul Delouvrier to the post of Delegate General for the Paris Region (see figure 4.4) – a newly-formed post resulting from the creation of a regional administrative structure for France, with the Île-de-France region consisting of the Paris Basin including rural land up to 100 km from the heart of the city. In 1966, Delouvrier added the post of Prefect to his responsibilities, with control over the budget of the whole region.[37] Delouvrier was actually forced from office in 1968, but by then he and de Gaulle had set Paris on a new course.

Delouvrier set about defining a new strategic vision for Paris and its region, to include mechanisms for its modernization and sustainable growth within a context of some decentralization from what was seen as the congested heart of the city to the under-provided suburban realm.[38] He borrowed Gravier's device of contrasting Paris with the French 'desert' (see page 41), but instead compared the capital city to the suburbs that surrounded it – the product of sixty years of unplanned development. However, Delouvrier's thinking also brought a change from old objectives of containment. Urban growth in the Paris region was now to be accepted and planned for, even in the face of criticism from those supporting provincial interests.[39] Major proposals included the creation of a series of new towns; new motorway construction; the establishment of the

Figure 4.4. Paul Delouvrier on the right, with a colleague.

RER network of cross-regional rail services; the addition of further suburban development poles to the one at La Défense envisaged under the PADOG; and the identification of protected zones within the built-up area.[40] The agglomeration was to be shaped at a large scale, to operate along a series of axes parallel to the Seine. As a hundred years earlier, Parisian planning came to the fore as an international model.[41] But although the capital city aspects of Paris were starting to play a role as a context, particularly to a Gaullist government keen to establish the image of France's global power status, the principal concerns were still those of the internal operation of the Paris agglomeration itself.

The urbanist and geographer Pierre George, in a theoretical essay clearly reflecting discussions about Paris, suggested that in planning for an efficient modern world capital one possibility was to remove from the inner city all the functions related to the servicing of the city itself and disperse these to a series of decentralized poles within the wider urban environment.[42] In their place the inner city would be given over to national and international functions. George's commentary can be read as an attempt to rationalize the suburban development poles created under the 1965 *Schéma Directeur d'Aménagement et d'Urbanisme* (SDAU).[43] In addition to service activities for the city, many of the 'capital city' functions of Paris have over the succeeding years been relocated to the suburban poles, including major offices (and ministries) at La Défense and in one or other of the new towns, and higher education and research institutes in the suburban Massy-Saclay axis in the south. Various fiscal measures and building regulations have encouraged such suburbanization including building height restrictions and higher taxes on inner-city office space.[44] However there has never been any suggestion that a new capital-city district should be created outside the historic core, and the vast majority of Paris's national- and international-level functions remained scattered within the city centre, as they have been for centuries.

The detailed urban plans for the City of Paris, published in 1980, made clear the aim of limiting inner-city employment growth.[45] Big firms, and government itself, had little choice but to decentralize any growth plans, particularly to La Défense and the new towns.[46] The only objective referring to the capital city role started by indicating the importance of preserving the character of Paris, with further mentions of the quality of urban space and of safeguarding the city's image.[47]

As well as its concerns with the Paris Region as a whole, the Fifth Republic introduced, effectively for the first time since the 1860s, a real interest in managed change in the inner city – the Ville de Paris itself – but within certain parameters. As with Haussmann and Napoleon III, this went beyond simply operational issues and embraced major prestigious projects to benefit the image of the city, and of France. Increasingly, the means for doing so has concentrated on the 'cultural' rather than the political or economic dimensions of a capital city role. Successive heads of state have steered particular Parisian projects, but it was under Mitterrand (1981–1995) that the prestige of France and of the Presidency, within an expanding European and global context, came to greatest prominence as an influence on capital city planning. As Ambroise-Rendu observed in 1987, even before all Mitterrand's *grands projets* had been initiated: 'Four presidents have done as much to the city as two emperors in the last century'.[48]

A number of sometimes complementary and sometimes competing forces have been at work in shaping strategic intervention within the City of Paris. These include capitalist land exploitation to secure higher rents;[49] technical solutions to perceived problems; the manipulation of the social composition of the population through renewal and related schemes;[50] preservation and landscape conservation; and the development of prestigious symbols for the aggrandisement of the reputation of the instigator (notably the President of the day), of the city, or of France as a whole.

From the very start of the Fifth Republic, decisions were taken that would transform parts of the inner city and bring massive demolition and reconstruction work. The removal of Les Halles markets is a classic example, where new powers were used to initiate action, but with long delays before outcomes could be agreed on.[51] The 1960s have been demonized as the period in which much of the old character of Paris was swept away,[52] but an obvious riposte would be that Paris had stood still for too long, and that a period of planned intervention was manifestly needed to improve the functioning of the city and to re-establish it as an efficient capital for a world power. But conservation was also on the agenda, with the 1962 'Malraux Act' providing for the creation of conservation zones (*secteurs sauvegardés*) within cities, and supplementing fragmentary legislation of 1840, 1913 and 1930 on historic sites.[53] Within Paris, the implementation of such zones has partly been seen as a further cause of social change at the neighbourhood level, particularly in the Marais, which was designated as a conservation zone in 1965.

In order to manage the strategic planning of the City of Paris, the Atelier Parisien d'Urbanisme (APUR) was created in 1967, its initial brief being to bring forward a *Schéma Directeur* for the city itself. As usual, however, such a task was not entrusted simply to Parisian interests: the funding of APUR was initially set at 42 per cent by central government, 42 per cent by the Ville de Paris, and 16 per cent by the Île-de-France Region. After 1978 this changed, with the Region refusing to support an organization that was devoted to planning within only one of its *départements*: state funding was reduced to 25 per cent with the other 75 per cent being borne by the Ville de Paris.[54] APUR proved an effective instrument for strategic thinking about the City of Paris, firstly under the Presidents of the Republic and later, after 1977, under Chirac as mayor. Its Director between 1968 and 1984, Pierre-Yves Ligen (see figure 4.5), has been hailed as 'Haussmann II' because of the breadth of his organization's activities.[55] Nevertheless, Ligen's achievements largely related to the functioning of his patch of the agglomeration (the inner city of Paris itself) as a great city rather than as the French capital.

Georges Pompidou, President from 1969 to 1974, came closest to a complete transformation of inner Paris, with his desires to modernize the city through the adoption of 'vertical urbanization' and the construction of high-rise buildings that had been precluded by existing building regulations.[56] He also promoted the conversion of the Seine embankments into vehicular expressways, through concern to adapt the city for the car. Had Pompidou lived longer the course of thinking about Paris might well have changed, towards the enhancement of greater functional and zoning concentration within the city, and the facilitation of car transport. Pompidou was less interested in the traditional image of the city centre than others before him, in some ways harking back to Le Corbusier's unimplemented 1925 *Plan Voisin*. A few weeks after Giscard d'Estaing's election as President of the Republic in the spring of 1974 he halted most of Pompidou's plans (at the cost of compensation to the developers), including the left-bank expressway.[57]

Figure 4.5. Pierre-Yves Ligen, Director of APUR, 1968–1984.

There was also continued rivalry over Parisian issues between the President and the municipal council[58] – until 1977 under the tutelage of a state-appointed Prefect, and after 1977 under its elected mayor. Under Giscard these tensions came to the fore. Giscard's ideas for Les Halles were opposed by the city's elected council, resulting in stalemate until the state withdrew its interest from the project in 1978. This was seen as a victory for the city over the state although, as a *quid pro quo*, the city council agreed to stay out of another of Giscard's projects – the Cité des Sciences et de l'Industrie at La Villette (figure 4.6). The struggles of the period 1974–1978 would probably not have occurred in a 'normal' city.[59]

The election of François Mitterrand as President of France in 1981 gave a particular boost to central government concerns with the planning

Figure 4.6. La Cité des Sciences et de l'Industrie.

of Paris for European and 'world city' roles. The 1980 draft SDAU for the City of Paris had already talked about 'reaffirming its influence as capital'.[60] Stress was placed on administrative functions (both nationally and for international organizations), education, culture, and on the needs for transport into the city from the airports, and the importance of the hotel sector. However, paradoxically, it was also under Mitterrand that steps were taken to boost decentralization in France, such that the state's hold on the fortunes of Paris were weakened.

During the early 1990s the wider comparative position became much more clearly articulated. The 'White Book', a consultation produced jointly by the regional and City of Paris planning agencies in 1990,[61] set the context in its preamble, stating its vision of the Île-de-France as the greatest European metropolis. This, and the 'Charter for the Île-de-France' (*la Charte d'Île-de-France*) of the following year,[62] referred explicitly to a context involving the pressures of globalization and the completion of the European Union's single market in 1992. The role of Paris on the international stage was seen as problematic: it was noted that the Paris region had been performing poorly in, for example, the attraction of foreign direct investment and the establishment of European headquarters of major American and Japanese firms.[63] In the 'Charter', important sections were devoted to the prospects of Paris becoming the European capital, with the main competitors seen as London (with its financial market five times that of Paris), Brussels (with its European institutions), and Berlin (which was, at the time, expected to benefit from German reunification). Accessibility questions were seen throughout the document in European, rather than simply French, terms, with access between Paris and Brussels, Frankfurt and Milan accorded as much (or more) emphasis as links to the French provinces.

Within the 'Charter', debates about the future capital city development of Paris are couched at the regional or agglomeration level, rather than in terms of the inner city – in which the major capital city functions remain largely concentrated despite some suburbanization to the suburban growth poles, mentioned earlier. Strategic thinking is still concerned with the contribution that the reorganization of the suburbs can make to the achievement of stated future goals, these now being conceived in international terms. Paris has never had the modern 'capitol complex' of major national and international functions characteristic of newer cities planned from the outset as capitals. Current thinking would lead to the further spatial diffusion of these functions, albeit within a limited number of development nodes scattered throughout the entire agglomeration. This actually represents a scenario of 'no change' from that operating over the period since the early 1960s.

In these circumstances the Presidential *'grands travaux'* or *'grands projets'* of the past twenty-five years have gained particular importance as tangible manifestations of capital city 'monumentalism'. Much attention has been devoted in particular to the activities of Mitterrand's years as President,[64] but several of his schemes were in the tradition of those started by his predecessors, or even constituted the fulfilment of their intentions (see figure 4.7). For example, a project for the completion of the city's principal axis at La Défense had been chosen by Giscard in 1980, but one of Mitterrand's first presidential actions the following year was to cancel what had been proposed as too small for the character of the site. What finally materialized in 1989 was a privately-financed building, with the major decisions all having been taken by the President.[65]

The relations of the major presidential projects to the capital city functions of Paris are interesting and complex. Economic and political spheres have taken second place to culture, with clear implications that one dominant construction of Paris's claim as a world capital lies in the cultural realm. Among other outcomes, presidential initiatives over the past forty years have given rise to three major new or enhanced art spaces (the Pompidou Centre,[66] the Musée d'Orsay, and the extended Louvre), a new opera house (the Opéra Bastille), a new site for the national library (at Tolbiac on reclaimed railway yards), a complex of museums, exhibition and performance spaces on the site of old abattoirs (at La Villette), and a cultural centre of the Arab World.[67] The Institut du Monde Arabe also had a political rationale, an example of France claiming an international relationship with a geopolitical region generally seen in limited terms of self-interest by other Western powers.[68]

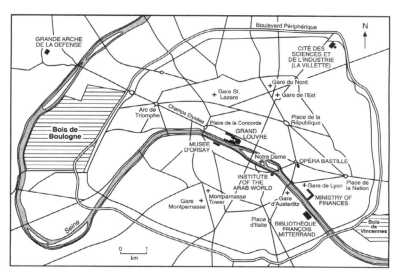

Figure 4.7. The *grands travaux* of recent presidencies.

Figure 4.8. La Pyramide du Louvre.

The dominance of cultural objectives can be seen in the 'Grand-Louvre' project, which entailed removing the Ministry of Finances from its prestigious city-centre palace to permit the extension and remodelling of the Louvre art museum (figure 4.8). The Ministry's new site at Bercy was the only major construction explicitly designed as a government building, although various ministry offices also now occupy part of the Arche at La Défense, along with a museum of the History of the Rights of Man – built to reflect the ideals of the French Revolution and opened in the bicentennial year.

These major projects are dotted around the city, some in prestigious sites, others not. Only La Grande Arche at La Défense (figure 4.9) makes a significant contribution to the established urban image of the city, providing a centre-piece to the development pole initiated in the late 1950s. Links into the overall urban infrastructure are not always perfect, for example at La Villette. On the other hand, the new national library has been provided with an additional *métro* line, and the Opéra Bastille occupies a space on one of the major historic intersections of the city – although not one that has previously been seen as a prestigious location. Only the Grand Louvre and the Pompidou Centre are located in the heart of the city centre. The projects are not intended to create an ensemble, but to provide animation to particular quarters and thus play a role for local development within Paris, with a particular emphasis in several cases on the historically poorer eastern quarters of the city in which they have been intended to stimulate regeneration. In some ways they continue a long evolutionary tradition in the city involving the creation of major monuments and national functions scattered throughout the built environment – as with the Arc de Triomphe, the old Opera House, and Les Invalides of the previous two centuries. However, their wider contemporary role in enhancing the overall reputation of Paris as an international capital of culture is also clear.

The very end of the twentieth century may have marked a turning point in the involvement of central government in major projects in Paris. The period of the presidential schemes appears to be at an end, with Chirac (first elected President in 1995, and re-elected in 2002)

Figure 4.9. La Grande Arche at La Défense.

showing less interest in making his mark on the city as president – having already done so as mayor of Paris between 1977 and 1995. Effective decentralization of planning powers to local authorities has now taken place. The Stade de France built at Plaine St Denis (just outside the northern boundary of the City of Paris) for the 1998 soccer World Cup was a 'capital city' project designed to showcase France on a world scale, and given greater symbolic importance through the victory of the French team. But the stadium's construction showed that such a project now required intensive collaboration and compromise between central and local state elements.[69]

Increasing concerns over Paris as a global capital have not, however, diverted attention from regional imbalances, the efficient operation of the city region and the detail of local planning policies.[70] The 1994 *Schéma Directeur* (see figure 4.10) envisages the population of the capital region as rising to around 11.8 millions by 2015,[71] and designates the prime zone for urbanization as the *département* of Seine-et-Marne to the east of the City. Not only is this area still relatively under-urbanized and under-developed (despite the presence of Marne-la-Vallée New Town) but it is also on the side of Paris that can be best connected to the rest of Europe through motorway, rail (TGV) and air links (from Roissy – Charles de Gaulle airport), thus according most with the international ambitions for the city. As in other capital cities, such international transport links have been the site of major recent investment with the airport and the TGV links being particular prestige projects as national gateways – although the final entries to Paris, through the ageing Gare du Nord and overcrowded motorways and rail links from the airport remain unchanged.

Continuing a theme developed in the plans of the 1960s, the future evolution of the region is seen in terms of polycentric urban space, with a number of regional growth poles and axes within the suburban realm. Four of these poles are to be 'centres of European reach' (*centres d'envergure européennes*). A fifth such designation is made of the City of Paris itself. The earlier consultation documents had drawn particular attention to the need to bolster the suburban poles in order to maintain Paris's 'pre-eminent international

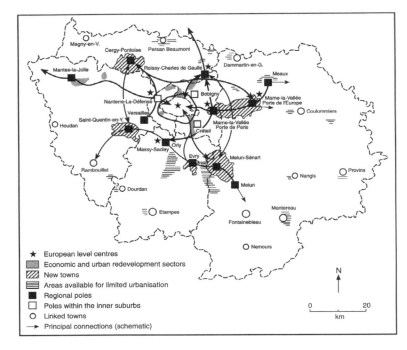

Figure 4.10. The 1994 *Schéma Directeur* for the Paris Region.

position'.[72] French planning since the 1960s, at a number of scales, has consistently utilized growth pole concepts as key elements in desires to reduce some of the great structural imbalances in all aspects of French life – between the capital and the provinces, and within the capital region between the city and its suburbs.

Conclusion

Planned intervention in France's capital city region was remarkably inconsistent over the twentieth century. Much of inner Paris remains largely as Haussmann planned it in the 1850s and 1860s. The following period was marked by inaction, with little strategic development to accommodate Paris to the twentieth century, to the demands of a rapidly growing population, or to the further enhancement of its capital and world city status. In the early post-war period the desire was still to rebalance Paris and France, but from the inception of the Fifth Republic (1958) onwards the aim has been to help Paris grow whilst seeking to ameliorate serious imbalances within the capital region.

Increasingly, attention has been paid to the need for the French capital to maintain its place internationally. Current thinking over strategic planning reflects the desire to bolster Paris's position as a major world capital, and one that can compete on the global stage – but especially with London and Berlin, seen as competitors within Europe.[73] Planners and politicians have not sought to create a functional zone of capital city attractions, instead placing new developments throughout inner Paris and thereby retaining an urban ethos that is largely unchanged for two centuries and that has given the city much of its international image. At the start of the twenty-first century, the attributes of Paris as a capital city are recognizably those that have emerged from over a millennium of evolution.

There are, however, clear signs of an increasing emphasis on culture as the sphere in which Paris has the greatest competitive advantage over its potential rivals, and state intervention has reinforced that line.[74] French politicians are acutely aware that Paris needs continued support to achieve their ambitions for it. Many of the issues to be faced in the Paris region are also the problems of the competitiveness and prestige of France as a whole. But that is in the nature of capital cities.

NOTES

1. Noin and White (1997), pp. 1–4.
2. Noin (1976); George (1998).
3. Hall (1998), pp. 937–938. Interestingly, the term has rarely been used in France.
4. Sutcliffe (1993); for Paris's influence on Latin American capital cities, see Almandoz (2002).
5. The city's functions as a seat of international organizations should not be overlooked: it houses the headquarters of the OECD and UNESCO among other bodies, as well as having been the original base of NATO. Norma Evenson has noted that 'to some observers, being the capital of France has long seemed too modest a destiny for Paris'. See Evenson (1984), p. 259. On the wider issue of a 'European capital' see Hein (2001).
6. Noin and White (1997), p. 11.
7. Sutcliffe (1970).
8. For example, see Hall (1997a); Sutcliffe (1993); des Cars and Pinon (1992).
9. Sutcliffe (1970), p. 321.
10. Readers wishing to consider the wider view of French twentieth-century planning as applied to the Paris agglomeration and region are advised to consult the following major studies on the topic: Evenson (1979); Gaudin (1985); Lacaze (1994); Pinon (2002); Sutcliffe (1970); Voldman (1997)
11. Rapoport (1993), p. 54.
12. For example, the Pompidou Centre or the new National Library at Tolbiac – where there has been extensive landscaping rather than city planning. Paris has had a series of building height regulations since 1607: see Evenson (1979), pp. 147–154; Bastié (1975), pp. 55–89.
13. Ambroise-Rendu (1987), p. 20. The role that pressure from the Paris populace had played in bringing down the Second Empire was overlooked, but the fact that they had done so was used to caution against the power of Paris in the future: on the former point see Price (2002).
14. Olsen (1986), p. 53.
15. Veuillot (1871) quoted in Marchand (1993), p. 124. For more extended discussion of the relationship of Paris to the provinces, especially during the Third Republic (1871–1940), see Cohen (1999) and George (1998).
16. Ambroise-Rendu (1987), pp. 193–194.
17. Lacaze (1994), p. 34.
18. Gravier (1947).
19. Berger (1992).
20. The spatial extent of the agglomeration, defined in terms of built-up area, grows constantly. Data shown in Figure 1 result from separate calculations for each year shown. See Noin and White (1997), pp. 19–23. The Paris Region, now known as Île-de-France, is wider than the agglomeration.
21. It is ironic to note that although the 1960s saw a change to planning for the future population growth of the Paris region, the annual rate of increase since 1962 has been slower than it was in the earlier post-war years.
22. This was carried out at the instigation of the state, not as a result of lobbying by local authorities: see Ambroise-Rendu (1987), p. 194.
23. Sutcliffe (1979), pp. 71–88.
24. The fall in property values after 1870 meant that the public authorities could complete some schemes without too much expense in terms of compensation for expropriations: see Sutcliffe (1981), p. 135.
25. Evenson (1984), pp. 273–274. The Basilica of Sacré Coeur in Montmartre (1871–1919) was an unpopular exception to this trend, representing, as it was argued to do, an atonement by the people of France for the 'sins' committed by the Second Empire (1852–1870) and the Paris Commune, see Harvey (1979), pp. 362–381.
26. See Evenson (1979), pp. 2–75 for discussion of

unfulfilled plans for Paris up to 1940; also Bastié (1984), pp. 174–181. On the garden cities, see Burlen (1987). Also Sellier and Brüggemann (1927). The garden cities were an interesting exercise in urban design, and showed French adaptation of planning ideas from elsewhere, but they were unrelated to the city's function as the French capital.

27. Hall (2002), p. 222; Sutcliffe (1970), pp. 104–112, 240–243; Evenson (1979), pp. 212–216; Evenson (1984), pp. 274–275. See also Fourcaut (2000).

28. Roncayolo (1983), p. 146.

29. See Evenson (1979), pp. 275 and 329–332. Some of the ideas in the section for regional planning for Paris probably underpinned the later Prost Plan of 1934. One of the French prizewinners in 1919, Jacques Gréber, although involved in various small-scale developments in France, is more notable for his plans for the Canadian capital, Ottawa: see Gordon and Gournay (2001), pp. 3–5; Gordon (2001).

30. Such as the Eiffel Tower (1889), the Grand and Petit Palais (1900), and the Palais de Chaillot at the Trocadéro (1937).

31. The presidential projects of the Fifth Republic (since 1958) have more recently involved major international design competitions, but again largely relating to individual buildings.

32. Bastié (1984), pp. 178–179; the Sarraut Law of 1928 was significant in curbing the worst excesses of suburban expansion.

33. Dagnaud (1983), p. 219.

34. Evenson (1984), p. 278; Marchand (1993), pp. 258–261.

35. White (1989), pp. 13–33.

36. Stovall (1990). See also Guglielmo and Moulin (1986), pp. 39–74.

37. Savitch (1988), pp. 100–106.

38. Ambroise-Rendu (1987), p. 21. A notable difference between the 1960 PADOG and the 1965 SDAU lay in the fact that the former envisaged 10 millions as a plausible population for the agglomeration, whereas the latter foresaw 16 millions: see Dagnaud (1983), p. 220.

39. Marchand (1993), p. 314; Thompson (1970), p. 216. Nevertheless, traditional official concerns over the balance between Paris and the rest of France were maintained – in particular a series of eight *métropoles d'équilibre* were identified by DATAR (the *Délégation à l'Aménagement du Territoire et à l'Action Régionale*) with the aim of counterbalancing Parisian growth.

40. See Noin and White (1997), chapter 4.

41. As Peter Hall has observed, 'If audacity is a criterion for merit in urban planning, then the Paris Schéma Directeur of 1965 must surely belong in some category by itself'. Hall (2002), p. 346.

42. George (1967), pp. 287–292.

43. The total number of new towns to be built had quickly been reduced from 8 to 5.

44. IAURIF (1993).

45. The successful restriction of growth in Paris can be seen in the fact that in the twenty years from 1975 to 1995, the increase of office floor space in the City of Paris was limited to 11 per cent, against 194 per cent in the inner suburban *départements*, 296 per cent in the outer suburbs, and 1542 per cent in the new towns.

46. DREIF (1995); Lacaze (1994), p. 59. Peter Hall has suggested that development in central Paris was held back in order to assist the growth of La Défense, but motivations were probably more complex, including the general desire to prevent the 'Americanization' of the Paris landscape. Hall (1998), p. 926.

47. SDAU (1980), pp. 74, 92.

48. Ambroise-Rendu (1987), p. 25.

49. See Godard *et al.* (1973) for the argument that this was the logic behind the large-scale early renewal schemes in the 13th *arrondissement* beyond Place d'Italie.

50. Arguably, this was a concern of Chirac's period as mayor of Paris – see Carpenter, Chauviré and White (1994), pp. 218–230. Ironically, by 2002 Chirac, as President of the Republic, found himself 'co-habiting' with a Socialist mayor of Paris, Bernard Delanoë. Chirac was not the first to manipulate the social composition of Paris in this way: Napoleon and Charles X had held similar views in the earlier part of the nineteenth century: see Rouleau (1985), p. 215. Lidgi (2001) has argued that Jean Tibéri, Chirac's successor as mayor after 1995, backtracked on some of Chirac's aims in this respect.

51. Michel (1988). This provides a detailed account of the evolution of the plans for the area.

52. Chevalier (1977).

53. Kain (1981), pp. 199–234.

54. Ambroise-Rendu (1987), pp. 264–267.

55. Michel (1988), p. 215.

56. 1967 saw a new general building height code introduced which enhanced the possibility of high-rise construction somewhat in some of the outer parts

of the City of Paris lying closest to the suburbs, but in 1974 restrictions were tightened once again in the inner city. See Evenson (1979), pp. 175–179. This period largely coincides with Pompidou's influence over the French capital.

57. Ambroise-Rendu (1987), pp. 22–23.

58. Chaslin (1985), pp. 12–13.

59. Chirac, as mayor of Paris, in 1978 declared that 'l'architecte en chef des Halles, c'est moi!': in practice that role fell to APUR. See Michel (1988), pp. 61–71, 243–244 on the whole episode. In view of more recent negative views of the Halles redevelopment as a failure, including plans in 2004 to reconstruct a large part of it, Chirac would probably not wish to be reminded of his earlier claim.

60. APUR (1980), pp. 92–96.

61. DREIF/APUR/IAURIF (1990). DREIF is the Direction Régionale de l'Equipement de l'Île-de-France; IAURIF is the Institut d'Aménagement et d'Urbanisme de la Région d'Île-de-France.

62. IAURIF (1991).

63. See Marchand (1993), pp. 373–376 for a discussion of how Paris has failed to gain foreign direct investment, and is being left behind as an international business centre. See also Berger (1992), p. 19. The most recent report on this issue was produced in July 2002 by IAURIF as *Note Rapide sur le Bilan du SDRIF*, No. 302.

64. Some commentators have argued that Mitterrand's projects were actually more about self-aggrandisement than about Paris as the capital of France – e.g. Woolf (1987), pp. 53–69.

65. Andreu and Lion (1991), pp. 570–80. See also Ambroise-Rendu (1987), p. 25 on Mitterrand's personal involvement in projects.

66. The full name is the Centre National d'Art et de Culture Georges Pompidou.

67. The only major sphere of culture that has not been provided for is that of orchestral performance, and this failure to provide Paris with a world-class concert facility has become an issue of some controversy.

68. The Institut has nevertheless had a chequered history, including the failure of certain Arab partners to provide financial support unless allowed to use the Institute for their own propaganda purposes; see Marchand (1993), pp. 350–351.

69. Newman and Tual (2002), pp. 831–843. Current discussions over the possibility of a new concert hall also illustrate this shift in balance away from the state: in November 2002 the Minister of Culture argued that the City of Paris, instead of the state, should contribute the bulk of the cost.

70. DREIF/APUR/IAURIF (1990), p. 25.

71. DREIF (1994).

72. DREIF/APUR/IAURIF (1990), p. 57.

73. On issues of wider competition over the 'capital' of Europe see Hein (2001)

74. Robert (1994) emphasizes the view that any future role for Paris as a 'world' capital is likely to depend on tourism, culture and congresses rather than on economic power.

Chapter 5

Moscow and St Petersburg: A Tale of Two Capitals

Michael H. Lang

In the twentieth century, Russia had two very different capitals; one, St Petersburg, the imperial capital, was modern and very European. The second, Moscow, was ancient and intensely Russian. The century opened with St Petersburg as Russia's capital and the seat of autocratic power of the Tsarist government and the royal court. However in 1918, after the Bolshevik Revolution, the new Soviet leader, Lenin, 'temporarily' moved the capital back to Moscow where it had been since the fourteenth century. The Revolution abolished both the Tsarist system as well as fledgling efforts to establish a Western style democracy, and imposed a Marxist-Leninist form of government. As a result, Moscow had a new and important status as capital of the world's first communist country, the Union of Soviet Socialist Republics or USSR. This chapter will give an overview of the history of the planning of these two great capital cities and the important role played by rulers such as Peter the Great and Stalin. The outpouring of Soviet plans and designs for turning Moscow into a model communist city will be assessed.

Both capitals were indelibly marked by the cruel hand of totalitarian rulers who effectively dictated their design and planning. Both capitals deserve attention because of their importance to modern planning history and the fact that Moscow with a population of 9 million and St Petersburg with 5 million are the two largest and most important Russian cities. Moscow in the first decades of the twentieth century, while no longer the capital of Russia, remained its dominant urban centre. If St Petersburg was the focus of court life and governmental ministries, Moscow retained many of its important ancient functions. The Tsars were still crowned in the Kremlin, the Russian Orthodox Church was based there and the industrial and trading functions remained. Both retained claim to the cultural heritage of Russia as represented in such notable museums as the Hermitage and Russian Museum in St Petersburg and the Pushkin Museum of Fine Arts and Tretyakov Gallery in Moscow.

St Petersburg: Origins

St Petersburg was an entirely new city, started in

1703 by the autocrat Peter the Great (1682–1725) as the new capital of the Russian Empire. He was Russia's first modern planner. His motto was 'For a new, ordered state, an ordered capital'.[1] He set the pattern of strict control of urban development by the state. Plans for Russian cities were mandated and had to be drawn up according to strict guidelines delineated in the Building Statute and the relevant sections of the Complete Collection of Laws as early as 1649. St Petersburg was to be a 'paradise' whose splendour was intended to surpass anything Europe had to offer. Many feel that he and the other Russian rulers who followed in his footsteps, particularly, Catherine the Great, succeeded.

From the outset, Peter was engaged in all facets of the city's design and planning including choosing its location. He is famously quoted as saying, 'The City will be here'. Pushkin's poem, 'The Bronze Horseman' made the emperor forever infamous as the flood-prone city's chief architect. Peter plotted its canals and even its main avenue, the famous Nevsky Prospect (1715). His edicts covered all matters connected with the city's design such as model houses for 'noble', 'wealthy' or 'common people'. Under his eye were laid out the famous Fortress of Peter and Paul, the Admiralty buildings with their gleaming, majestic spires, the Twelve Collegia (governmental ministries) and the signature ensemble surrounding the Summer Gardens. The latter was to set the tone for the design of the city for the next one hundred and fifty years.

Thus, Peter controlled the essential aspects of the city plan that was based on the three ray street plan focused on the Palace Square (figure 5.1), the Admiralty buildings on the embankment of the Neva River and the Fortress of Peter and Paul on the facing embankment. In contrast to more organic Russian cities, it was to be a geometric, orderly and controllable city of straight streets and canals along the lines of Amsterdam. But Peter wanted his city to surpass anything in Europe: 'his city would soar like an eagle: it would be a fortress, a port, an enormous wharf, a model for all of Russia, and at the same time a shop window on the West'.[2] It must be noted, however, that Peter's paradise was built by the conscripted labour of serfs and prisoners, many of whose lives were sacrificed to the task.

Eventually, St Petersburg duplicated all of the functional elements common to a capital city of a major country and which had previously existed in Moscow. Universities, scientific institutes, art museums, theatres, concert halls, history museums and libraries were constructed and placed in prominent settings. As a result much of the social, artistic, intellectual and cultural life of Russia became centred on the new capital.

St Petersburg at the start of the twentieth century was a capital marked by stunning contrasts between rich and poor. Much of the squalor was due to the quickening pace of industrialization during the century's first decade. Fuelled by the demand for labour, St Petersburg's population in 1913 reached 2,125,000, which was nearly four times its population in 1864. Grappling with feverish population growth and a

Figure 5.1. General Staff Building on Palace Square opposite Winter Palace and Hermitage Museum (1754–1762).

poorly organized housing industry, St Petersburg for all the architectural grandeur at its centre, had the most poorly housed and serviced population of any capital city at the start of the twentieth century. To this were added the agonies of World War I and staggering losses at the Front. It was a situation that in October 1917 helped lead to the storming of the Winter Palace (figure 5.2) and the Bolshevik Revolution.

Figure 5.2. Winter Palace and Hermitage Museum.

Several city planning movements flourished in St Petersburg and Moscow just prior to the Revolution. The court architect, Leontii Benois, teamed with his brother Alexandre and other artists to return to the classical aesthetic, to introduce historic preservation and establish a city planning profession.[3] Ivan Fomin, a Benois apprentice, developed plans in the classical tradition for undeveloped areas of the capital. After failing to secure support from the Duma, Leontii Benois led a team that prepared an independent plan in 1910. This plan was influenced by the 1909 Greater Berlin Plan, and started the approach towards comprehensive planning of an industrial metropolis.[4] Other Russian planners were active in the international movement advocating garden cities and de-signed a number of garden city inspired communities.[5] Although the Duma authorized new planning statutes in 1916, World War I stalled implementation.

Moscow: Origins

The planning of Moscow was quite distinct from that of St Petersburg. Moscow is one of Russia's oldest cities. Founded in 1147, Moscow developed as a walled city with the Kremlin at its centre. As it grew, additional protective walls were erected further out and roads were extended toward the gates, aimed toward distant cities such as Tver and Smolensk. Thus was set the original radial concentric pattern of the city. But the plan as such was an organic plan, the product of a multitude of *ad hoc* decisions taken by various authorities and private parties each pursuing their own interests. The importance of Moscow as a centre of Orthodox Christianity cannot be underestimated; after the fall of Constantinople in 1453, Moscow was proclaimed as the third Rome ('and there shall be no fourth'). A ring of beautiful walled monasteries marked the periphery of the city. Population grew, fuelled by trade and the growing wealth of local merchants and nobles as well as immigration from various European nations.

The walled Kremlin has long been the heart and soul of Russia and of Moscow. Rebuilt in 1495 by Ivan III, this ensemble of buildings and spaces continues to inspire wonder and awe due to its opulent designs and monumentality. Its gigantic brick walls were topped by eighteen towers and five gates, some as high as 76 metres (249 feet). Inside its huge grounds were scattered the palaces of the Tsars, ancient churches, historic monuments, arsenal and government buildings. The Tsar's family and the imperial court were housed in monumental facilities in the Great Kremlin Palace (1837), the Tarem Palace (1635), and the Faceted Palace (1485). The centrality of faith to the Russian state was manifested in the Patriarch's Palace (1656), seat of the head of the Russian Orthodox Church. Governmental administrative functions were housed in the Senate (1790). The public realm was represented by the paved vastness of Red Square, guarded by Resurrection Gate. In front of the gate was the large marker designating the centre of the Russian Empire where Russians come to have their pictures taken. Here too is St Basil's church (1561), Kazan cathedral (1637) and the trading rows (figure 5.3).[6] Also in the Kremlin complex were Alexander Gardens (1821) containing the obelisk commemorating 300 years of Romanov rule (1913) and the tomb of the unknown soldier (1967). These elements comprised the heart of an inspiring capital complex that was almost without peer.

Moscow saw a comprehensive plan drawn up in 1775 but little of it was ever implemented. Indeed, several plans were prepared over the years but little was done even after the opportunity for rebuilding afforded by major fires in 1773 and 1812. One noteworthy exception produced the ensemble that included the Bol'shoy Theatre and the square in front of it.[7] By the start of the twentieth century, Moscow was still a unique blend of Eastern and Western architectural styles. Small, wooden buildings with gingerbread detailing around the windows jostled with the imposing, stone mansions of the nobility; myriad Byzantine style churches adjoined modern buildings designed by foreign architects.

Planning the Capital for World Communism

A new epoch dawned after the Bolshevik Revolution and the relocation of the capital back to Moscow. This epoch was epitomized by Moscow's unique status; it was once again both Russia's capital city as well its most economically dominant city. But most importantly, it was now the showcase city of an emergent new country based on communist principles laid out by Marx

Figure 5.3. Contemporary view of Red Square with GUM department store (replacing trading rows) on left and St. Basil's Church in background.

and Engels. Communism's entirely antithetical attitude to capitalist principles, specifically its requirement of state ownership of the means of production and support for a worldwide proletarian revolution, ensured that what occurred in Moscow would reverberate around the world.

The Bolshevik seizure of power and the immediate nationalization of the land cleared one of the major obstacles hindering effective urban planning. Indeed, it was largely for this reason that city planners were among the few segments of educated society to welcome the Revolution. During the 1920s and 1930s it occasioned a tremendous outpouring of exciting and radical plans, designs and proposals for the new Moscow. The fever to produce the new communist capital travelled outside of Russia and attracted the attention of a host of famous architects and planners such as Le Corbusier, Frank Lloyd Wright and Mies van der Rohe. A number of constructivist buildings, such as the Izvestia building and the Zuyev Club were built, but most were not due to lack of resources during the civil war in the 1920s (figure 5.4). The future seemed full of fantastic possibilities; a 'city on springs', a 'cosmic city', and 'horizontal skyscrapers' were all mooted.[8]

Impassioned debates raged about the new proletarian society that was being formed and all aspects of its new social and economic relationships. As a result, modern Moscow became emblematic of the long sought alternative to capitalist cities and the economic system that gave rise to them. And as the capital of the USSR, it was to be much more than Paris or London, it was supposed to be a workers' paradise, nothing less than the perfect egalitarian city of the future. As a result, its leaders understood that modern Moscow had to be developed as both an evocative and inspiring capital, as well as an urban environment that offered a superior way of life to its citizens. Thus, the adequate provision of new housing forms became as central to its planning as did monuments, boulevards and governmental ensembles.

Moving the capital proceeded chaotically. By 1918, Lenin, his family and his close associates were ensconced in the Kremlin and soon the huge Kremlin complex was filled with military and administrative offices. As a result, other government officials and their staffs, fresh off the train from St Petersburg, engaged in a competitive struggle for space in the streets and precincts near the Kremlin. There were no purpose-built buildings available. Instead, hotels, offices and mansions vacated by departed corporations or

Figure 5.4. Monument to the Third Internationale by V. Tatlin, in Soviet abstractionist *avant-garde* style, was also designed as a functional building proposal with revolving internal volumes suspended on cables, in studio, Petrograd (Petersburg), 1919.

the wealthy were commandeered; for example, the old Nobles' Club became the House of Unions. In what would become the most hated symbol of communism, the Cheka (KGB) took over a former insurance building on Lubyanka Square, a few blocks from the Kremlin.[9] The shortage was such that Lenin was forced to use the Bol'shoy Theatre to hold communist party meetings. Moscow's city council or Duma was soon superseded by the Moscow Soviet. This new local administration came to be dominated by the central government.

Stalin's Socialist Capital

The years after the Revolution were difficult ones for the USSR. It was only after the death of Lenin in 1924 and the ascendancy of Joseph Stalin (1879–1953) that various 'socialist reconstruction' projects were realized. Most of these projects are noteworthy because of their monumental size. All required Stalin's approval. Often he would drive through the city at night with his bodyguards inspecting building projects and issuing detailed instructions. He was held in such fear that one building was erected with a disjointed façade; the architects had submitted two versions for his approval and he mistakenly had approved both. Rather than risk approaching him again, they built both versions.[10]

That Stalin was the master planner of Socialist Moscow was beyond question. According to Lazar Kaganovich, Stalin became involved first with the need to improve city infrastructure and services and '. . . Comrade Stalin kept enlarging the boundaries of the discussion until it got to the desirability of a general plan for the rebuilding of the city of Moscow'. To carry this out, a special Central Committee Commission was established. Stalin reportedly was an active participant in all sessions, listening to expert presentations and making suggestions (figure 5.5). The result according to Kaganovich was '. . . old Moscow becoming Stalin's Moscow'.[11]

Under Stalin, Moscow saw the destruction of many of its oldest buildings, particularly churches, in violation of Lenin's pronouncement

Figure 5.5. Late 1940s painting of Joseph Stalin making map designations with members of the Politburo. Members include Kaganovitch, (standing on the far right) and Khrushchev, (standing at door).

calling for the protection of all ancient buildings. Stalin also expanded the official government policy of atheism. As a result, many of the signature religious buildings in and around the Kremlin were demolished. In the Kremlin, a monastery and a convent were destroyed in order to build the Presidium (1929), the headquarters of the executive arm of the Soviet parliament. Later, during the Khrushchev era, a large modern glass and steel office building, the Palace of Congresses (1961), was constructed within the Kremlin for communist party conferences. In Red Square, the ancient Kazan cathedral was demolished, as was the Resurrection Gate. The monuments of modern Russia were added: Lenin's massive tomb, the graves of John Reed and other selected heroes of the Revolution and World War II. Much of the destruction in Red Square was designed to facilitate the massive display of military might during the May Day parades that became emblematic of the Soviet state (figure 5.6).

Most symbolic of Stalin's new Moscow was the demolition in 1931 of the huge cathedral of Christ the Redeemer, off Red Square. It was removed to make room for a monumental Palace of the Soviets, which was to be the centrepiece of the Soviet capital. This was to have been a colossal building with a tower soaring 315 metres topped by a statue of Lenin 100 metres high – three times the size of the Statue of Liberty in New York harbour. Its conception was based on the cold war tensions and the quest for international dominance. It was no accident that it was to be taller than the Empire State Building and larger in volume than the six biggest New York skyscrapers combined (figure 5.7). The Palace was never built due to site problems and a municipal swimming pool replaced what was to have been the world's largest building set within the world's largest plaza.[12]

In the years following the Revolution, several notable civic projects were built, including

Figure 5.6. Painting of May Day military celebrations in Red Square by K.F. Unon, 1942.

Figure 5.7. Winning entry of the Palace of Soviets by B. Yofan, V. Gel'freikh, and V. Shchuko, 1932, set the tone for Soviet Architecture into the 1950s.

the Volga navigation canal and the Moscow underground. The latter, in particular, stirred civic pride due to its heroic scale and rapid construction and the lavish decorations of the stations.[13] The stations were illuminated with massive chandeliers with the walls decorated with ceramic tiles, and artwork commemorating historic events (figure 5.8). The Moscow underground, bus and trolley systems were part of Stalin's vision for an urban, high-density, metropolis, like New York. A similar, but smaller, system was constructed in St Petersburg.

Monumental boulevards were another Stalinist initiative. Soviet planners straightened and widened many of the radial streets leading to the Kremlin. A high price was paid. Architectural historians decried the fate of classical Moscow, much of which was lost by road widening and housing developments that affected ancient streets such as Tverskaia (Gorgii) in the 1930s and the new Kalinin Prospect that cut through the old Arbat district in the 1960s. When informed of opposition, Stalin instructed that demolitions be conducted at night. These boulevards constitute an unwavering commitment to monumental style planning traditions. This commitment is all the more remarkable due to the low per capita car ownership that prevailed in the USSR in the 1930s.

Whole sections of Moscow were rebuilt according to Stalin's notions of civic design, he ordered that the boulevards be lined by

Figure 5.8. Moscow Underground Station.

large apartment blocks so as to complete the monumental panorama viewed as one rode in toward the Kremlin (figure 5.9). The apartment blocks were designed in a neoclassical style, built around large, open communal courtyards, often referred to as a 'superblock'.[14] These imposing and commodious apartments were reserved for favoured members of the Communist Party, the military, sports figures and the like.[15]

In addition, Stalin worked to rid Moscow of its vestiges of the 'large village', demolishing numerous one-storey wooden houses near the city centre. This was controversial since a number of these districts represented the older picturesque

Figure 5.9. Stalinist neoclassical architecture.

Russian style. Many average Russians bemoaned the loss of their 'wooden Moscow', but, of course, they were not consulted in such decisions. More telling was the strong role of the Communist Party and its ideological approach to urban design, a generalized and shifting approach often called 'socialist realism'.[16] Many existing squares were enlarged and lined with new civic buildings crowned with portraits of revolutionary heroes. Large statues and monuments to revolutionary heroes, literary and artistic figures completed the ensemble.

Perhaps the most visible symbols of the Stalin era were the seven monumental skyscrapers, which dominate the skyline to this day. They were built after World War II under Stalin's orders to show off to visiting dignitaries. The buildings' bulky wedding-cake style represented his personal design preferences. The skyscrapers housed several government ministries, hotels and apartments. The towering design for Moscow University, another Stalinist project, was even more lavish. Set on a commanding hillside site, with the city below, it continues the Russian tradition of dominating design ensembles.[17]

Stalinist planning approaches, such as wide boulevards and large blocks of flats serviced by an underground transit system, were later used in St Petersburg's suburbs and many other cities in communist Eastern Europe. In contrast to Moscow, St Petersburg's historic central precincts were preserved and completely rebuilt after the devastation of the 900-day siege suffered during World War II.

Socialist Reconstruction

Kaganovich's 1931 monograph, *Socialist Reconstruction of Moscow and other Cities in the USSR* is perhaps the best statement on planning Moscow under Stalin.[18] Kaganovich was an old Bolshevik and confidante of Stalin who organized the building of the Moscow underground. His book contained his report on the planning of Moscow and the resultant resolution that was passed by the Central Committee of the Communist Party in 1931 (figure 5.10).

Kaganovich's report was hardly a stirring call for planning a new capital; rather, it featured a frank assessment of Moscow's problems, especially housing.[19] He knowledgeably described the need for comprehensive city planning as an

Figure 5.10. The cover of L.M. Kaganovitch's report on city planning featured an aerial view of Moscow with both old style housing and modern multi-storey housing developments. Superimposed was the long shadow of Lenin's figure, arm imploringly outstretched, hand pointing to the future city.

integral part of the process of building the new 'proletarian' socialist capital. His report gave a clear overview of the comprehensive planning needed and the goals to be achieved in such areas as housing, streets, sanitation, energy, transportation, education, health services, etc. Missing, however, was any significant analysis of Moscow's capital functions, either symbolic or practical, and how they would be maintained or reconfigured under socialism. Indeed, the Kremlin was not even mentioned.

Instead, Kaganovitch focused on housing since he felt this was the crux of the matter to chart the course of the new communist capital. He advocated strict limits to growth by prohibiting new industrial development and called for a capital city set within a planned hierarchical system of cities. This approach was used to prepare comprehensive plans for Moscow and St Petersburg in 1935.[20]

Throughout the communist period, broadly conceived garden city thinking played a surprisingly powerful role in the planning process. While a number of interesting communities were built along these lines, by and large the garden city notion was rejected for Moscow.

Lev Mendelevich Perchik, like so many communist officials, imbued the 1935 plan with both patriotic and propagandistic ramifications:

> Every clause . . . speaks of only one thought, one desire: to improve in every way and to enhance to the utmost the well-being of the toiling masses of the Red capital of the glorious socialist fatherland, to make Moscow a city worthy of its great title-capital of the USSR . . . new Moscow – Soviet Moscow – is a world centre, a flourishing socialist city, the international capital of the workers and toilers of all lands, it is the dream city of all who are oppressed and exploited.[21]

The 1935 Moscow Plan

All of this broad strategizing found expression in the 1935 plan for Moscow which demanded a halt to all experimentation in planning and moved to establish principles common to all socialist cities (figure 5.11). These were:

- limited city size;
- state control of housing;
- planned development of residential areas (the superblock and micro-region);
- spatial equality in the distribution of items of collective consumption;
- limited journey to work;
- stringent land-use zoning;
- rationalized traffic flow on a hierarchy of new roads;
- extensive green space (parks and city greenbelt);
- symbolism and the central city (May day parades);
- town planning as an integral part of national planning.[22]

Post 1935 Planning

Moscow's planners drew up comprehensive plans in 1971 and 1989 that continued and developed the thrust of the 1935 plan. The 1971 Moscow Genplan put the focus on the establishment of eight 'town regions' with populations from 650,000 to 1,340,000 within Moscow, each a self-contained, serviced, sub-centre of the city, akin to the borough form of government in New York or London.[23] Another notable feature was the plan's emphasis on public transport development rather than highway development.[24]

The 1989 plan called for extensions to the city's infrastructure such as new subway lines and other public works. It predicted that Moscow's population would move upward to 9.5 million in 2010. Garden city thinking was again represented by the proposal to develop a 'system of satellite cities' which would catch the overspill

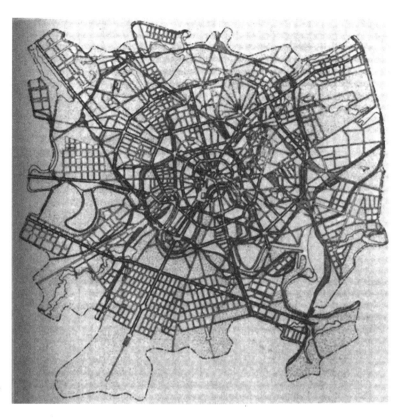

Figure 5.11. Moscow Master Plan, 1935 by L.M. Kaganovitch.

population. The greenbelt was to be expanded. All polluting factories were to be relocated. However, the changing political climate under Gorbachev and Yeltsin led to the abandonment of this plan.[25]

Failures of Communist Planning

The failure to cap the population growth and the failure to provide adequate housing accommodation were the two most signal failures of planning Moscow. All of the plans for Moscow up to the present, called for limiting its population. Moscow's planning apparatus, for all its size and supposed authority, lacked the power to zone or otherwise control the development of the city.[26] Most of the major building activities such as housing, day care facilities etc. were undertaken by individual industrial enterprises, often with the enforced labour of political prisoners in the Stalinist era. These enterprises reported directly to central ministries and functioned independently from city plans, much like public authorities and special districts in America. It is a situation that persists today although the situation is changing due to privatization of land and the influx of large domestic and foreign development interests eager to put up large projects.[27]

Clearly, the failure to control the growth of the population had a deleterious effect on the integrity of the greenbelt around Moscow. The 1935 plan had established a 10 km wide swath of open space around the entire city.[28] The effect on the city plan was significant but not catastrophic because the growth came in the form of high-rise flats rather than low-density sprawl. As a result, Moscow still has a clear, if ever-expanding, city boundary with the surrounding countryside. In the 1990s, for the first time, the population of the city dropped, falling back from just over 9 million to 8,881,000 in 1993.[29]

Housing policy is another signal failure of socialist planning. Poor housing conditions were one of the factors that precipitated the Revolution so it appears strange to an outsider that even today Moscovites remain so minimally housed. There are a myriad of reasons for this, but can be traced to the initial decision to favour the development of the military industrial complex at the expense of the consumer/workers' social and physical needs, as well as the rampant demand fuelled by industrial growth.

After Stalin's death, Krushchev's new housing policy was aimed at the masses (figure 5.12). Using industrialized housing methods, production rates rose impressively, but quality was quite low. To this day these units are referred to as 'Khrushovey', combining his name with the term for slum. Today, virtually all new housing projects on the outskirts of Moscow are built of ferroconcrete, giving many districts the sterile appearance of 1960s-style American high-rise housing projects; a Russian version of Le Corbusier's Radiant City. Another problematic area for the Soviet capital has been the signal lack of adequate and convenient commercial and service facilities.[30]

Post-communist Moscow

Following the fall of communism in 1990, Moscow, under an activist Mayor, embarked on a major development programme to reverse

Figure 5.12. Painting entitled 'Wedding on the Street of Tomorrow' by Y.I. Pimenov, 1962 shows 'Khruschovey' industrialized housing.

many Stalinist deprivations. The capital has seen a remarkable programme of reconstruction and restoration of historic structures such as the Victory Memorial, the Gostinny Dvor (guest's court), Kazan Cathedral and Resurrection Gates at Red Square to name a few.[31] A wider programme of historic preservation and revitalization includes museums and nine railway stations dating from Tsarist times.[32] Clearly, much remains to be done given the years of neglect. 1990 saw the designation of Moscow's Kremlin and Red Square as UNESCO World Heritage sites. New monuments have also been constructed, notably a controversial statue of Peter the Great. It is quintessentially Russian in its conception, its grandiosity and its many hidden meanings.

Strenuous efforts to engage in more open, participatory planning have led to a wider dialogue about Moscow's future.[33] New strategic plans for both Moscow and St Petersburg have been drawn up following Western participatory approaches. One offshoot of this has been the growth in active citizen participation and even protests regarding road plans and historic preservation issues.

Housing construction has emerged slowly from its centralized administrative pattern. Downtown has seen the development of large, new post-modern hotels and office and apartment buildings, often funded by foreign consortiums. These have impinged on such historic areas as the Arbat, a major tourist attraction near the Kremlin.

The fall of communism opened a new chapter in Moscow's history; one that will allow it to build on the many successes of communist planning and the 1935 plan by adding the private sector as a player.[34] One of the most notable new projects has been the five-level underground shopping centre at Manezhnaya Ploshad near the Kremlin's walls. Extending some five floors underground, it provides a modern, attractive shopping venue and food court. A joint venture built along Western lines, it is covered by a beautiful park, fountains and statues.

Conclusions

Tsarism and constitutional democracy were overthrown in St Petersburg while the new Communist state was forged in Moscow where it was, in turn, overthrown. Clearly the Revolution of 1917 is central to the modern history of Russia's two capitals; the forces it unleashed profoundly influenced their planning and design. But the modern history of these two cities also reflects the duality of old and new and Eastern and Western cultural traditions inherent in all things Russian.

In St Petersburg, this duality can be seen in the Soviet government's decision to rebuild completely the former imperial capital including its outlying imperial palaces after the World War II destruction. Since the fall of communism, official support for the former capital has strengthened and the city's original name was restored. The remains of the last royal family were interred with the other Tsars in the Peter and Paul Fortress in a religious ceremony attended by President Yeltsin. Recently, President Putin, who is from St Petersburg, held several meetings with world leaders in the city, lending credence to rumours concerning the possibility that the capital of Russia might once again return to St Petersburg. Whatever the case, these actions form part of a strengthening process of official sanction for the monuments, ceremonies and observances associated with the former monarchy and its imperial capital.

Historians can recognize that planning in these two capital cities was influenced by several important factors. First was the personal influence of the autocratic authority of the Tsar or the various communist dictators, particularly Stalin and Khrushchev. Second was the role played by the many pre-Revolutionary architects and planners who remained in Russia to help build the ideal communist city. They were instrumental in establishing and maintaining a planning movement that was based on blending foreign notions with Russian design traditions to meet the needs of the new Socialist society. Also important was the competition with the West, which led to the decision to develop a high-density, urbanized society showcased in Moscow and St Petersburg. The results were based on the interplay of these forces. The fall of communism has radically opened up the decision-making process. This, and the reintroduction of a private market, will presage major changes. It will be interesting to see how these changes influence the future course of planning these two great Russian capitals.

NOTES

1. French (1995), p. 18.
2. Volkov (1995), pp. 10–14.
3. Starr (1976), p. 225.
4. Enakiev's 1912 book *Tasks for the Reform of St Petersburg* was an influential text.
5. Lang (1996) p. 795.
6. Supplanted by the monumental GUM Department store in the eighteenth century.
7. French (1995), p. 19.
8. French (1995), p. 39; Colton (1995), pp. 215–218.
9. Colton (1995), pp. 99–100.
10. Colton (1995), p. 325.
11. Colton (1995), p. 252.
12. Colton (1995), p. 332.
13. 'The subway of the revolution is a revolution in subways.' Perchik (1936), p. 59.
14. French (1995), p. 37.
15. The perimeter design of these signature buildings was influenced by New York City public housing projects designed by Frederick Ackerman. Ackerman was a passionate advocate of perimeter garden court style housing and an opponent of high-rise slab and tower in the park styles. Interestingly, Ackerman is known to have met with Soviet architects during his long career. Lang (2001), p. 150.
16. Parkins (1953), p. 108.
17. Colton (1995), p. 329.
18. Also important is Perchik's *The Reconstruction of Moscow*, published in 1936 and which celebrated the planning and construction accomplishments of Stalin calling him, 'the great architect of socialist society'. Perchik (1936), p. 72.
19. Kaganovich (1931), pp. 12–13.
20. Parkins (1953), p. 71.
21. Perchik (1936), p. 41.
22. Bater (1980), pp. 28-29; Parkins (1953), pp. 37–41.
23. The administrative micro-regions were never implemented.
24. Colton (1995), p. 523.
25. Colton (1995), p. 719.
26. French (1995), p. 83.
27. Colton (1995), p. 735.
28. French (1995), p. 88; Colton (1995), p. 482.
29. Colton (1995), p. 758.
30. French (1995), p. 110.
31. While generally welcomed, the cost of some of these undertakings has been criticized, particularly the reconstruction of the Cathedral of Christ the Redeemer at a cost of some US $200 million. Vinogradov (1998), p. 104; Glushkova (1998), p. 120.
32. Vinogradov (1998), p. 105.
33. Colton (1995), p. 726.
34. Luzkov *et al.* (1998), p. 66.

Chapter 6

Helsinki: From Provincial to National Centre

Laura Kolbe

In the planning history of every city, there is a moment of transition into 'the modern'. In Helsinki, that moment came in 1899, when a city plan competition was arranged for the Töölö district near Helsinki's geographical centre. Although a tumultuous preparation process preceded this first competition in the field, it was a definite breakthrough in the history of Finnish city planning. The new planning ideals of the early 1900s were obvious in Nyström and Sonck's winning scheme: a picturesque and landscape-adapted urban street network and an intimate atmosphere accompanied by architectural effects.[1]

Helsinki's development before 1914 was similar to that of many other medium-sized capitals in continental and northern Europe. At that time, the city was a part of the Russian Empire and can be compared with many similar cities in the other European empires.[2] An administrative tradition of civil servant rule, an economic structure still oriented towards agriculture, a lack of capital, and slow industrial and infrastructure development caused the urbanization process to begin late in Finland. The driving force towards modernization consisted not of a weak civic society but of bureaucratic cadres and enlightened civil servants. The situation changed when in the 1870s and 1880s Helsinki started to grow. The pressures of new forms of production, new political, social and national groups, increasing trade and traffic, and new urban life styles started to influence planning. Helsinki developed gradually in the final quarter of the nineteenth century to become the cultural and political centre of the country, the real capital of Finland. The Töölö competition coincided with the peak of the take-off and, in one stroke, it launched Helsinki's bureaucratic town planning onto the modern day European stage.[3]

Provincial Capital: Imperial and Centralist Planning

We cannot possibly grasp the scope of this change unless we know the historical roots of centralized rule in Finland. In a large, sparsely inhabited and peripheral country like Finland, the central government has been an active founder of towns and cities. Many historic cities were established by the crown for military, administrative, trade or educational purposes. Public control exercised by a professional bureaucracy is well established and has been in existence at least since the seventeenth century. The ambitions and investments of the central power have strongly influenced the development of Helsinki. It is not an old bourgeois trade town, but came about as the result of political projects. The change in town planning policy that occurred around the year 1900 should be seen in light of the change in the relationship between the state and the citizens. It implied a gradual effort to transfer planning from the central power to the local level, from the civil servants to the citizens. In this sense, too, Helsinki is an interesting case to study.[4]

In 1812, Helsinki was made capital of Finland by decree of Russian Emperor Alexander I. Finland had been separated from its old mother country Sweden in 1809 in the wake of the great wars in Europe, and annexed to the Russian Empire. The Grand Duchy of Finland immediately dealt with the capital city question. During the period (from about 1200 to 1809) when Finland was a part of Sweden, Stockholm had been the capital of the Finns, and administrative, economic and cultural contacts took place directly between Stockholm and the provinces. In the new situation, Finland's development was influenced by St Petersburg, at the time capital of Russia and an expansive centre of power on the Baltic Sea rim. Due to national and military considerations Helsinki – located only 400 km from St Petersburg – was made capital of the Autonomous Grand Duchy of Finland.[5]

The emperor pronounced the genesis of the new capital, and the state took its planning and construction – and finance – in hand. Becoming an Imperial Russian city after being a Royal Swedish town implied an ascent in the urban hierarchy. In 1812, the Russian Emperor appointed Helsinki's first town planning authority under the name of the Reconstruction Committee. Johan Albert Ehrenström, a famous military engineer born in Helsinki and a connoisseur of European arts and culture, was invited to head this committee. This marked a clear change of policy, since the old Helsingfors had grown very slowly without a city plan.[6]

Helsinki became a means of expression for the Russian imperial power. Ehrenström's town plan was completed in 1812, to be confirmed by the emperor in 1817. It showed the classical town planning ideal of European princely towns and a regular, functionally and architecturally proportioned urban structure. Construction had begun a year earlier in 1816, when the Reconstruction Committee hired Berlin-born Carl Ludwig Engel, an architect who had also worked in Reval (Tallinn) and St Petersburg. 'L'Empéreur a gouté et apprové le Plan. Dans son exécution il seroit un des monuments glorieux de son règne', Ehrenström later commented in one of his letters. The new city's identity as a provincial government centre was backed by an architecturally and ideologically suitable Imperial style.[7]

The Reconstruction Committee finished its work in 1825. The world could see that Helsinki was able to compete in beauty with the most prominent European capitals. Monumentalism and classicism expressed the spirit of this centrally governed garrison, and administrative and university town. After the Vienna peace treaty of 1815, a regular and hierarchical classicism conveyed a message of political conservatism, continuity, order and stability. The city centre, where the imperial and urban dimension met in the stone buildings of the central Senate

Square, had been chosen for the upper strata of society, while the suburbs with their wooden buildings were of lower status. Monumentalism expressed itself in both the Senate House (today the National Government Building) and in the main edifice of the university on the opposite side of the square. The large Lutheran Cathedral on the northern edge of the square dominated the panorama (see figure 6.1).[8]

Figure 6.1. Imperial Senate Square depicting the historical and classical spirit of Helsinki.

Up until the 1850s, the town plans, enlargement plans and park plans, etc. of the surveyors and engineers were fitted into a square grid street pattern. The responsibility for the planning had been transferred to the City, but the Emperor still confirmed the plans. The railway, which terminated at what was then the northern rim of the city, influenced the city's development. The major western road Västra chausséen (later Henrikinkatu and Mannerheimintie) had been sited through this area and the wedge-like area between road and railway has been a problem to the growing city even to the present day. The railway system demonstrated the importance of the State's goals and investments to the whole nation and turned Helsinki into a main port for export and a real capital with connections all over the country. This began to change the originally imperial character of the city centre. The Railway Square became a second monumental square, which was spatially expressed by the union of bourgeois capital activities, a national awakening and urban modernism. The square was edged by businesses, the Ateneum Art Museum, the College for Industrial Arts (1887), the Finnish-language National Theatre (1902) and two of the most famous hotel-restaurants, the Fennia and the Seurahuone.[9]

During the imperial reign, the right to make urban plans was monopolized by the authorities much in the fashion of centrally governed countries. The Senate, i.e. the highest State authority, also undertook many initiatives, such as encouraging Helsinki to build a water supply network (in 1862), and promoted the construction of stone buildings (instead of wooden) by granting state loans. Yet, planning proposals still had to be confirmed by the Emperor. Distinction, hierarchy and fire safety were still apparent features in the urban plan and construction rules for Helsinki that came into force in 1875.[10]

Tradition and Change: Municipal and Bourgeois Urban Planning

A central issue here, when we talk about capital cities, is the relationship between the state, the municipality and public opinion. It is obvious that at every stage of the modern urban history of Finland these players have a relationship when it comes to capital planning and the image of the number one city of the nation. Towards the end of nineteenth century an effective system of municipal administration was created (see figure 6.2). The 1875 reform of the municipal administration marked a change in urban planning. The city's administrative court became the central administrative body.

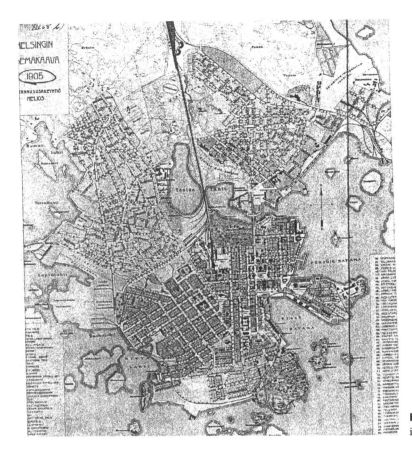

Figure 6.2. Helsinki plan drawn in 1905.

The new city council, consisting of enlightened and influential men of the community, approved the urban plans and construction rules. The shift from the central to the local level expressed itself in people's attitudes. The idea that a town should be planned before it grew – not after – gained ground.

The influence of municipal authorities on urban planning and the formation of a national profession of urban planners implied a first phase in the modern era. The planning competition for the Töölö district mentioned earlier was a result of this change in custom. The trend was backed up by Helsinki's development towards a major city, by population growth and by an increasingly varied social fabric. In the 1910s when its population exceeded 100,000 inhabitants, Helsinki joined the international metropolis category. The city started expanding to the north – the first industrial district – and to the west, where some more industries and a port came about. The Paris-style Esplanadi Park with its twin boulevards and large areas of stone housing were expressions of bourgeois wealth and growth of capitalism in Helsinki (see figure 6.3).[11]

The modernist turn towards local urban planning was seen in the years 1906 to 1908. The introduction of architect education in Finland meant the birth of modern urban planning. In 1908, the city's new authority for municipal planning, the Urban Planning Committee, was assigned to reform and make new urban plans, improve public transport between the central and the peripheral districts, and take measures for drawing up a master plan. During just a few years, many technical offices were created. The models were Sweden's new and progressive law on urban planning (1907) and Stockholm's land purchase policy. Both promoted modern suburban planning and building. Urban planning shifted from engineers to architects. For the first time since Engel's days, architects as a cadre got the opportunity to influence construction and planning in Helsinki.[12]

When Helsinki's first city planning architect Bertel Jung took up office he determined that a master plan for the city should be drawn up in the spirit of Vienna's metropolitan plans by Otto Wagner and the 1910 urban planning exhibition in Berlin. Jung's first comprehensive master plan for Helsinki was inspired by the Berlin exhibition and was based on population forecasts, as was then the practice in central Europe. Jung recommended high construction coverage and population density for the historical parts of Helsinki, which in turn prompted transport arrangements resembling Stockholm's local railways. Because Helsinki's land purchase policy was still weak, Jung's master plan was never confirmed nor carried out.[13]

Figure 6.3. Esplanadi (the Esplanade), first proposed in 1812. Now a park and an elegant boulevard connecting the old harbour area and Market Place to the city centre.

A third modernist and democratic phase was seen between 1908 and 1914. Helsinki had become the cultural and political centre of the new country, a real capital. The introduction of the one-chamber parliament in 1906 coincided with the birth of working-class and bourgeois values, the national awakening, a golden era of arts and culture and the change in urban planning. In line with European models, planning the capital city underlined technological modernity, aesthetic dimensions, urban intimacy and organic growth instead of regularity and ready-made patterns.[14]

But municipal support for planning was weak, so enlightened private companies hired skilled young architects to plan large residential districts. The land companies combined capital with technical skills and urban architectural visions from designers like Eliel Saarinen.

The Republic of Finland: Capital City Planning and a New Urban Centre

The First World War shattered the old world, destroyed cities and gave birth to new national states. Helsinki remained the natural capital

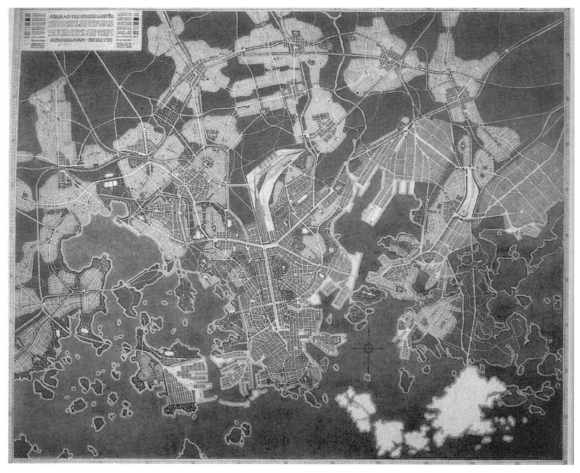

Figure 6.4. *Pro Helsingfors* proposal for a master plan for Helsinki, 1918. Drawn by architects Eliel Saarinen and Bertel Jung.

HELSINKI: FROM PROVINCIAL TO NATIONAL CENTRE

Figure 6.5. Kuningasavenue (King's Avenue), 1918 proposal by Eliel Saarinen. Aerial perspective in an ink wash.

when Finland separated from Russia and became an independent republic in December 1917 in the shadow of the World War and the Russian Revolution. A bloody civil war in the spring of 1918 divided the nation and interrupted social reforms for some time. The master plan proposal *Pro Helsingfors* made by Saarinen and Jung together in the spring of 1918 marked the beginning of a new era (see figure 6.4). The plan was commissioned by commercial counsellor Julius Tallberg, a business magnate and a major force in the background of municipal policy. Tallberg's commission must be seen as the last display of strength of the 'enlightened bourgeoisie'. In his preface, Jung wrote that the city's officials trapped in the exhausting treadmill of everyday matters would not have been able to produce a similar comprehensive plan.[15]

In the background of the 1918 plan we find the theoretical population goal: according to the lowest forecast, Helsinki would have 370,000 inhabitants in 1945. The metropolitan atmosphere in the capital of an independent state was manifested in the extension of the central business district by filling Töölönlahti Bay and moving the railway station northwards to Pasila. A new central road, the Kuningasavenue (King's Avenue) united the old and the new railway station neighbourhoods and symbolized Helsinki's political role as a national centre (see figure 6.5). The avenue's name also reminds us of the monarchist dreams nurtured among certain bourgeois circles in Finland in the Spring of 1918. The name was later changed into Valtakunnankatu (Nation Street). Kuningasavenue's wide transversal roads to the east and west formed an urban backbone for the city and shifted the emphasis from the old imperial centre. Housing, manufacturing and part of the harbour functions spread outside the city boundaries along well-developed suburban railways.

Pro Helsingfors was the only modern master plan proposal made in Finland in the 1910s that followed international trends. Although it did not have a lawful mandate, Saarinen and Jung's principles influenced the development of the capital throughout the twentieth century (see figure 6.6). Among the major problems to be solved in the 1920s and 1930s were the matter of the railway station, the housing problem and the planning of a new city centre. According to the outline made by city officials in the 1923 official plan, the Töölönlahti Bay was filled, but the resistance of the state railways prevented the transfer of the main railway station. The city centre problem remained unsolved, although a solution was sought in a 1925 competition for a comprehensive city plan. The winner, architect Oiva Kallio, suggested that the area should be developed in the spirit of urban monumentalism and historical-classical dignity. In much the same way as Saarinen and Jung, Kallio emphasized the city centre as a space for both public functions and housing.[16]

The symbolic value of the Töölönlahti Bay area grew when the planning of the most important political building of the new republic, the Parliament House, started. The new unicameral building needed premises commensurate with its new functions. Above all, there was a need for a chamber where all 200 deputies could assemble

Figure 6.6. Helsinki skyline, 1931.

in plenary session. That marked the beginning of a building project that went on for nearly twenty years. The alternative solutions were either to extend the 1889 House of Estates (which was built close to Senate Square to accommodate the non-noble estates) or build a completely new home for the parliament. From 1907 to 1930 the Parliament met in rented premises.[17]

The planning of the Parliament House and its location were the result of an architectural competition held in 1923. The building, designed by architect J.J. Sirén, was completed in 1930. National and local interests were combined in the key issue, the location. The real planning work before the competition was in the hands of the Parliament, but the City of Helsinki controlled the land use. The areas around the old administrative city centre and the central coastal park area (Tähtitorninmäki) close to the old harbour were the favourite locations of many architects. The city was not willing to give up park area or allow more buildings in the Senate Square area, but it was prepared to sell state land in the Töölönlahti Bay area. The city planners had already reserved space there for public and cultural buildings. The rocky site of the parliament building was spacious, undeveloped and hilly and thus had the potential to create a ceremonial effect.[18]

It was now possible to plan a new institutional symbol and democratic city centre by moving the political power away from the 'Russian and imperial' Senate Square. This was realized only in part. In the planning competitions the architects had a free hand to place the building within the city structure. Monumentality caused some problems. It was difficult to combine the

Figure 6.7. The new Parliament House by J.J. Sirén was completed in 1930.

over-dimensioned and idealized mass of the Parliament House with its modest surroundings. The result was a lonely stone castle with strong roots anchoring it to a rocky outcrop (see figure 6.7). Security and stability were sought with the aid of the architectural form and material. Steps at the front of the building linked it to the main street area. However the square planned by Sirén and other architects for the front of the building was never completed. The whole composition speaks the same 'national' language as the old administrative buildings – that of Classical architecture.[19]

The economic depression of 1930, however, hampered all larger urban design and construction in both the old and new city centre, and the planning of the Töölönlahti Bay area only resumed after 1945. A 1949 competition endorsed the idea that the Töölönlahti area should be preserved as a green area and the main railway station be kept in its present location. In the 1954 plan the Kamppi district, west of the station and Parliament, was planned as the new administrative centre. The planning of this area was in the hands of the city and the role of the state was marginal.[20]

With the 1952 Olympics (first planned for in 1940), Helsinki joined the exclusive club of Olympic cities. A joint committee of national and local governments and sport groups planned the games. The Olympics activated planning and left a permanent mark on the streetscape and also introduced functional transport arrangements and sports facilities. The Olympics consolidated Helsinki's position as a capital both in Finland and abroad. This sports nationalism, together with local and national forces, turned Helsinki into a modern sports city.[21]

The Nation State: Metropolisation in the Capital City Planning

Finland fought two wars against the Soviet

Figure 6.8. Architect Alvar Aalto describes the 1964 Helsinki Centrum plan to President Urho Kekkonen and leading urban politicians.

Union, in 1939–1940 and 1941–1944. Although the country was not invaded nor was Helsinki destroyed in the extensive 1944 bombardments, the war years were a definite turning point. By 1945 the population had risen to over 300,000, and migration to Helsinki grew steadily. Its land area grew five-fold when the suburban zones were annexed to Helsinki in 1946 by decree of the national government. This annexation brought urban planning in Helsinki into a new era, where capital, regional, metropolitan, suburban and traffic planning could be seen as a whole. The 1960s saw suburban expansion within a planned structure.

The role of local planning was strengthened in 1964, when the city's urban planning office was founded and a political urban planning board appointed. Municipal planning and long-term municipal economic planning were consolidated. Since 1964 the local authorities have played an exceptionally strong role in urban and capital city planning in Helsinki. This role is based on the city's strong land ownership, estate policy and its investments in basic infrastructure.[22]

The tasks of the new city planning department included the implementation of Alvar Aalto's monumental plan (presented in 1964) for the city centre (see figure 6.8). The City of Helsinki had assigned the development of the plan for the centre to Aalto in 1959. This commission also included the Pasila area, the first future urban centre along the railway. Aalto's proposal presented a dense urban structure in Kamppi and a monumental approach around Töölönlahti Bay, thus emphasizing Helsinki's position as a capital. The area in between the railway station and the Parliament building was covered by a terrace and the shore of Töölönlahti was edged with a row of buildings for cultural purposes (music hall, opera, theatre, museum etc.) (see figure 6.9).

Figure 6.9. 1964 Helsinki Centrum plan model by Alvar Aalto.

The plan soon caused much political debate. At the end of the 1960s the new generation of planners and architects heavily criticized Aalto's efficient traffic arrangements and his idea to sacrifice the green shores of the Töölönlahti Bay and Kaisaniemi Park areas. Although the City Council had approved Aalto's plan as a basis for further planning, only two buildings included in it have been built to date: the Finlandia Hall by the bay and the Sähkötalo (Electricity House) in the Kamppi area.[23]

After 1959, regional planning and master planning became mandatory. The largest projects in Helsinki developed the Itä-Pasila in the 1970s and the Länsi-Pasila area in the 1980s into a concentration of workplaces, housing, public administration and offices in the spirit of La Défense, outside Paris, and for the same reason – deflecting modern growth away from the historic core. This commercial construction was complimented by large-scale community construction throughout the metropolitan area.[24]

From 1945, the state, the municipalities and various civic organizations had been interested in the planning of the capital city. Anxieties about the changes caused by rapid growth and the criticism expressed after 1968 by the new Left were directed against the ideals of efficiency, in favour of historical values. During the post-war rebuilding era, many wooden and stone buildings from the nineteenth century were destroyed or converted for other functions in the name of business and efficiency. By the 1950s, however, certain civic and heritage organizations had expressed their concern about the old milieu. In 1952, the City of Helsinki declared the Senate Square and its surroundings the historical centre of the city, which should be preserved. [25]

Since the 1970s, the urban policy of Helsinki has been characterized by the maintenance of a strong capital city centre. Preservation of historic buildings became of the highest importance. Meanwhile, the problem of the Töölönlahti Bay area has remained unresolved. Not even the 1985 all-Scandinavian competition managed to produce a permanent solution for this republican forum. Construction has been started here and there, one property at a time, including the Museum of Contemporary Art on the Mannerheimininaukio (Mannerheim Square).

Conclusions: The Spirit of Urban Planning

Many of Helsinki's strongest meanings are embodied in waterways. The capital city of Finland was planned by the Baltic Sea. The river, the ocean waterfront areas, the bays, shores and coastlines, as well as the isthmus site have, to a varying degree, figured prominently in the historical development of the city. The sea has played a role in shaping the city's capital and symbolic image, as well as its spiritual urban essence. The historic centre, located on the narrow peninsula, is linked to the sea in an exquisite fashion and its neo-classical waterfront façade is a well known symbol of the capital city. Extensive harbour and industrial areas express the economic vitality of the city. With the rapid industrialization in the early twentieth century, land was reclaimed from the sea for harbours and dockyards. Also suburban planning has moved along the coastline.[26]

Helsinki has a particularly rich shoreline and very varied spaces linking the city and the water. The new urban waterfront projects will reinforce Helsinki's capital image as a maritime city. Shore area planning has become timely due to the rearrangement of harbour and industrial areas and oceanfront zones have become desirable residential and recreational areas. Even the official residence of the President of Finland, Mantyniemi was planned in 1994 on a rocky pine promontory inside a narrow western bay. The plan of the building is shaped into a long broken sweep facing the sea.

Helsinki, the seat of government in Finland, is of quite recent origin. It was built in the nineteenth and twentieth centuries, which were 'the centuries of capitals'. There is nothing mediaeval or feudal in Helsinki's development. The first phase of planning took place under the special circumstances of Russian rule, yielding a city of order and dignity. Engel's city plan created the white, architectural image of neoclassical parts of central Helsinki. The city still retains rather low roof heights and any vertical element is highly visible in the townscape. Helsinki today is no longer bound by this neoclassicist framework. In the past one hundred and fifty years, alternative urban and planning approaches have been explored and a unique capital city has been constructed.

NOTES

1. Brunila and Schulten (1955); Nikula (1931).
2. Blau and Platzer (1999).
3. Klinge and Kolbe (1999).
4. Kervanto Nevanlinna (2002).
5. Åström (1957), pp. 42–58.
6. The old Helsingfors (Swedish for Helsinki), founded in 1550 on the northern shore of the Gulf of Finland, had failed its basic task of competing with old Hanseatic Tallin on the opposite side of the Gulf. Numerous fires, wars between Russia and Sweden, occupations, plagues and scarcity had labelled the evolution of the town. Its transfer in 1640 to what today is central Helsinki on the isthmus did not improve the situation. See Klinge and Kolbe (1999), pp. 7–15; Blomstedt (1963); and Stenius (1969), Maps, pp. 70–82.
7. Åström (1957), pp. 42-58, Blomstedt (1963) pp. 258–264, Klinge and Kolbe (1999), pp. 24–32. See also Lilius, (1984), pp. 9–11 and Hall (1986), pp. 69–75.
8. Eskola and Eskola (2002).
9. Lindberg, and Rein (1950); Kervanto Nevanlinna (2003), pp. 83–91.
10. Åström (1957), pp. 129–141; Hall (1986), pp. 72–75.
11. Kuusanmäki (1992).
12. Kuusanmäki (1992), pp. 159–162. Åström (1957), pp. 178–192. Nikula (1931), pp. 150–151.
13. Nikula (1931), p. 191, Kuusanmäki (1992), pp. 174–196. Also Sundman (1991), pp. 524–527. Helsinki's first city planning architect Bertel Jung (1872–1946) belonged to the polyglot and learned generation of architects who reformed Finnish urban planning. Before 1914, Jung travelled to many places in Europe and studied matters of contemporary urban planning. In 1915 he described the importance of the stimuli and models he had picked up during these trips: first the German, then the British and the American school. At the same time, he kept his eyes open for developments in the rest of Scandinavia. Young Finnish architects interested in urban planning including Jung himself and Eliel Saarinen, Lars Sonck, Gustaf Strengell, Harald Andersin and Otto-Ivar Meurman felt that international shows and congresses were important. This was seen especially in capital city planning.
14. Nikula (1931), pp. 102–109; Åström (1957), pp. 174–176.
15. Kuusanmäki (1992), pp. 8–15. See also Korvenmaa (1992). Architects played an important part in the consolidation of the national identity and Finland's international position. As a profession anchored in the ideological core of the middle class, architects became an expression of social and aesthetic transformation. Internationalization supported the construction of the Finnish identity; interaction was the forum in which a national character best showed its worth. Architects in Finland were just as familiar with Camillo Sitte, Joseph Stübben and Werner Hegemann as with Ebenezer Howard's or Raymond Unwin's works.
16. Kolbe (1988). The vision was of a suburban community that was geographically delimited and socially varied and planned for; an enlightened working class and middle class that put European reform ideas into practice. An Anglo-Saxon element was definitely the trust in private action and in the strength of a natural, intimate, anti-aristocratic and anti-urban (i.e. 'picturesque') community planning. Direct influences from English garden cities and Raymond Unwin's ideas – and from Swedish villa communities – could be seen in many of the Helsinki plans.

17. Kolbe (1988), pp. 150–160.

18. Jung (1918). In Saarinen's and Jung's plans, Helsinki expanded along the suburban railways and along separate 'exit roads' and the local road network to the east and west. Parts of the areas were reserved for manufacturing, whilst others became villa-like suburban zones or densely inhabited urban centres of housing.

19. Sundman (1991), pp. 78–82.

20. Hakala-Zilliacus (2002), pp. 311–322.

21. Hakala-Zilliacus (2002), pp. 85–90.

22. Hakala-Zilliacus (2002), pp. 91–100.

23. Kolbe (2002), pp. 86–88. Kervanto Nevanlinna (2002), pp. 194–196.

24. Nikula (1931), pp. 278–283. The classical tradition was also seen in the suburbs. The garden town Käpylä was built partly by the municipality just north of the centre. The initiative had been made in 1910 (before Finland had become independent) by *Yhdistys Yleishyödyllisen Rekennustoiminnan Edistämiseksi*, an association for the promotion of construction for the common good. After a long period of war and crises, Käpylä became a guinea pig area for municipal construction. The buildings were designed by architect, Matti Välikangas. Käpylä manifested the new social housing ideals of that day, with spaciousness and closeness to nature. New housing areas that lay close to the centre predominantly had an integrated streetscape.

25. Kolbe (2002), pp. 96–98. Also Raatikainen (1994), pp. 7–36.

26. Kolbe (2002), pp. 221–256.

Chapter 7

London: The Contradictory Capital

Dennis Hardy

London has throughout its modern history been an undisputed capital. In the twentieth century, in spite of enormous changes in its context and functions, it maintained and even enhanced its historical dominance. Yet there is in this tale of success an inherent contradiction, namely, that London's continuing supremacy occurred in spite of, rather than because of, political intervention and planning. On the contrary, during the past century official efforts were directed less to encourage its inherent dynamism than to restrain it. The purpose of this chapter is to explore this essential contradiction, between London's continuing dominance and status as a capital city and a marked absence of official policies and schemes to promote it. London, it is argued, has evolved in this period as a Global Capital by default.

This account of London is in three sections – the first summarizing the immense changes that transformed the city in the twentieth century; the second reviewing the nature and extent of public intervention in relation to its capital status; and the third questioning why London's continuing dominance as a capital has not been fully reflected in its architecture and civic design.

The Tale of Two Cities

London at the end of the twentieth century was, in many ways, a very different city from that which witnessed the ending of the Victorian era. When Queen Victoria died, in 1901, after a reign of sixty-four years, London was in every sense an imperial capital. Britain ruled a vast empire that included great swathes of territory on every continent except Antarctica, with a population of more than 400 million people, and London was at the heart of it.[1] At the late Queen's funeral, representatives were brought from all parts of this far-flung Empire to march behind the coffin, brightening the streets of the capital with their exotic costumes and skins of every hue. For many of the onlookers this was the first time they had seen non-white races, other than in books recounting the exploits of missionaries and explorers.

Figure 7.1. The City of London in the early twentieth century, looking east towards the Royal Exchange.

In Westminster, the Houses of Parliament at the start of the twentieth century took decisions that had a bearing on the lives of people across the world; and in nearby Whitehall civil servants ensured that the extensive colonies were governed according to British administrative traditions. A short distance to the east, in the City of London, merchant banks financed investment in distant railways and mines, irrigation schemes and plantations, while in the same square mile the headquarters of trading companies conducted business with every part of the Empire (figure 7.1). Still further east, the huge complex of docks that had been constructed successively from the start of the nineteenth century showed in tangible form what a long history of global contacts meant: with a constant stream of ships bringing in tea and spices from the Orient, ivory and cocoa from Africa, wheat and refrigerated meat from the Americas and timber and furs from the Baltic (figure 7.2).[2]

Nor, as Jerry White illustrates in his seminal book on London in the twentieth century,[3] was the significance of all of this divorced from the minds of ordinary people. When news arrived in May 1900 of the Relief of Mafeking (a rare piece of good news in a desperate war fought to protect British interests in South Africa), White describes the extraordinary scenes of jubilation. It was, he suggests, as if 'every one of its 6.5 million citizens who could crawl from their cots or hobble from their beds' found their way onto the streets of London.[4]

London was then not only pre-eminent within its own country (larger than twenty-two of the next largest British cities put together)[5] but also a world city. The American historian, Jonathan Schneer, describes it in 1900 as the 'imperial metropolis',[6] a world city without equal. This dominance could be measured not simply in volumes of trade and business but also, claims Schneer, less quantitatively in its imperial architecture and cultural ambience.[7] And to the contemporary, H.G. Wells, who himself hailed from a south London suburb, his city (a constant source of fascination to him) was unquestionably 'the richest, largest, most populous city' that the world had ever seen: 'immense, vast, endless!'.[8]

Figure 7.2. Contrasting views of London's fortunes as a port: with bustling activity at the start of the century compared to empty wharves in the 1970s.

Consider the situation, in contrast, at the end of the twentieth century. The Houses of Parliament were still to debate world affairs, but no longer as influential leaders of an imperial power. The Empire had long been dissolved, and Britain had yet to settle on a new identity, whether as a part of Europe, an adjunct of the United States or leader of a phantom Commonwealth. If the nation was finding it hard to divest itself of an imperial role, the same was true of the capital. London, no longer the hub of an empire, was carving out for itself a new world role – a role that is most vividly illustrated by the transformation of the square mile of the City itself, the capital's inner sanctum. This latter zone has, historically, been seen as the commercial heart of the capital, and in spite of its small resident population (totalling just 6,000) it remains a separate administrative authority with a remarkable degree of autonomy.

If only in terms of the appearance of the City of London, the contrast with the past is striking. The marble buildings and mahogany furnishings of traditional banks and trading companies that characterized the imperial era have largely disappeared. In their place is the glass and steel of towering modern blocks that house the offices of world corporations. In one of the few public interventions that has had a decisive influence on the fortunes of the capital, Margaret Thatcher's government in the 1980s engineered a radical process of financial deregulation. In what is popularly termed the 'Big Bang', a myriad of closed shops and quaint practices were replaced overnight by a new, totally computer-based system. Instant communication along internet highways was in stark contrast to the sedate pace and style established in a past age, when representatives of trading houses, wearing silk top hats and black frock coats, delivered messages by hand. The Big Bang was remarkably successful in enabling London, as a financial centre, to compete effectively across the world. American, Japanese and European banks all sought a foothold in the capital, contributing in no small way to a new boom in office building within and close to the City. London had once again established itself as a leading financial market in the global economy, linked through networks to New York, Tokyo, Frankfurt and other major centres. Because of its financial strengths, it was in the 1990s, in the words of Jerry White, firmly installed 'in the very top drawer of *world cities*'.[9]

Inevitably, this magnitude of change was not to be confined to the City boundaries. Just to the east, the great docks of a former era now offered a valuable land bank to accommodate new development. Although formerly the link to its sprawling Empire, by the end of the century the traditional trade had dried up and with it the upstream docks themselves. Coinciding as this did with the decline of associated industries that once lined the waterside – such as engineering, ship-works and flour refining – the way was now open for a significant shift eastwards of the financial centre. Great complexes of office blocks appeared in the redundant docks, together with shiny new buildings housing innovative media and information industries (figure 7.3). And, as if to cement its new role, the former docklands were also to provide the City with its own urban-based airport as well as waterfront housing for office workers.

Perhaps nothing better illustrates London's transformation to its new role as a global city than the modern dominance of air traffic. It is hard to imagine that at the height of the imperial era the relatively few who travelled abroad did so largely by ocean liners. Many of these would have left from the London Docks themselves, their time of departure governed by the tides. In contrast, air traffic has burgeoned beyond almost anyone's expectations. The capital is now served by a ring of airports that in 2000 recorded an annual throughput of 116m passengers, with the prospect of a doubling of this total over the next twenty years.[10] London Heathrow itself is the busiest airport in the world, and a mere look at the departure screens is sufficient to illustrate the complexity of the modern global networks in which London figures in such a significant way. Nor are these routes merely to serve travellers to other parts of the world. Traffic is two-way, with visitors coming to the capital for a variety of reasons including the modern phenomenon of tourism. Only 1.5 million tourists visited the capital in 1960, in contrast with 13.5 million forty years later.[11]

Figure 7.3. Contemporary photo of Canary Tower with new towers alongside, signifying the emergence in the former Docklands of a modern counterpart to the City of London.

Geographical and social change, too, reflects the inherent dynamism of the city over the past century. On the face of it, the population of the capital increased quite marginally (from 6.5 million to a little over 7 million) but this belies a massive redistribution from the core to the suburbs, which mark the enlarged boundaries of the modern city, as well as the effective regionalization of London into the surrounding Home Counties. Each day, a large percentage of the workforce commutes from this surrounding region into Greater London (nearly half a million alone using the overground rail system).[12] Nor is it geography alone that has changed. London has always attracted immigrant groups, but never on the scale witnessed in the second half of the century. In a sense, the Empire turned in on itself, and huge numbers from former colonies came to settle in the capital of the Mother Country. By the end of the twentieth century, one in three of the capital's population is classified as ethnic minority, and in some of the capital's thirty-three boroughs there is now a majority of non-white residents.[13] Amongst the most recent incomers are large groupings from the Balkans, the former USSR and central Asia. London, the international city, has itself become multi-racial.

This transformation from an imperial to Global Capital was dramatic but unquestionable. Yet, how much of this was due to conscious planning and policy support?

London the *Bête Noire*

In 1951, the postwar Attlee government was responsible for the first major celebration of peacetime, the Festival of Britain.[14] Located mainly on the south bank of the Thames, opposite the seat of government itself, the Festival, although ostensibly a celebration of nation, was a golden opportunity for the capital to present itself in a new, postwar role. It was unthinkable that the event should be staged anywhere but in the centre of London, and millions flocked to it from all parts of the country. There had been nothing like it since the Great Exhibition at the Crystal Palace, one hundred years before, held then to celebrate the reign of Queen Victoria and the splendour of the British Empire. And as if to add to the potential of the capital to promote itself, the main exhibits on the South Bank were augmented by the living example of rebuilding in the East End, with the model scheme of Lansbury on the visitors' map. London, came the message, was remaking itself, a Phoenix quickly arising from the ashes of wartime destruction (figure 7.4).

Figure 7.4. Tower Bridge with smoke rising around it following a night of enemy bombing. Extensive damage to the city during the Second World War was to provide an important stimulus for post-war replanning.

But the Festival of Britain was a notable exception. Beyond that, successive governments did little to advance the cause of the capital, and, indeed, did rather more to seek to erode it. For, while London remained the undisputed capital, it had for long attracted its own opposition, in the form of visionaries and reformers who campaigned relentlessly against 'the great wen'.[15] In 1890, William Morris encapsulated the thoughts of many when he described a dream of London, 'the modern Babylon of civilization',[16] disappearing altogether. To Morris, London embodied all the ills of industrial capitalism, forcing its unwilling citizens into a life of misery. With the overthrow of capitalism he envisaged a situation where its people would freely leave the city for small towns and villages in the countryside. His views were shared not only by fellow revolutionaries, such as Peter Kropotkin,[17] but also by more moderate reformers.

Ebenezer Howard was one such reformer, a gentle individual who simply believed that, even without a political revolution, people would be happier in small towns than in the amorphous capital.[18] His remedy was the creation of garden cities that he foresaw would naturally attract both businesses and individuals, instead of locations in London. Indeed, the final chapter of his seminal book on garden cities is devoted to a discussion of the future of the capital, following the establishment of garden cities. Because of the pull of the latter, London as it had become would slowly disappear. At that point, however, in a seeming concession to big city advocates he acknowledged an opportunity to reconstruct it on totally different lines. As a direct result of Howard's book, the Garden City Association, was formed in 1899 with the specific aim of campaigning for his ideas. In addition to the positive promotion of garden cities, the Association took on a wider brief of encouraging good planning, which included the containment of London.[19]

Nearly four decades after the garden city campaign was launched, the hitherto outlandish idea of dismantling the capital received a qualified seal of approval, when no less than a Royal Commission recommended that it was time to take action. The Barlow Committee (named after the chairman, Sir Montague Barlow) advocated a system of national planning, sufficiently strong to curb the continuing growth of London while at the same time diverting some of this growth to other parts of the country in economic need.[20] Published in 1940, the Barlow Report (as it was popularly known) proved to be an influential backcloth for the no less influential wartime *Greater London Plan*, prepared by one of the leading town planners of the day, Patrick Abercrombie.[21] Taking his cue from Barlow and, in turn, the long campaign of the Town and Country Planning Association, Abercrombie set in place the building blocks that were to shape London's postwar future. No more growth was the watchword, with the introduction of a green belt to set the limits of development to its 1939 extent, and new towns beyond that to enable the decanting of some of London's 'excess' population.

The *Greater London Plan* (1944) was, in fact, just one of three strategic plans that set the framework for postwar development (figure 7.5). In the previous year, Abercrombie's work with the London County Council's chief architect, J.H. Forshaw, led to the publication of the *County of London Plan*, covering what was effectively the inner core of Greater London, minus the square mile of the City itself.[22] This last omission was filled by the City of London Corporation, which, after its first submission was rejected for failing to be sufficiently visionary, returned with a very different plan in 1947.[23] In this latter case, the planners responsible were Charles Holden and William Holford, but although they responded to the challenge to create a plan for the financial heart of a world city their narrow geographical remit limited its chances of success.

In the event, it was the *Greater London Plan*, with its aim of curbing London's growth,

Figure 7.5. 'How should we rebuild London?' The question was the title of a book, published in 1945, by the campaigner, C.B. Purdom. Illustrations were by Oswald Barrett (known as 'Batt'), with this one conveying a sense of grim determination to turn the various plans current at the time into reality.

that exercised most influence. Successive governments adopted this mantra of restraint; people and firms were induced by various means to leave the capital, and even the national government eventually relocated some of its offices to other parts of the country.[24] But there was an ambivalence about the process: a recognition of the plan to decentralize yet also an unwillingness to create an effective structure to make this happen. It was not until 1963 – with the formation of the Greater London Council and its remit to govern the whole of London within (and in parts including) the Green Belt – that an attempt was made to match metropolitan plans with a corresponding government structure. The experiment was weakened at the outset by the fact that the GLC was politically divided and was challenged at every turn by its constituent thirty-two London Boroughs and the City of London Corporation, sufficiently powerful in themselves to impede metropolitan policies. There were also sharp political differences between County Hall (the home of the GLC) and national government across the river at Westminster. The production in 1969 of an updated plan for London, the *Greater London Development Plan*,[25] was immediately undermined by a mixture of political conflicts and a failure on its part to reflect a changing public mood. A claim that the plan was designed 'to maintain London's position as the capital of the nation and one of the world's great cities'[26] proved empty rhetoric.

The Greater London Council was, from the outset, never far from political controversy, and it was this that eventually spelt its downfall. Quite simply, the single-minded Prime Minister, Margaret Thatcher, could no longer suffer the taunts from across the river of a radical leader of the GLC, Ken Livingstone, and in 1985 duly abolished the Authority. Not for the first time in its history, London was seen by central government as being just too important to trust to its own devices. And not for the first time London was left without an over-arching system of government, a situation that left different factions free to pursue their own interests. In something of a power vacuum, the City of London Corporation (fearing the emergence of a rival commercial centre in the former docks) competed with the London Docklands Development Corporation to match the supply of new office space. But with a downturn of the economy in the early 1990s, much of this in both areas remained empty; in 1992 the monumental docklands office development on Canary Wharf, for instance, recorded a vacancy rate of 40 per cent.[27]

More recently, the election, nationally, of New Labour in 1997 brought renewed hopes of a more unified system of government for the capital, with the promise of an elected Mayor and new agencies to support the office. Ken Livingstone was returned with an impressive majority as the people's choice. The prospect of making an impact on the lives of Londoners has, however, been seriously diminished by vindictive politicians in national government who continue to resist serious attempts by Livingstone to impose his own will.[28] Yet, in spite of his limited powers, Mayor Livingstone and the Greater London Authority have, to date, produced what amounts to only the third overarching plan for London in its modern history. *The London Plan*, as it is called, published in draft form in June 2002 and finalized in 2004,[29] marks an important change of direction, in which new growth is heralded as a sign of the capital's vitality. The challenge becomes one of providing for this growth by reversing what is seen as the chronic under-performance of London's physical and social infrastructure. His vision is not to restrain London but to develop it as 'an exemplary sustainable world city'[30] (figure 7.6).

The record of strategic planning over a longer period, however, shows that London has not enjoyed in the twentieth century the level of support bestowed on many other capital cities. Unlike Paris, it cannot demonstrate in the past century prestigious *grands projets*; unlike New York, it has not benefited from successive mayoral initiatives; and unlike Beijing it lacks the dynamism to re-make itself in a modern image. It has, in its recent history (prior to the appointment of a Mayor at the end of the century), quite simply missed the kind of strong leadership that a capital city of this stature might have expected. That it has retained its place as a great world city is hardly a reflection of good government; instead, in the words of Michael Hebbert, London is a great city 'more by fortune than design'.[31]

But if official support has been lacking, and if fortune alone cannot be seen as a full explanation, there must be something else to account for London's continuing world capital status. Its population is once again on a rapidly increasing curve, with a booming economy and an expected addition of 700,000 people over the next fifteen years to a total of 8.1 million.[32] London 'stands

Figure 7.6. Restraint and growth: Sir Patrick Abercrombie, the foremost architect of mid-century plans for the containment of London, and Ken Livingstone, Mayor of London at the start of the present century and champion of a more densely populated capital.

today in a unique position in the world economy . . . [it] has not had such a taste of pre-eminence since 1900 and the experience is unfamiliar'.[33] Clearly, there have to be factors other than official plans and policies to account for this.

The simple answer is that people want to be in London, to share in the capital experience. Most influential are corporate and institutional decision-makers, who see political and commercial advantage in doing so. Historically, this has long been the case. First, merchants and traders organized themselves into Guilds to protect and promote their own interests, influencing governments but not relying on them. Even today, the great Livery Halls in the City of London stand testimony to centuries of wealth accumulation in the capital. Industrialization in the eighteenth and nineteenth centuries, saw, at first, decentralization to the points of production. But, later, amalgamations led to centralized head offices, and the need for associated finance houses to organize investment (not only in British industry but also across the world) encouraged co-location in London. In the twentieth century, the process continued, surviving the demise of heavy industry and a corresponding rise in the service sector. More recently, the emergence of the information economy, although seemingly less governed by the need to be in one place, has simply reinforced past patterns of concentration.

Nor is the attraction of London confined to decisions taken in corporate boardrooms. None of this would work if successive generations of individuals were not prepared to brave the difficulties of living in a big city. London, in spite of its massive drawbacks, so eloquently described by William Morris and a long line of later critics, continues to exercise an irresistible pull on the very people who can make a city work. Young people are either lured to the capital for their higher education (the student population at any one time is more than a quarter of a million),[34] and stay on after, or decide after graduating elsewhere that life in the capital is preferable to provincial mundanity.[35] They see quite rightly that there is greater job choice, higher salary levels, and the buzz of the capital's social and cultural life. When they settle down and take responsibility for a family some of this allure palls, but by then, if they move away, they have made their own contribution to innovation and change in the capital.

And to an indigenous population must be added successive waves of immigrants to Britain, many of whom have seen London as the obvious starting point in their cultural journey. Before the twentieth century, first the Huguenots and later the Irish made their own social and economic contributions to the capital; followed by Jews from eastern and central Europe, who settled first in the East End. The twentieth century itself has witnessed even greater diversity, to create a truly cosmopolitan city. While some immigrants have more readily assimilated than others, there can be little doubting their part in revitalizing areas of the London economy and strengthening international links.

In the early years of the twenty-first century, it was becoming easy to feel complacent. London, surely, had risen to the many challenges of poor government and limited support in the previous century and was more strongly placed than ever to continue its role as one of the world's great capitals. Yet, on two consecutive days in July 2005 any tendency to a newfound sense of optimism was severely jolted. First, on 6 July Londoners were delighted to learn that their city had been chosen, against the odds, to host the 2012 Olympic Games. That evening crowds gathered spontaneously in Trafalgar Square, to celebrate; there was a warm feeling that London was a good place to be. Then, in the following morning's rush hour, terrorist attacks on the public transport system changed everything. No-one could question the fact that the terrorists had targeted London because of its symbolic status as a capital city. The awful lesson was that in the years to come there will be a continuing price to pay for capital status, not only in London but elsewhere across the world.

Capital Icons

The dominance, over the years, of London as a capital, it is argued above, owes little to political support and positive planning. Endorsing this view, the planning historian, Thomas Hall, has noted that by the end of the nineteenth century, in contrast to most other European capitals, London had grown without an overall plan.[36] In musing over this inherent contradiction, there is one further conundrum: why have there been so few attempts in the past century to express the dominance of London in its architecture and civic design? Where are the symbolic representations of its capital status that one might reasonably expect?

The conundrum is even more intriguing when one takes account of earlier examples of iconic schemes. Many of these are a product of the monarchical history of the nation, in which successive rulers (before and after the erosion of their powers in the seventeenth century) sited key strongholds in and around the capital: the Tower of London and royal palaces at Greenwich, Hampton Court and, later, Buckingham Palace (with its ceremonial Mall) in the centre of the city. The Houses of Parliament themselves date as a place of government from the eleventh century, while the neighbouring Westminster Abbey symbolically reflects the proximity of Church and State. Further east, St Paul's Cathedral rises above even much of the modern City skyline, one of the few enduring products of an earlier phase of rebuilding. To the many tourists who come to London these are all remarkable symbols of a great capital. But such symbols are also entirely pre-twentieth century, and perhaps what is really remarkable is how little of significance has been added since then.

Some gestures were made in the years before the First World War, when London was at its imperial zenith. At that time, parts of London were re-made in a fashion befitting an imperial capital: government offices along Whitehall, prestigious quarters of the West End, and individual buildings in the City. Although planned in isolation, they shared a style of architecture that was variously described as imperial, baroque and pompous, but reflecting, in populist terms, what people thought the capital of an Empire should look like. 'We have nothing as yet to compare with at least a half of London's magnificence', mused a contemporary traveller from New York.[37]

There was also a larger scheme: the great sweep of neoclassical buildings along Kingsway, around Aldwych and into the Strand, a project conceived in Victorian times but officially opened by the monarch in 1905. It was heralded at the time as the 'largest and most important improvement ... in London since the construction of Regent Street in 1820', and in the view of Schubert and Sutcliffe, 'its scale and appearance put it squarely in the lineage of Haussmann'.[38] It was undoubtedly a notable feat of urban engineering and it displayed an unusual sense of coherence, but as an imperial symbol the end result was hardly of world significance. The idea of the Strand as a link between Westminster and the City was sensible but the outcome was certainly not monumental, since the Kingsway itself (running from north to south) bears little relation to any of the main axes that define the capital. The scheme may well have been in the lineage of Haussmann, but Haussmann it was not.

Only towards the end of the twentieth century were there signs of a renewed engagement with symbolism. The celebration of a new millennium offered a unique opportunity for a bold statement of the capital's continuing role as a world centre; if earlier opportunities had been missed this was a fresh chance to say something about London in a post-imperial, post-modern world. There was no shortage of monumental projects, sufficient to encourage the view that 'millennial London [had become] a spectatorial space, a touristic theater even for its workday citizens'.[39] Another critic has even ventured that suddenly 'self-deprecating old London [is] bent on transforming itself into a city that has not just the civic grandeur of Paris, and the stylish bustle of Barcelona, but the glamour of Manhattan as well'.[40] The British Museum, the Science Museum, the Royal Opera House and Tate Modern were all cultural centres that attracted major renewal schemes. But there were three, in particular, that were designed primarily with the millennium in mind: the Dome, the Wheel (popularly known as the London Eye) and the Millennium Bridge (figure 7.7). All three are sited alongside the Thames, and each is in its own right an impressive structure.

As an icon feature The Dome (designed by Richard Rogers) is certainly impressive in appearance, like a mighty igloo, sited on former industrially damaged land by the Thames at Greenwich. It is several miles downstream from the centre of the city, but enjoys the kudos of a location astride the mean time meridian. The idea was the brainchild of the pre-1997, Conservative government, but when New Labour came to power in 1997, instead of abandoning what already seemed to many a hare-brained idea,

Figure 7.7. The Dome, the London Eye and the Millennium Bridge are three new riverside landmarks introduced to celebrate the Millennium.

the Dome was embodied as its own centre stage millennium project. It was intended to demonstrate the achievements of Britain in the previous millennium, but to many it simply took the form of yet another vacuous theme park. Visitor numbers in its first year were well below targets, with many Londoners simply exercising their individual right not to visit something that they had not asked for in the first place. Massive subsidies failed to make it viable and within the year it had closed. To add insult to injury it proved difficult to find a buyer, and only a scheme in which it was effectively given away with a vague promise of a share in future profits enabled the government to cut the unwanted tie.

In contrast, the 'Eye' took its place in the very centre of London, on the banks of the Thames alongside the former County Hall. Funding came wholly from the private sector, and the brand name of British Airways added to public confidence in the project (even though its opening was delayed by several months for technical reasons). In spite of the traditional nature of the concept – big wheels have been popular in pleasure gardens and exhibition sites across the world for more than a century – it, like the Dome, is an impressive structure. Some 135 metres at its highest point, the Wheel consists of thirty-two glass-covered capsules, each large enough for twenty-five passengers. Unlike the Dome, however, there is nothing obscure about its purpose, in that it does no more or less than offer visitors an opportunity to see the whole of the city. Londoners can pick out where they live in relation to everywhere else, while tourists with a single view can locate all of the popular landmarks. The appeal of this simple pleasure has proved to be enduring. It cost taxpayers nothing and was there to use or not. In the event, it has been enormously popular – with 3.5 million visitors in its first year of operation, and a larger figure in the following year.[41] The original expectation was that it would be a temporary landmark for the millennium but, as a result of its popularity, planning permission has been extended to enable it to remain for at least twenty-five years.

The third symbolic creation was the Millennium Bridge, an elegant structure designed by Norman Foster with the assistance of the sculptor, Antony Caro. It was eagerly awaited as the first new crossing of the Thames in a century, and offered its users spectacular views of St Paul's Cathedral as well as a useful route to the new Tate Modern on the south bank. Like the wheel its very simplicity made it immediately popular with the public. Yet, although it was restricted to pedestrians, it had to be closed within days of its opening because of an unfortunate propensity to wobble. Extensive engineering works had to be undertaken before it could be used again, at which point it once more attracted large numbers of users.

There is something faintly ironic about London's uncharacteristic flirtation with *grands projets*: one has proved to be an outright failure and two have succeeded only after embarrassing delays. But there is also something fitting about the experience, revealing a certain uneasiness if not a cultural scepticism of symbols for their own sake. London has never fully committed itself to the kind of bold civic statements that many other capital cities – like Paris, Berlin, Rome and Vienna – have done. Perhaps this is because of its own ambivalent status as the capital of a nation that opposed slavery while building an Empire, that is neither republican nor wholly monarchist, that despises autocracy while at the same time preserving an aristocracy, that is never comfortable with too much political intervention, that is essentially reformist but with a political revolution in its history, and that wishes to be seen as modern but is embedded in its own heritage. There is nothing new in the contradictions of its latest icons; it is simply that embodied in these is something of the modern history of London itself.

NOTES

1. The phrase *Heart of the Empire* was, in fact, popularized through a book published in 1901 with that title, edited by the Liberal journalist and politician, Charles Masterman.
2. For more details on the trade at this time see Hardy (1983).
3. White (2001).
4. *Ibid.*, p. 4.
5. *Ibid.*, p. 5.
6. Schneer (1999).
7. *Ibid.*, p. 10.
8. White (2001), p. 4.
9. *Ibid*, p. 211.
10. Greater London Authority (2002), p. 181.
11. White (2001), p. 212.
12. Each weekday, 466,000 workers travel by train alone into central London. Greater London Authority (2002), p. 183.
13. Greater London Authority (2002), p. 145.
14. A good account of the Festival of Britain is to be found in Banham and Hillier (1976).
15. The phrase was originally coined by William Cobbett, and used in his critical commentary on England *Rural Rides* published in 1830.
16. William Morris, *News from Nowhere*, first published 1890; see 1970 version, p. 55.
17. Kropotkin's most influential anti-urban tract was, perhaps, *Fields, Factories and Workshops*, first published as a book in 1899 and reissued with annotations by Colin Ward in 1985.
18. Howard, Ebenezer (1898); revised and republished as *Garden Cities of Tomorrow*, (1902).
19. For a comprehensive history of the organization and its campaign, see Hardy (1991). The Association became the Garden Cities and Town Planning Association, and later the Town and Country Planning Association.
20. The outcome was eventually published as the *Report of the Royal Commission on the Distribution of the Industrial Population*, London: HMSO, Cmd 6153, 1940.
21. Abercrombie (1945).
22. Abercrombie and Forshaw (1943).
23. This episode is recorded in Cherry and Penny (1986), pp. 136–141.
24. 66,500 Civil Service jobs were relocated from the capital between 1962 and 1989. Hebbert (1998), p. 133.
25. Greater London Council (1969).
26. *Ibid.*, Written Statement, p. 10.
27. Hall (1998), pp. 924–925.
28. This relationship is well illustrated by the case of funding to renew London's ailing underground rail system. Against the wishes of Mayor Livingstone the government has imposed a costly private/public partnership deal that will bring few benefits to Londoners for many years.
29. Mayor of London (2002).
30. *Ibid*, p. xii.
31. This is the sub-title of Michael Hebbert's book (1998).
32. Greater London Authority, (2002), p. 15.
33. Hebbert (1998), p. 150.
34. Higher Education Funding Council for England (2002), p. 129. See also Buck *et al.* (2002).
35. Greater London Authority (2002), p. 143.
36. Hall (1997*a*), p. 91.
37. Howells in White (2001), p. 11.
38. Schubert and Sutcliffe (1996), p. 116.
39. Levenson (2002), pp. 219–239.
40. Sudjic in Levenson (2002), p. 229.
41. Figures obtained directly from Press Office, British Airways London Eye.

Chapter 8

Tokyo: Forged by Market Forces and Not the Power of Planning

Shun-ichi J. Watanabe

The Tokugawa shogunate, which had been in power in Japan for two hundred and sixty-five years, came to an end in 1867 when political control was transferred to new rulers acting in the name of the young Emperor Meiji. Known as the Meiji Restoration, this marks Japan's birth as a modern nation. In 1868 the Emperor made an official journey from the imperial capital, Kyoto, to Edo, the city 500 kilometres to the east that had been the seat of the Tokugawa shogunate.[1]

The Emperor took up residence in Edo Castle, making it the Imperial Palace. There was much protest from the people of Kyoto, so the government decided not to make the change of capital to Edo official. However, a strategic decision was made for the Emperor to stay in his Edo residence, making the city the *de facto* capital, and the name was changed to Tokyo, meaning 'Eastern Capital'. Thus Tokyo, which has never been legally declared as the capital of Japan, became the seat of government at the beginning of the modern age.[2]

At the time, Japan was still a developing country whose economy was based on agriculture. Tokyo was located in the middle of the fertile Kanto plains; the central location in the nation was beneficial as well, and the port of Yokohama, a base for trade with the Western world, was only 30 kilometres away.

The city's population started expanding in the mid-1600s, reaching over 1 million in the late 1700s, making it the largest city in the world.[3] However, the departure of former Tokugawa officials and their retainers after the Meiji Restoration reduced the population to roughly half by the early 1870s.

The Meiji government tried to rebuild the castle town of Edo into the modern capital city of Tokyo. Of all the major world capitals today, Tokyo has the unusual history of being one of the only ones to have been developed as a modern capital city considerably after it had first become a big city.

As a capital city at the centre of Japan's efforts to modernize and Westernize, Tokyo represented the country's most important site for exchange with foreign countries. It became the showcase for such novel modern innovations as gas lamps, railroads, and brick-built districts. National railroad and telegraph systems connected the entire country to Tokyo. Many of the future leaders of the country migrated from the provinces to Tokyo, seeking jobs in the capital's important sectors, including government, military, business, and education. The city started to grow rapidly as a huge political, economic, and cultural centre.

Meiji leaders sensed the power and newness of Western cities and viewed them as symbols of civilization. They wanted to transform Tokyo into a showcase as grand as the Western capitals. Following a fire that destroyed large parts of the Ginza district in the heart of the city, the construction of the Ginza Brick District Project was overseen by the British architect Thomas Waters from 1872 to 1877. Shortly thereafter, the German planners Hermann Ende and Wilhelm Böckmann were invited to Japan in 1886 to develop a plan

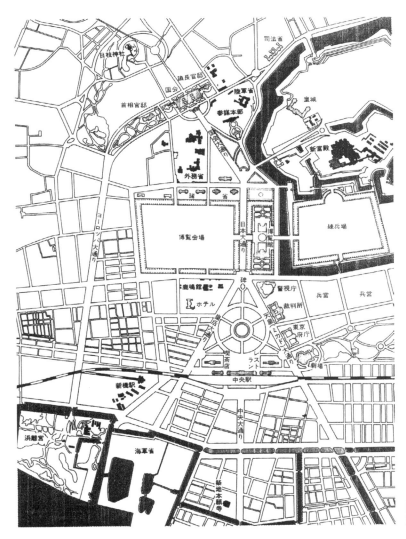

Figure 8.1. Government Quarter Project by Wilhelm Böckmann, 1886.

for the Government Quarter Project (see figure 8.1). The image the city was pursuing was the equivalent of Berlin as capital of Prussia, or the Paris envisioned by Baron Haussmann.

To the Meiji leaders, the destruction of the traditional built environment was a desirable symbol of modernization. In contrast, Edo Castle, now the Imperial Palace, was well preserved as a vast, kilometre square area of land in the midst of the urban core. For the Japanese, symbolism of the imperial system was not an overwhelming built environment, as in other nations, but the carefully secluded void creating a natural sense of solemnity. This rule treats the Imperial Palace as 'forbidden' space even to subways, and leaves the rest of Tokyo's built environment less symbolic but more functional as in ordinary large cities.

In 1888, to reconstruct the city as the seat of the Imperial government, the first planning legislation, known as the Tokyo Urban Area Improvement Act, was enacted.[4] The target area for reconstruction was Tokyo's urban core. The Tokyo Urban Area Improvement Commission was formed to discuss the plan, which obtained Cabinet approval and then was implemented by the Tokyo Prefecture. The Commission was under the jurisdiction of the Home Ministry, which, in the seventy-five year period until World War II, had great political influence due to its control of local government, police, urban planning, and building control.

The Japanese Planning System

Modern Western urban planning, which emerged at the end of the nineteenth century, aimed at total control of ever-growing industrial metropolises. Its goals were, first, to control urban structure at the large scale by planned provision of urban infrastructure and by planned development of suburbs and, second, to create comfortable urban spaces at the small scale through land-use controls and urban design.

In terms of urban structure, Meiji leaders wanted to improve the infrastructure of the urban core, but did not include the idea of controlling the urban structure of the entire city. In terms of urban space, they succeeded in building only limited areas in the urban core in Western style as symbols of civilization, but most of the city was left untouched, remaining a dense sea of overcrowded wooden residential structures. For a long period thereafter, urban planning in Japan dealt with individual urban facilities and buildings as separate entities through direct provision and indirect regulation. An explanation why they did not have a more expansive vision lies in the very different urban physical and political conditions in Japan in contrast to Western Europe and North America.

The Western approach to urban planning started with preparations of plans by professional planners and development of legal mechanisms to implement such plans. In contrast, planning did not even exist as a profession in Japan. It was accepted that urban growth was a natural process, and that the aim of planning was to mitigate incrementally the accompanying problems. The scale and speed of urbanization were great, but sufficient controlling power was not given to planners to combat urbanization.

An Emerging Metropolis, 1880s–1920s

After the initial post-Restoration decline, Tokyo's population began to increase again, passing Edo's peak by the mid-1880s. By 1900, the population was a little less than 2 million, and the urban area reached a radius of 7 to 8 kilometres from the Imperial Palace. To its west, the urban area expanded beyond the borders of Edo, spilling into other towns and villages. This suburbanization, which began about fifty years later in Japan than in major cities of the West, signified Tokyo's growth into a metropolis. As the city set out on a course of industrialization and modernization, urban growth took place in a regulatory vacuum without any of the land-use controls of modern urban planning.

Looking at Tokyo's urban core at the beginning of the twentieth century, we find that the pre-existing urban form of the castle town of Edo had almost disappeared as a result of reconstruction projects by the Tokyo Urban Area Improvement Commission. It was common to place modern facilities in the urban core, replacing the traditional built environment (except for the castle) instead of creating new precincts outside. Along with the construction of Tokyo Central Station in 1914, there was a modern office district and a new government building district. In 1920, the construction of the National Diet Building started on former army land about 500 metres south of the royal moat (see figure 8.2). The Diet's monumental façade refrained from facing the Imperial Palace but faced Tokyo Station. The location and orientation were exactly as in Böckmann's Government Quarter proposal some thirty years before.

The Diet Building marked this spot as the new centre of the capital of the modern nation. If one stood nearby at the main intersection in Hibiya and looked around, to the north-east one could see the newly developed Hibiya office district, to the south-

Figure 8.2. National Diet Building, 1920.

east the traditional Ginza shopping district, to the south-west the fairly well planned Kasumigaseki government quarters, and to the north-west the Imperial Palace. All of the most important areas of the capital were located within a mile.

This new age of expansion brought innumerable problems. Tokyo was a city of great population density despite its inadequate infrastructure. The streets that resulted from urban improvements were incomplete – covered with dust during dry periods and full of muddy pools in wet weather. Streetcars were always full, and a network of elevated trains or subways had yet to be built. Industrial pollution affected the city and suburbs, sewage systems were undeveloped and only the urban core had water supply and electricity.

The government tried to respond to the pressures associated with Tokyo's rapid urban growth. Especially after World War I, major cities in Japan suffered frequent urban rioting by industrial workers and the urban poor, and the security of the capital became a serious political issue. The Home Ministry, which had focused on agricultural areas, turned its attention to the hitherto unexplored urban areas. This led to the introduction of modern urban planning ideas then being developed in Western Europe and North America.

The City Planning Act of 1919

The 1919 City Planning Act became Japan's first planning law for the entire area of all major cities, not just the urban core of Tokyo. It institutionalized two important planning tools: (1) land readjustment programmes (*Kukaku Seiri*), adapted from German models; and (2) zoning system, adapted from US models.

The Act, basically designed for the rapidly expanding capital city, was a highly centralized system controlled by the national government with little power given to prefectures or municipalities. This planning system was uniformly applied to other Japanese cities despite their situations being quite different from those of Tokyo. This new social technology of planning was carried out directly by the national government through the City Planning Section of the Home Ministry. The Tokyo City Planning Commission, which succeeded the Tokyo Urban Area Improvement Commission, was also placed within the Home Ministry and was chaired by the Deputy Home Minister.[5]

For urban design, the 1919 Act expected that planning would first enable orderly development of urban land through land readjustment, and then building control would place new buildings under proper control. The urban image was still vague, but Shigeyoshi Fukuda, the city's architect, published a detailed vision for Tokyo's future.[6] The Tokyo metropolitan area had a population of about 3 million in 1918 and

Figure 8.3. New Tokyo Plan by Shigeyoshi Fukuda, 1918.

was expanding beyond its boundaries. Fukuda estimated that this population would grow to 6.76 million fifty years in the future. He concluded that considerable land-use planning would be needed to accommodate the growth in an area within an average commuting time of one hour. Fukuda depicted a hierarchical system of the urban core and sub-cores and an overall land-use plan; he presented a plan for a network of urban infrastructure including subways, elevated trains, and major thoroughfares (see figure 8.3). This master plan proposal was probably the first master plan in Japan. It demonstrated considerable reflection on the city's urban structure, but barely touched on the urban space in the city.

The 1919 Act aimed to solve urban problems in existing overcrowded urban areas by creating orderly suburban development and efficient transport systems. As a result of private railroad company development activities, provision of transportation to the suburbs succeeded. However, the urban area that emerged lacked order because of weak land-use controls. The 1919 Act thus succeeded in influencing urban structure but failed to improve urban space adequately.

The Capital under Reconstruction, 1923–1935

At lunchtime on September 1, 1923, the Great Kanto Earthquake of magnitude 7.9 hit the Tokyo and Yokohama areas. Fires raged for three days, burning 3,600 hectares of land in the urban core, or 46 per cent of Tokyo's land area. The toll of the earthquake and the 134 subsequent fires was a loss of 70,000 lives, or 3 per cent of the city's 2.3 million inhabitants. The homes of 1.55 million people, or 67 per cent of the population, were lost. This calamity was Tokyo's first large-scale natural disaster as a modern city. The government temporarily put aside discussion of national urban planning policy and focused its efforts on reconstruction.

The Home Minister Shinpei Goto, whose efforts had provided the impetus for the creation of the 1919 Act, became director of the Imperial Capital Reconstruction Agency created directly under the Cabinet. Fearing loss of public confidence, Goto quickly declared that 'the seat of the Emperor should not be moved from Tokyo'.[7]

Goto succeeded in passing the Special City Planning Act and a reconstruction budget of 46.8 million yen (or US$23.4 million). During reconstruction, the planning of Tokyo was transferred from the Home Ministry to the Reconstruction Agency. Over 6,000 specialists were brought to the agency from around the country.

Reconstruction took seven years to complete, and the final expenditure was 82 million yen. The main projects undertaken were land readjustments that provided roads, bridges, parks, and neatly designed blocks and lots on the destroyed land. At first, landowners protested strongly against land readjustments because part of their land was to be taken for public use, but in the end approximately 80 per cent (3,000 hectares) of the burned land was renewed by land readjustment. The whole reconstruction programme can be evaluated as a great success in terms of the immensity of the undertaking, the development of planning and design methods, and the training of many planning experts.

The Reconstruction Agency did not release any master plan that illustrated a desirable urban image as the reconstruction goal.[8] However, the focus of their efforts was clearly to make the city more earthquake and fire-resistant, and to modernize it. Above all, reconstruction destroyed the castle town of Edo and created the modern capital of Tokyo. It brought broad and straight streets in a more or less gridiron plan to the urban core, the long-time dream of planners and residents. In contrast, because of the lack of land-use controls, disorderly developments occurred in suburban areas, where many people moved after the earthquake. It was not until 1925 that the first zoning regulation was put into effect in the Tokyo City Planning Area covering the city and surrounding suburban municipalities stretching 16 kilometres from the centre.

The Wartime Capital, 1935–1945

Japan became increasingly militaristic in the 1930s and officially entered World War II in December 1941, and urban planning became linked to the war effort. The primary concern for Tokyo was air defence, and the 1919 Act was amended in 1940 to include it as a planning goal. In response to American bombing raids, which began in April 1942, air defence was sought through the use of fire-resistant construction methods, establishment of parks and open spaces, and forced evacuation of residents. In the context of war, urban planning no longer involved the building of the city, but was concerned with its dissolution.

   Tokyo's Greenery Plan, under preparation between 1932 and 1939, had established an 'open space' zoning classification. With the outbreak of the war in 1941, however, parks and open spaces changed their function from environmental preservation to air defence. Thus, the Air Defence Greenbelt was designated in Tokyo in 1943. Such a measure was taken only in Tokyo and its region due to the importance of the national capital.

As Japan expanded militarily into East Asia, Tokyo had become not only the capital of the country, but also the military capital of East Asia. However, there were no significant architectural or planning efforts undertaken in Tokyo to symbolize this new status. By the time Japan surrendered in 1945, compulsory evacuation and bombing of the capital had reduced Tokyo's population from 7 to 3 million.

Rebuilding the Bombed Capital, 1945–1960

World War II ended on August 15, 1945. General MacArthur, in charge of the Allied Occupation Forces, established his General Headquarters in the former Daiichi Life Insurance Building. It was located well within a mile of the Diet and Imperial Palace and reflected its imposing form in the royal moat, as if watching the Imperial Palace as the true ruler. MacArthur found that Tokyo was the perfect location for a command centre to control Japan, and he governed the defeated nation by using fully the existing Japanese bureaucratic system.[9]

Tokyo sustained the worst destruction and casualties of the 120 cities damaged during the war. A total of 852,000 houses were destroyed, 88,000 lives were lost, and 15,900 hectares of land were burned, covering approximately 28 per cent of Tokyo's land mass.[10] The housing problem was severe, as houses had been lost due to air raids and compulsory evacuation and also housing was needed for citizens returning from abroad after the war.

Progress on land readjustment was exceedingly slow due to the acceleration of inflation, shortage of experts, and opposition from residents. Only 1,200 hectares of land (6 per cent of the originally planned area) were actually treated, and the rest of the area left unplanned. Another problem in conducting land adjustment activities was the lack of an official master plan to regularize and systematize planning efforts. The city's leaders were facing urgent problems needing immediate resolution, and did not have time to contemplate 'images' of how they wanted their capital to look. MacArthur was eager to reform institutions for a democratic and anti-military Japan, but was not eager to create a new planning system or urban image for Japanese cities. Furthermore, the new constitution declared the Emperor to be a mere symbol of the State without any political power. As a result, Tokyo did not seek to create any positive urban image that would be associated with such power.

In terms of urban structure, reconstruction planners gave up any hope of having total control of the urban area, as such a goal was unrealistic under these conditions. However, their reconstruction efforts succeeded in rebuilding limited but strategically important areas such as the urban core and areas surrounding major train stations.

In terms of urban space, they succeeded in rebuilding wide major thoroughfares, but buildings facing them were poorly landscaped, probably because severe post-war economic conditions did not allow private construction companies to take advantage of the opportunity. The construction of fireproof, high-rise buildings on the major thoroughfares would have to wait until Japan's period of rapid economic growth that started in the 1960s. While post-war reconstruction stopped short of large-scale modern building efforts, it is fair to say that it provided the foundation that enabled the subsequent rapid growth of Tokyo.

The Capital in the Economic Boom, 1960s–1970s

After gradual recovery in the 1950s, Japan entered a period of high economic growth in the 1960s. During this period, the country completed its conversion from an agricultural economy to an industrial one, resulting in highly accelerated migration from rural farming areas to large cities, especially Tokyo.[11]

Tokyo's Ward area[12] had a population of 2.78 million at the end of the war in 1945; the population rapidly increased to 6.7 million in the next ten years. By that time, Tokyo's metropolitan area started to spread over the Kanto plains into other parts of Tokyo Prefecture and the three adjacent prefectures. The metropolitan population amounted to about 15 million. By the 1970s, the metropolitan area had spread even further, with the population reaching about 30 million.

In Tokyo's core, any remaining available land, including gardens, was used to build small-scale wooden rental apartments to address the housing shortage.[13] During this period of rapid growth, rather than planning determining urban form, economic factors precipitated growth, resulting in a lively but disorderly urban space.

Tokyo, in addition to drawing some of the principal administrative functions from Osaka and Nagoya, became an increasingly important national headquarter that managed the national economy and connected it with the international market. The capital function developed there was not only politico-administrative control over the nation but also societal control through the 'iron triangle' composed of political, administrative, and business sectors.

The keywords in the era of rapid growth were 'information' and 'international'. Tokyo's symbol of 'information' was the 333-metre Tokyo Tower built in 1958. One could view the sweep of the Kanto plains from its observation deck. It embodied Tokyo's role as the cultural capital sending out information about the social fashions, values, and ethos of the day as the national centre of TV networks rather than the political capital that directly controlled the nation.

The 'international' symbol was the Tokyo

Olympics of 1964. This event showed the country and the world that Japan had recovered from the war. The government took advantage of the event to build the Capital Expressway, the first urban expressway in Japan, and to improve major thoroughfares. Previous restrictions on building height were replaced by building floor area ratios, which allowed the construction of skyscrapers. The former water purification plant in the Shinjuku sub-centre was renewed as a high-rise district. Thus, Tokyo's urban structure and urban space were dramatically transformed, providing a physical setting that would allow further high growth.

The era of rapid growth also signalled a new phase of urban visions, with large-scale development proposals like Kenzo Tange's Tokyo Plan 1960 (see figure 8.4). It proposed a high-rise city of residences (for 10 million people), office buildings, and expressways in Tokyo Bay. On the periphery, the Tama New Town Plan, accommodated 300,000 people in 3,000 hectares of undeveloped hilly land about 30 kilometres west of Tokyo Station.[14] Based on a well-coordinated master plan, the urban structure of the new town as a whole has been appropriately controlled; innovations in the design of multi-family and single-family dwellings, residential space, and

Figure 8.4. Tokyo Plan 1960 by Kenzo Tange, 1961.

open space have led to much better designed urban space.

The strain of very rapid growth in the 1960s led to increased social pressures in the 1970s. Overcrowding, environmental pollution, noise disturbance, building disputes for sunlight exposure, housing shortage, and long commutes, as well as the oil shock of 1973, were among the issues confronted. The 1968 City Planning Act proved inadequate in protecting the environmental quality of residential areas.[15] The political mood shifted from a conservative one that supported rapid growth for big enterprises, to a progressive one that valued decentralization and a high quality of life and environment for residents and workers.

This shift was clearly symbolized by the successful 1971 re-election campaign of Tokyo's progressive Governor Minobe. He announced a concept of 'Open Squares and Blue Skies for Tokyo' in a plan proposal calling for citizen participation (to discuss many issues in the 'open squares') and control of the environment (to regain 'blue skies') but did not envisage any concrete urban structure or urban space.[16] What is remarkable about the election is that, for the first time, urban policy for the capital city had become the central issue of a campaign.

In 1975, public elections for the heads of Tokyo's wards were restored after twenty-three years. This resulted in a large shift in administrative responsibilities for urban planning to each ward. From this point, each ward, as an independent municipality, started to develop know-how in administering urban planning. Decentralization went into full swing. Responding to this political mood, citizen movement groups consequently sprang up advocating various kinds of community building, or '*machizukuri*'.

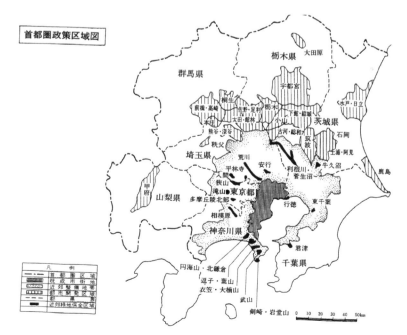

Figure 8.5. Capital Region Improvement Plan, 1958.

Tokyo in the Capital Region, 1950s–1990s

Tokyo's rapid growth in the post-war era became a serious national issue. The problem was not the city's role as capital, but rather the over-concentration of the nation's political, economic, and cultural activities. The problem grew from city to regional scale, so the government enacted the Capital Region Improvement Act in 1956, regulating the area roughly extending in a 100-kilometre radius from Tokyo Station (see figure 8.5). Two years later, the nationally appointed Capital Region Improvement Commission announced the first Capital Region Improvement Plan. It became the first statutory urban regional plan in Japan and was amended almost every decade, with the fifth plan in 1999.[17]

The capital region planning that started to combat Tokyo's growth problems ended up as a weak regional mechanism to co-ordinate roughly the location of such regional infrastructure as expressways and railways and that of population, urban settlement, and regional green space. The notion of a planned effort to build and maintain the capital itself was weak; the visionary approach to capital planning was lacking. In its place were a series of efforts that minimally addressed the pressures created by continued growth of the city.

Later amended in 1976, the third Capital Region Improvement Plan adopted the policy to designate Yokohama and other strategic large cities outside Tokyo's Ward area as Business Core Cities and to encourage their development as regional sub-centres. The plan assumed that the population and national functions, if seen at the national scale, would continue to be concentrated in the capital region itself. It also recognized a new stage of development in which business functions, in addition to residential, industrial, and research functions, were to be decentralized from Tokyo to sub-centres within the region.

However, this policy of intra-regional decentralization was in some way ambivalent, as the plan also allowed business functions to continue to locate in the Ward area as well as the newly developed Bay Area sub-centre in Tokyo's port area. Here again, the power that determined the location of various urban functions that create urban structure and urban space was not the power of the government or planning but the power of the market.

'Global Capital' in the 'Bubble', 1980–2000

With the globalization of the Japanese economy in the 1980s, the strong yen invited foreign financial organizations to establish operations in Tokyo. Tokyo, located in a time zone staggered from those of New York and London, ranked along with the major stock markets and financial activities, justly earning the status of a Global Capital. However, these developments were not the result of a concerted planning effort, but rather the happenstance results of strong economic markets and subsequent financial investment. The central management functions of finance and information continued to concentrate heavily in Tokyo, resulting in increased social and political pressures.

In the mid-1980s, enormous demand for office space led to abnormally inflated land prices, which first started in the urban core of Tokyo, spread to the surrounding areas, then to suburban areas, and finally to other large cities all over Japan. The 'bubble' economy saw giant speculative office projects in the urban core and established residential communities cleared for redevelopment. The face of Tokyo changed considerably, but in 1991 the bubble burst, and four years later land prices dropped to one-quarter of their worth at the height of the bubble.[18]

From a planning perspective, the key projects of the bubble era were the Bay Area Sub-centre project and the plan for Relocation of the Capital Functions from Tokyo. The Tokyo Bay area had been a favourable site for land reclamation projects as far back as the Edo period (1600–1868). The Tokyo prefectural government began large-scale reclamation projects as part of the improvement of Tokyo Port, starting with a teleport in 1985. The 1987 Bay Area Sub-centre Plan included 115,000 jobs and 44,000 residents, covering 448 hectares of land in Tokyo Bay. The prefectural government created the landfill areas, provided the infrastructure, then leased building lots to the private sector at high prices. However, with the collapse of the bubble economy in 1991, demand from the private sector disappeared, and the project became a financial drain for the city.

Second, the relocation of the capital functions away from Tokyo has been a frequent public issue since the 1960s. With a fairly simplistic idea, some argued that growth had made Tokyo too big and too dominant, and that this problem could be solved through relocating the capital functions. There was, however, little discussion of feasibility. The topic was reintroduced suddenly in the early 1990s, when the extreme concentration in Tokyo came to be seen as one of the most serious domestic problems. Along with the rise in land prices came a worsening of living conditions; also, with Tokyo's sapping of various functions from all over Japan came a loss of economic vitality in the rest of the country.

Recognizing these growing problems, the national Diet enacted the Act of Relocation of the Diet in 1992. The idea was to relocate the Diet itself rather than the entire central government and Imperial Palace. It is amazing that such an important enactment as this went on with almost no intensive public debate. Three years later, in 1995 an investigation committee recommended that the new city with the Diet and other necessary functions would require 9,000 hectares of land and would accommodate a maximum population of 600,000. The committee also announced that construction should start no later than 2000. The next year, in 1996, a commission was formed to choose the site. Many areas applied, and it took three years to narrow down the candidates to two. Tokyo Prefecture, which until then had been fairly complacent, began to become vocal in its protest. And in the current post-bubble era, timidity has replaced bold planning, and the project has hardly moved forward. Now people regard it as unrealistic.

The capital problem of a nation should be a great concern not only to the people and government of the present capital but also to the entire nation. Yet the rationales for the relocation have been changing from time to time and are not very convincing. They have ranged from its contribution to projected increases in domestic consumption, to safety concerns in times of natural disasters, to renewal of the mood of the time, and to the benefits of decentralization. The issue has been dealt with very politically. The Act passed the Diet so easily because the bubble economy and post-bubble troubles have distracted the people and government. There may have been hidden envy towards Tokyo's material prosperity by the provinces and the hidden desire of the people of Tokyo to run away from serious urban problems.

Tokyo's Future Perspectives, 2001 and Beyond

In concluding this chapter, it is appropriate to

address briefly the future perspectives for Tokyo in the twenty-first century. As more and more people were disappointed by the failure of the government sector, the expectation for the market and voluntary (or citizen) sectors grew around the turn of the twenty-first century.

One example from the market sector is the Urban Renaissance Program pushed by Prime Minister Koizumi since 2002. It facilitates urban renewal by private developers by deregulating planning and building controls. The type of urban image being sought approaches that of Le Corbusier's 1930 plans for megastructures with large open space areas. However, many critics believe that megastructures will lead to degradation of residential quality of life, and may invite another bubble of land inflation.

For the voluntary sector, expectations that focus on the role of citizens have risen, especially since the 1990s. Citizens as free and independent individuals began to work in local affairs and to participate in local government through '*machizukuri*' or 'community building' efforts. They envision a city built on a human scale in incremental, small-scale construction projects. With citizens' new desire to participate actively in urban planning, they are developing the necessary skills and tools to contribute to planning processes. Tokyo in the twenty-first century will likely be a mosaic of well-designed and efficient post-modern structures along with small-scale residential and commercial mixed-use developments that create a very comforting environment.

Finally, it is helpful to draw a comparison between Tokyo and other capital cities. With the exception of part of the Meiji era, Tokyo has placed very little emphasis on plans for its urban structure and urban space as the capital of Japan. Rather, because Tokyo was the political, economic, and cultural centre of the country, the past hundred years have been a time of rapid population growth and accompanying urban problems. While it may be a slight exaggeration, one could say that the history of Tokyo was that of a large city rather than the history of a capital city. The history was determined far more by the natural forces of the market than by the power of planning.

As we embark on the twenty-first century, Japan is entering a phase of negative population growth. Can we easily change the previous Japanese planning models that have been premised on continued population growth, into a new planning paradigm that is not premised on growth? Can we design the various functions of the capital city into a proper physical environment? Can we remake the Tokyo that has been a convenient place for business and government into a truly liveable city for its residents?

These are questions we need to ask as we move forward to consider our vision for the future of the capital city of Tokyo.

NOTES

1. During the Edo period, Edo was the *de facto* political capital, Kyoto (which had been the imperial capital since 794) remained the *de facto* cultural capital, and Osaka was the *de facto* commercial capital. See Sasaki (2001).

2. For an overview of Japanese city planning, see Ishida (2004) and Watanabe (1993). For the planning history of Tokyo, see Fujimori (1982), Ishida (1992), Ichikawa (1995), Jinnai (1995), Koshizawa (1991), Tokyo Metropolitan University (1988) and Watanabe (1980, 1984, 1992). For general histories of Tokyo, see Cybriwsky (1991), Ishizuka and Narita (1986) and Seidensticker (1983, 1990).

3. Smith (1979), p. 51.

4. At that time there was no word for 'urban planning' in Japanese and the word 'urban improvement' was the closest. See Fujimori (1982).

5. In other prefectures the commissions appointed their governors as chairs. Building administration was

carried out through the police bureau of each prefecture under the jurisdiction of the Home Ministry.

6. His article, article 'Shin-Tokyo (New Tokyo)' was published in 1918 with the assumption that an urban planning law would be in place to regulate the city's development. See Suzuki (1992).

7. See Tsurumi (1976), p. 587.

8. Just prior to the earthquake, Goto had invited the American political scientist Charles A. Beard to tour Japan and lecture on city planning, a topic that was still unfamiliar to the Japanese. The earthquake occurred immediately following Beard's departure, and Goto invited him back to Japan for advice on reconstruction planning. Beard's proposals (see Beard (1923)), based upon 'peace-time planning', were too idealistic and hardly suitable, according to the skilled specialists of the agency. The reception for Beard's proposals was poor, and he returned discouraged to the United States the following year.

9. The Occupation Forces, working to transform Japan into a democratic and decentralized nation, abolished the Home Ministry in 1947. The Ministry of Construction was established in 1948 as the national organization that oversaw the country's planning, building, and housing activities.

10. Ishida (1992), p. 143.

11. Along with Tokyo, their destinations also included such cities as Nagoya and Osaka, in the so-called Pacific Belt or Tokaido Megalopolis, stretching southwestward from Tokyo. New expressways and bullet trains linked cities within this zone in a high-speed transportation network.

12. At the turn of the twentieth century, *Tokyo-shi* (City of Tokyo) consisted of 15 *Ku* (wards), and the rest of *Tokyo-fu* (prefecture) was divided into *Gun* (county) and *Shi* (city) areas. Annexations in 1932 created one of the world's largest cities. Currently, each of Tokyo's 23 *Ku* corresponds roughly in function to the municipalities of other prefectures.

13. In the suburbs, areas that did not have the necessary urban infrastructure (roads and water supply) witnessed disorderly development, or sprawl, although the patterns of growth were different from sprawl in North America.

14. The plan was later amended to provide housing for 370,000. Ground was broken for the project in 1969, and expansion is still underway at the present.

15. The 1968 Act was the first full-fledged legislation to address issues of land-use controls. It mainly aimed to prevent urban sprawl through a new development control system, and more detailed land-use categories.

16. In contrast, the Liberal Democratic candidate Hatano proposed a concrete plan to invest 4 trillion yen over a five year period for the redevelopment of Tokyo's urban core, including railroads, thoroughfares, and high-rise housing. This plan shared Tange's vision to some degree.

17. The first plan designated a regional core, a greenbelt (not realized), and satellite urban development areas, for new towns and improving existing cities. Many dormitory communities were created, but the only self-contained new town was Tsukuba Science City, which, located 30 kilometres north-east of Tokyo, attracted dozens of ministry laboratories from Tokyo.

18. The economic failures that resulted have led to ten years of stagnation referred to as the 'lost decade'.

Chapter 9

Washington: The DC's History of Unresolved Planning Conflicts

Isabelle Gournay

The equation: 'Washington + planning = L'Enfant 1791 + McMillan 1902', although justified from a historical and aesthetic standpoint, obscures growth patterns experienced by the city and its suburbs over the last century. The District of Columbia (DC) had 280,000 inhabitants in 1902 but major suburban growth occurred during the Depression and New Deal. By 1950, the National Capital Region (NCR) was home to 1,752,248 people, with 46 per cent (802,178) living in DC and the rest in the Maryland and northern Virginia suburbs.[1] In 2000, with only 572,059 inhabitants, DC ranked twenty-first among US cities and accounted for less than 12 per cent of the region's total population of nearly five million. Although Washington still represents the epitome of Peter Hall's Political Capitals on a global scale, the percentage of federal employment within the DC limits has decreased since 1900.[2] Yet, it is undeniable that the city's and region's livelihood depend on the administrative sector and service industries.

L'Enfant's plan inaugurated modern capital city planning,[3] putting in sharp focus the relation between political and physical networks. Its layout exemplifies how, to quote Stephen Ward, the United States 'gave back to Europe the notion of the city-wide master plan and the grand approach to urban landscape design'.[4] The Frenchman's vision of widely set apart 'centres of decisions' for the legislative and executive branches allowed the Federal district to stretch several miles and have room to grow.

L'Enfant's firing by George Washington foreshadowed how the twentieth-century planning history of the US capital was one of unresolved conflicts and endemic tensions. Three major explanations come to mind. First, the notion that Washington belongs to all US citizens, rather than to its inhabitants, is fixed in the national psyche. Accordingly, its planning enacts rituals of nationhood and symbolizes democratic ideals and international power. Enhancing the beauty of the ceremonial core took precedent over improving surrounding neighbourhoods because it catered to both patriotism and the tourism industry.[5]

A second explanation was (and remains) the imbalance between the city's demographic, economic and cultural significance and its political stature. Inherent to all political capitals, this disparity is particularly marked in a country as vast and ethnically diverse as the United States. Washington appears more cosmopolitan than other North American cities of similar size, with the seat of the Organization of American States (housed since 1910 in a magnificent building near the White House) and of post-World War II institutions such as the World and InterAmerican Banks and International Monetary Fund. The region is now home to a growing percentage of newcomers hailing from Latin American and Asian countries, but it has not entirely relinquished its Southern parochialism.

A third source of conflict is 'taxation without representation'.[6] Before 1967, there was no responsible local governance, only a board of commissioners consisting of two civilians and a member of the Army Corps of Engineers who administered the District. Responsible local governance was established in two stages. In 1967, President Lyndon B. Johnson appointed a city council and a 'mayor-commissioner', the African-American lawyer and housing official Walter E. Washington. This was followed by mayoral and council elections in 1974, in which Walter Washington was elected Mayor.[7] Washingtonians also gained some local governing power in 1964, when citizens were first allowed to cast their votes in Presidential elections. Although statehood is still denied, DC does have a non-voting delegate in Congress, whose role is to lobby on behalf of Washingtonians.

The preparation of plans for DC, and their implementation continue to depend upon 'feudal' annual congressional appropriation, an inadequate system given the transient and volatile nature of US legislative and executive leadership. For example, Congress uses its power to write legislation that bypasses local agencies and affects districts beyond the Federal precinct. For instance, Congress's designation of the entire Georgetown area as an historic district made gentrification official and triggered the exile of a large Black community.

DC planning is also affected by the District's financial insecurity, which influences the provision of public services. Its tax base is reduced by two factors: first, federal agencies are tax-exempt and second, like in many metropolitan areas in North America, most persons working in the District pay taxes in their place of residence in the suburbs.

Autocratic governance made little room for private citizens' participation. A group that carried a certain weight was the Committee of the 100 on the Federal City, founded in 1923 as a force of 'civic conscience' filling the gap between 'parochial self-interest and national politics'.[8]

In the late 1950s, Washington became the first large US city with a majority Black population. Consequently many conflicts affecting its planning have related to enduring racial inequalities, with a clear geographic divide. Districts west of Sixteenth Street (the north-south street on axis with the White House) are predominantly affluent and White while neighbourhoods further east are generally much less privileged and home to a large African American population.[9]

The need to cope with the opposing demands of the central city and suburbs resulted in constant shuttling between downtown plans and broad-based regional schemes. Ideological tensions caused other problems when profit-driven boosterism clashed with idealism, or technical expertise with partisan politics. Institutional or personal rivalries pitted federal agencies against District commissioners, mayors and citizen groups. Infighting occurred among agencies with ill-defined and fragmented responsibilities and, within agencies, between political appointees and professional staff. Conflicts of expertise added to the problem, especially out-of-town engineers opposing local infrastructure planners and high-profile architects for whom 'fixing DC' became an ego trip.

The McMillan Plan

The first major capital city plan of the twentieth century is usually named after its political champion, Michigan Senator James McMillan. As chair of the Senate Committee for the District of Columbia, he forged a behind-the-scenes alliance with the American Institute of Architects, park advocates and the Washington Board of Trade, a powerful organization in the absence of an elected city council. Members of the Commission he sponsored – architects Daniel Burnham and Charles McKim, sculptor Augustus Saint Gaudens and landscape architect Frederic Law Olmsted Jr – were exceptionally talented, charismatic and energetic. Retelling how McMillan shrewdly played the 'park card' to gain Congressional approval, Jon Peterson has convincingly stressed the plan's 'extraordinary breadth and complexity for its day'.[10]

Rarely did a plan arrive so much 'at the right time and in the right place'. The United States was beginning its ascendancy as an international power boasting political, economic and military pre-eminence. Drastic measures were needed to house a fast-expanding federal bureaucracy. With Theodore Roosevelt at the White House and other anti-trust politicians in power, it became possible to implement a key element of the plan – the removal of the railroad from the Mall. The terminal was relocated north of the Capitol grounds, in Union Station, Daniel Burnham's magnificent gateway and transportation hub.

Washington was becoming a cosmopolitan 'winter resort' for retired industrialists turned congressmen. Improvements had already taken place: in the early 1870s, trees had been planted by 'Boss' Alexander Shepard along L'Enfant's extra wide streets, land reclaimed along the Potomac, and height limitations (reinforced in 1910) implemented to protect the visual dominance of the Capitol dome. Given the overall mediocrity of other projects triggered by Washington's centennial as government seat, the moment was undeniably opportune for Burnham's cohort. They worked fast, touring Europe soon after the Commission was authorized, and preparing plans on the voyage home.

The Commission published a well-documented report and succeeded in building public support for their ideas. Their public relations and representation methods set important precedents, such as before and after models of the Mall that illustrated the contrast between mid-nineteenth century picturesque ideals and those of the City Beautiful Movement (see figure 9.1). Mounting exhibitions, organizing conferences with proceedings, giving cocktail parties, publishing lavish brochures, providing copy for journalists became part of the Washington planning ritual.

Although divided between its democratic intentions and subliminal 'imperialism', the McMillan Report remains an impressive and pragmatic 'working document'. For instance, in pre-air conditioning days, it wisely made provision for recreation in the shadow of the monumental core, to alleviate the hardship of heat waves. Inspired by precedents which the Olmsted firm prepared for Boston in the 1890s, the vision for an outlying park system was particularly influential. The north-western linear system following Rock Creek would 'foster urban renewal, transforming the creek's environs from marginal housing and scattered industries into dignified and prosperous residential neighborhoods'.[11] The system of stream valley parks generated by tributaries of the Potomac and Anacostia Rivers was extended miles beyond the District's limits in Maryland, an asset for residential districts that was fully exploited by planners up to the late 1970s.[12]

The McMillan Report illustrated many European planning landmarks and its proposals were derived from these precedents. 'Creative borrowing' was justified by a clear intention to 'restore, develop and supplement' L'Enfant's vision.[13] To Burnham and his friends the literal, almost 'naïve', aesthetic borrowings, practically inevitable in the charette climate that presided during the preparation of the drawings, played second fiddle to the general, long lasting 'character' they wanted to impart. Old World precedents were domesticated in the Mall, the plan's central spine. For example, in the design for the pool reflecting the Lincoln Memorial, a cross-shaped scheme similar to Versailles' Grand Canal was abandoned for much simpler geometry. This clear-cut dialogue between natural and man-made landscapes is purposely more abrupt than in Europe and a highly distinctive and successful trait of American urban design.

Extended westward, the Mall, stripped of many of its trees, assumed added visual significance. Although temporary buildings marred a large section for decades,[14] the Mall's great swath of lawn was host to rallies, demonstrations and is indelibly related in the popular psyche with Civil Rights marches. It was prolonged eastwards by a 'Modern Acropolis' centred on the Capitol and encompassing the Library of Congress, completed in 1897. The 1902 plan recommended that the square surrounding the Capitol include a new building for the Supreme Court and monumental office structures for the Senate and House of Representatives. Subsequent plans attempted to link the capitol complex with the new Union Station (see figure 9.2), some spearheaded by the architect of the Capitol, whose authority could compete with that of the planners.[15]

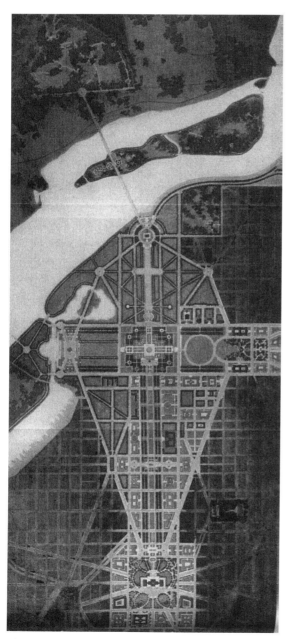

Figure 9.1. The 1902 Senate Park Commission (McMillan Commission) Plan.

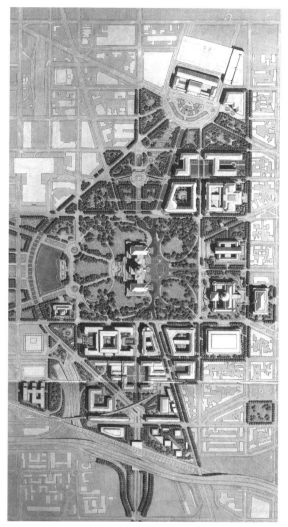

Figure 9.2. 1981 Master Plan for the development of the Capitol grounds by George M. White, the then Architect of the Capitol.

A more controversial aspect of the McMillan Plan was how it focused on the federal precinct, neglected surrounding districts and ignored the nagging presence of slum 'alley dwellings' in the shadow of the Capitol, the flip side of a city otherwise known for sheltering a prosperous, white-collar population. However, the assumption advanced in the McMillan Report that private developers would emulate the design consistency of the monumental core was not groundless. Affluent Northwest DC features some of the grandest and best-planned commercial and residential districts in North America developed under the aegis of the Board of Trade and designed by talented architects, many hailing from New York City.[16]

The powerful vision of the McMillan Plan provided a morphological and stylistic template for government and private architecture inspired by neo-classical paradigms, both French and American (in particular Jacques-Ange Gabriel's Place de la Concorde and Robert Mills' Treasury Building). No one can deny that Washington's ceremonial and administrative core, realized over the entire course of the twentieth century, is the grandest of all City Beautiful schemes. However, the national and international, direct and indirect, impact of the McMillan Plan is one of its few facets that remains little studied.[17]

The McMillan Plan implied considerable public spending but despite meagre initial resources and Congressional infighting, it was implemented to a remarkable degree over the century, especially since 1930. Its vision has lost none of its symbolic power, with a Mall inhabited by memorials and lined by government-run museums of national and international stature. The Museum of the American Indian was completed in 2004 on the last buildable space on the Mall. Prestige office space has gravitated around the Mall, catering to the administrative needs of the world's most powerful country, as well as national and international lobbies.

To the present day, the L'Enfant and McMillan Plans are safeguarded by the Commission of Fine Arts, a design review board approved by Congress in 1910. Its members are presidential appointees and primarily design professionals but its first chair was Charles Moore, who had been Senator

McMillan's personal secretary and edited the 1902 report.[18] The CFA's geographic purview has extended to encompass all new construction adjacent to the monumental core and vistas envisioned by L'Enfant.[19]

Charles Moore's major triumph at the CFA was the colossal office grouping known as the Federal Triangle. This 70-acre urban renewal operation was halfway between the White House and the Capitol. It was intended as a visual continuation of the government group centred on the White House. The triangle was already mapped out in the 1902 plan (but essentially for municipal use) and implemented in response to a dramatic space shortage in the late 1920s. Its political champion was the banker turned Secretary of the Treasury, Andrew Mellon (see figure 9.3), who also provided funds for the adjacent National Gallery of Art, completed in 1941.[20] He was advised by a board of prominent out-of-town architects chaired by Edward H. Bennett, Burnham's associate for the 1904 San Francisco and 1909 Chicago plans.[21]

At the heart of the Triangle, Bennett designed a Great Plaza that was never implemented, remaining a parking lot for over seventy years (see figure 9.4). Instead, a vast office building named after Ronald Reagan was completed in 1998, expounding a satisfactory urban design strategy but stodgy elevations. Concentration of Federal employment at such a scale was, and still is, controversial from the standpoint of traffic congestion, since Washington has one of the highest ratios of automobile commute in North America. But the Federal Triangle, one of the greatest achievements of Beaux-Arts inspired compositional methods (in plan and elevation), offers one of the most powerful and dignified statements of bureaucracy in any capital city.

The Interwar Period

A new phase in Washington's history began in 1926, after Congress was lobbied by the Washington Board of Trade to create the National Capital Park and Planning Commission (NCPPC) to centralize federal planning activities

Figure 9.3. Secretary of the Treasury Andrew W. Mellon viewing model of the Federal Triangle project at the Corcoran Gallery, 1929.

Figure 9.4. The 1902 Senate Park Commission (McMillan Commission), *Bird's Eye View of General Plan, from Point taken 4000 feet above Arlington*, rendering by F.L.V. Hoppin.

in the District of Columbia. The following year, opportunities for regional planning efforts opened, as a sister organization was created for the Maryland suburbs.[22] Until 1952, the NCPPC performed functions related to park acquisition and comprehensive regional planning for both the federal and local government. The Commission included ex-officio District and congressional members. Its chair and four public members were presidential nominees, including Frederick A. Delano (uncle of Franklin Roosevelt) and influential landscape architects and planners such as Olmsted Jr and Harland Bartholomew. Olmsted was the only remaining member of the 1902 McMillan Commission on the NCPPC but he had renounced the City Beautiful approach and led American and DC planning towards City Efficient objectives.[23]

The NCPPC suffered from understaffing and a weak political mandate. In its advisory capacity it could not prevent the construction of the Pentagon on intended parkland. Its most enduring inter-war legacy was the scenic landscaping of the Potomac River banks as recommended by the McMillan Commission, including the national highway to Mount Vernon, George Washington's home.

During the New Deal, priorities shifted from comprehensive to short-term planning objectives, such as the creation of the new town of Greenbelt, MD and construction of National Airport near the monumental core. After the war, planning initiatives were rather similar to those in other large US cities. Washington was at the forefront of urban renewal, since the NCPPC was authorized to designate urban renewal areas and adopt plans for them as early as 1945. National legislation enabled the NCPPC and its new implementation arm, the DC Redevelopment Land Agency, to receive direct federal funds and bypass the customary congressional approval process.

1950s Radicalism: Downtown Urban Renewal and Metropolitan Planning

In 1950, the NCPPC issued *A Comprehensive Plan for the National Capital and Its Environs* as a response to regional planning studies.[24] Engineer-planner Harland Bartholomew, the leading City Efficient consultant, prepared the plan.[25] This extensive, well-documented publication set practical goals, accepted low-density suburban development and proposed significant urban renewal of the downtown. Although it was well regarded by the planning profession at the time, the plan underestimated the growth of federal employment, was only advisory and left room for interpretation of its policies. Its symbolic role at the national level should not be discounted, however: by intervening with such gusto in DC the Federal government indicated it was willing to play a greater role in urban renewal planning.

A large portion of DC's Southwestern Quadrant was rebuilt and gentrified in an effort to combine objectives related to slum clearance, 'modern housing', waterfront redevelopment, freeway construction and, last but not least, the erection of new federal office buildings in close proximity to the Mall. Key players included planner Bartholomew, New York mega-developer William Zeckendorf and (opening a breach in DC planning's gender gap) architect Cloethiel Woodward Smith who, with her local colleague Louis Justement, set general planning directions in 1952. Although L'Enfant's street pattern was significantly altered (a trend also in the McMillan Plan) and too few amenities were planned, the modernist residential district was built at an appropriate scale. The quadrant had ingenious housing typologies by Smith and other local 'Young Turks' such as Charles M. Goodman.

Further east, on the south side of the Mall, renewal took a 'Burnham redux' turn, with out-of-towners' 'ego trips': L'Enfant Plaza (1960–1973) by the Zeckendorf/I.M. Pei team[26] and Marcel Breuer's office buildings for the Departments of Housing and Urban Development (1968) and Health, Education and Welfare (1976). The large-scale redevelopment of another area adjacent to the Federal Precinct, Foggy Bottom, gave birth to the infamous Watergate multi-use complex (1967), Washington's first Planned Unit Development (PUD), on the site of an old gas tank.

Paradoxically, the Highway Act passed by Congress in 1956, which altered so many city centres, had little impact on DC. In the 1960s, the combined efforts of Black and White civic leaders managed to stop new bridges and cross-town highways and the National Capital Planning Commission (NCPC) reviews helped delay and sidetrack freeway development. Only 10 miles of Interstate highway ran through the District of Columbia in 2003.[27] A single ring road – the 495 portion of the Interstate system – opened in 1964. The expression 'inside the Beltway', popularized by the media to differentiate political *cognoscenti* from the rest of America, also conveyed a new planning reality.

On the other hand, freeway construction played a determinant role in shaping the suburbs. Fear of atomic attacks and sheer lack of space were two major reasons why much federal employment migrated to the suburbs, which in turn experienced a phenomenal growth. In addition to 'White flight' from the District, many middle-class African-Americans started moving eastward to Maryland's Prince George's County by the mid-1960s. Suburban growth problems created political pressure for better regional planning. In response, the 1952 legislation reorganized the NCPPC as the National Capital Planning Commission (NCPC)[28] and created the National Capital Regional Planning Council (NCRPC).[29]

Because the various jurisdictions had conflicting goals, metropolitan planning, now viewed as a dynamic process, was a thorny issue. The NCPC was reinvigorated in the 1950s by planner William

Finley, and attempted to curb sprawl with its 1961 *Policies Plan for the Year 2000*.[30] This diagrammatic proposal focused on European-style 'finger plan' growth, with radial corridors of new town centres separated by undeveloped natural wedges[31] (see figure 9.5). Both ideas were implemented, but at a reduced scale. The new town of Reston in Virginia was located next to the new Dulles International Airport, designed by Eero Saarinen, a gateway as magnificent as Burnham's Union Station. The new town of Columbia in Maryland strengthened the notion of a Baltimore-Washington metropolitan corridor.[32] The federal government tried to direct this suburban growth with the NCRPC and the 1950 and 1961 plans, but eventually retreated to plans to revitalize DC.

The 1960s: Back to the City

Although rapid transit was mandated in the National Capital Planning Act of 1952, plans for the 103-mile Metrorail system were only adopted in 1968. Metrorail planning took place throughout the 1960s, with groundbreaking in 1969 and its first section opened in 1976. Designed by Chicago's Harry Weese, the downtown stations have dim, vaulted interiors, where exposed concrete is endowed with a 'sublime' character reminiscent of Piranesi's engravings. Because of the complex and highly politicized decision-making process which presided over Metro's implementation, its lines, which hardly extend beyond the Beltway, focus on servicing downtown office locations

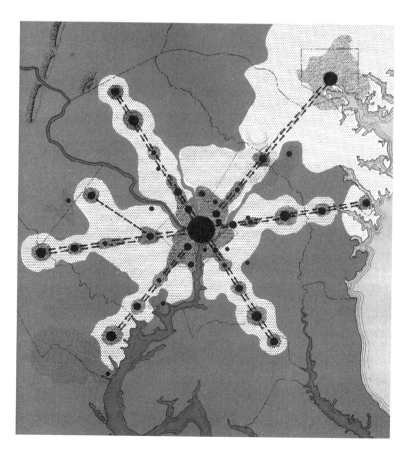

Figure 9.5. The 1961 'Finger plan' for the metropolitan region in the year 2000.

and do not address cross-town travel or serve ethnic neighbourhoods as adequately as they should.[33] Large-scale development near outlying metro stations, where government agencies have relocated (Silver Spring in Maryland, Ballston in Virginia), have benefited the suburbs much more than DC's eastern and southern districts.

Major downtown planning started in the late 1950s and the area was rescued from neglect during the Kennedy/Johnson years, a watershed as significant for Modern federal architecture and urban design as Teddy Roosevelt's presidency was for City Beautiful ideals. The northern, non-federal side of Pennsylvania Avenue, which is the parade route leading from the Capitol to the White House, was in poor condition, with dingy liquor stores and many boarded-up commercial fronts opposite the Federal Triangle. Patrick Moynihan, working together with planner Frederick Gutheim, initiated the planning process that helped revitalize the 'Avenue of the Presidents'. Their 1962 report presented 'Guiding Principles for Federal Architecture' that rejected reliance on tradition and clearly stated the need to avoid 'an official style'.[34] This effort led President Kennedy to appoint the first Pennsylvania Avenue Development Corporation, headed by International Style architect Nathaniel Owings of New York. It drew up a grandiose plan for the area in 1964 (see figure 9.6).

Historic preservation came of age during the 1970s, and the plan proposed in 1977 by

Figure 9.6. Model of the 1964 scheme designed by Nathaniel Owings for the Pennsylvania Avenue Development Corporation.

the newly created Pennsylvania Avenue Development Corporation expounded a vision less destructive of the urban fabric. Dismayed by the way the mammoth FBI Building (C.F. Murphy, 1967–1972) shunned passers-by, PADC developed specifications for massing, setback, and lot coverage that would unify both sides of the Avenue while encouraging pedestrian traffic. It was able to attract the Canadian Embassy (Arthur Erickson, 1981–1989) to the never-completed Municipal Center. It acted as a companion piece to I.M. Pei's 1978 East Wing of the National Gallery of Art.[35]

In addition to Pennsylvania Avenue initiatives, Jackie Kennedy, during her husband's presidency, helped safeguard historic landmarks around the adjacent Lafayette Square, which the McMillan Plan had slated for demolition. President Kennedy also helped NCPC leadership to pass from professional planners to local activists, nominating Elizabeth Rowe as the chairperson.[36] As a result, NCPC began intervening in riot-torn inner-city districts, with apparently limited success.

In conjunction with the State Department, the NCPC's proposed 1967 Comprehensive Plan included studies for an International Center grouping chanceries and international agencies, which were encroaching on upscale residential districts. A similar federal enclave was eventually developed, further away from the downtown, as a 'theme park' for diplomats.

Since Home Rule

The Home Rule Act of 1973 established a dual, complex, and sometimes counter-productive planning mechanism. It required NCPC to consolidate federal and district 'elements' in a single comprehensive plan[37] *and* the District government to develop its own Comprehensive Plan.[38] In addition, it established Advisory Neighborhood Commissions to represent citizens' interests. In 1984–1985, after a decade of *ad hoc* zoning decisions, the city's Office of Planning submitted its Comprehensive Plan for NCPC review of Federal interests and then to the City Council for approval. Every two years, the City's Office of Planning provides a formal report enabling ward representatives to review and approve amendments to its Plan's local elements. Currently, the Office is promoting targeted 'development projects', most significantly the Anacostia Waterfront Initiative (which expands on the vision of the McMillan Plan for a rather neglected area with great scenic potential) and advocates building housing in close proximity to the Federal precinct.[39] However, the natural course of speculative construction and rehabilitation has been favouring the downtown district immediately north of Pennsylvania Avenue, a renaissance that NCPC tries to monitor closely. In particular, the Eighth Street Corridor originating at the National Archives is being revitalized and a number of privately founded museums are moving into the area.

In the foreseeable future, the downtown urban decay partly caused by 1968 riot damage should recede further away from national symbols. A major economic downturn could slow this trend and the decay seems impossible to eradicate totally because many neighbourhoods have achieved only skin-deep prosperity. The metropolitan landscape of uniform blocks, which is more European than American in character, is being developed between Eighth Street and Union Station. But this growth, which DC needs so much from a fiscal standpoint, may have detrimental long-term consequences. 'Megastructures' such as the new convention centre adjacent to Mount Vernon Square, continue the trend towards closing of original streets, therefore further compromising the integrity of L'Enfant's Plan. Height regulations are not in jeopardy within DC, but the skyline secured from tall DC buildings has been marred by high-rise development in Roslyn, VA, just across the Potomac River from Georgetown.

The NCPC has become a mature agency whose focus has reverted to the symbolic core. Its latest urban design plan was released in 1997. According to its alluring brochure, *Extending the Legacy* is a 'dramatic departure from past federal plans', which 'eliminates obsolete freeways, bridges and railroad tracks that fragment the city' and 'reverses decades of environmental neglect'. It favours public-private partnerships as an alternative to congressional red tape. Waterfront development through buildings and recreational activities is once again promised, but the suburbs are hardly mentioned. Reviving a project it first issued in 1929, NCPC proposes visually to re-centre monumental Washington on the Capitol, which would facilitate the spread of future memorials and encourage economic growth and physical improvements on East and South Capitol Streets. Entirely advisory, *Extending the Legacy* is meant to re-focus federal and public attention on the planning and design elements proper to capital cities. It is not a comprehensive but a 'framework' plan: its watercolour renderings (see figure 9.7), undoubtedly conceived as pendants to those for the McMillan plan, are unlikely to exercise much impact. The architectural ideas, which tend to play a strong supporting role in successful DC plans, are far from compelling.

The NCPC's new planning focus is demonstrated by the *Comprehensive Plan for the National Capital*'s division of responsibility between the federal and local governments. The federal elements include:

* the federal workspace;
* foreign missions and international organizations;[40]
* transportation;
* parks and open space;
* the federal environment;
* preservation and historic features; and
* visitors.[41]

The federal government is no longer attempting to dictate regional land-use planning policy, and it now recommends supporting Smart Growth principles and local and regional planning objectives. Its regional planning activity is focused upon transportation, parks and environmental issues relating to well-defined federal interests for its workplaces, land holdings and visitor experiences.[42]

The District of Columbia is responsible for preparing the local elements of the comprehensive plan, including:

* economic development;
* housing;
* environmental protection;

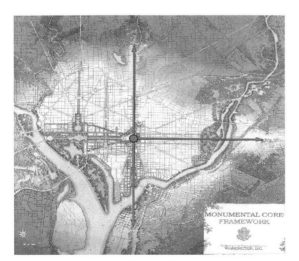

Figure 9.7. The 1997 *Extending the Legacy* plan, showing South Capitol Street as a new gateway to the city after the proposed removal of the Southeast/Southwest Freeway.

- transportation;
- public facilities;
- urban design;
- preservation and historic features;
- the downtown plan;
- human services;
- land use; and
- ward plans for the District's eight wards.[43]

This division of planning interests reflects Washington's status as a mature Political Capital within Peter Hall's typology. The federal government is no longer solely responsible for the comprehensive planning and development of the national capital region, and the feudal division of powers from the nineteenth century is gradually evolving into a more collaborative approach. But recent events have demonstrated that this division of planning responsibilities does not quite cover federal interests in monuments and security.

The proliferation of commemorative monuments is a major concern of NCPC and the Commission of Fine Arts. This design issue, with major land-use and ideological implications, is addressed in a Memorials and Museums Master Plan that encourages their dissemination beyond the monumental core.[44] Maya Lin's Vietnam (1982) and Lawrence Halprin's Franklin Delano Roosevelt (1999) Memorials are ambitious but unobtrusive designs with a profound understanding of broader planning agenda. On the other hand, many justifiable concerns have been raised about Friedrich St. Florian's World War II Memorial (2004) on the Mall, just west of the cross axis of the Mall, as it alters the grand, simple, east-west vista traced by L'Enfant and reinstated in the McMillan Plan.[45]

Finally, the 1997 *Legacy* plan did not address homeland security issues. Since the September 11, 2001 attack on the Pentagon, the century-long process of demonstrating political might and national unity by planning and building a dense but confidently open ceremonial city is being challenged. NCPC has addressed the heightened security issues quite rapidly, with a National Capital Urban Design and Security Plan approved in late 2002.[46] The design project by Michael Van Valkenburgh Associates to create a safe 'pedestrian-oriented public space' in the section of Pennsylvania Avenue fronting the White House was completed in November 2004. However, tourists will no longer be able to stroll into the Rotunda of the Capitol, whose entire east grounds (designed by Olmsted Sr) were excavated to create a secure visitor reception centre. These new measures to protect the Capitol contradict the spirit of symbolic and physical accessibility that mattered so much to L'Enfant and to the artisans of the McMillan Plan.

NOTES

1. The period of greatest demographic decline was in the 1980s. In 1950, Greater Washington had 1,815,150 inhabitants; in 2000 5,428,254. A rebound of 20,000 is expected in the next decade.

2. Federal employment in DC was 42 per cent in 1965 and is currently less than 29 per cent (US Bureau of Labor Statistics).

3. Gutheim (1977); Reps (1967) and Gillette (1995) are useful surveys of Washington's planning history.

4. Ward (2002). L'Enfant's map stopped at Florida Avenue and the outlying districts emerging in the late nineteenth century were planned without regard to its systematic street grid.

5. With 17.6 million domestic visitors in 2000, after a peak of 19.8 million in 1998.

6. This Revolutionary War slogan is emblazoned upon all new District of Columbia vehicle licence plates.

7. WAMU's DC Politics Hour is one of the most entertaining (but also disheartening) local public radio broadcasts. In reaction to the flamboyant but administratively inept Marion Barry (1979–1990) and (1995–1998), the present mayor, Anthony Williams, is

virtually an 'a-political' technocrat. Members of the City Council seem to thrive on perpetual dissent.

8. Striner (1995), p. 1.

9. Gillette (1995).

10. Peterson (1985).

11. Davis (2002), p. 128.

12. The National Arboretum along the Anacostia River in North East DC is also a magnificent, but insufficiently patronized, open space.

13. Moore (1902), p. 35; 'Creative borrowing' comes from Ward's (2002) classification framework for diffusion of planning ideas.

14. There are few structures as permanent as 'temporary' government office buildings. The Mall was filled with World War I era 'temps' almost as soon as it was re-vamped, see Reps (1967) pp. 169–172; Gutheim (1977), pp. 150–151 and 229–232. Some of these remained into the 1960s. Canberra had similar problems with its temporary offices of the 1920s, and the last of Ottawa's World War II temporary buildings was still in place in 2002.

15. The planning, design and development of the capitol complex has been the responsibility of the Architect of the Capitol's Office since 1793. The original Capitol was a multiple-use building, but new structures were built for the Library of Congress (1897), the Supreme Court (1935) and offices for the Representatives and Senators. See Allen (2001). The crest of Capitol Hill is now host to less architecturally distinguished but equally grand Congressional and Library annexes.

16. Longstreth (2002). Urban corridors achieving outstanding planning and architectural distinction are Connecticut Avenue, Massachusetts Avenue (also known as 'Embassy Row'), Sixteenth Street (on axis with the White House), all in the Northwestern Quadrant.

17. For instance, the plan was analyzed by Patrick Abercrombie in the second issue of *Town Planning Review*, see Abercrombie (1910). The plan regained its iconic status with post-modernism and triggered the imagination of Léon Krier. See Krier (1986).

18. Moore also edited Daniel Burnham and Edward Bennett's *Plan of Chicago*. See Burnham and Bennett (1909).

19. Kohler (1996), p. 74. In 1930, the Shipstead-Luce Act extended CFA's duties to advising the DC government on street façades of private buildings in a large section of the Monumental Core and on properties bordering Rock Creek Park.

20. Tompkins (1992). In fact, Moore made life miserable for Bennett.

21. In compensation for their pro-bono planning services, each architect designed a Triangle building.

22. The Maryland National Capital Park and Planning Commission was spearheaded by good government reformers led by E. Brooke Lee. It manoeuvred around obsolete and sometimes corrupt local political machines. See Gutheim (1977), pp. 206–210.

23. City Efficient planners advocated comprehensive planning, transportation systems, regional planning and zoning. See Olmsted (1911); Ford (1913); Peterson (1996).

24. In particular that of the New York Regional Plan Association. See Chapter 19 in this volume.

25. Bartholomew (1950).

26. Montreal's Place Ville Marie by the same team was initiated in 1958.

27. Levey (2000).

28. The missing 'P' relates to the 1952 withdrawal of the park function, granted to the National Park Service. In the 1960s, the NPS commissioned Skidmore, Owings and Merrill to prepare a master plan for the Mall. S. Dillon Ripley, who chaired the Smithsonian Institution for two decades, oriented the plan toward recreational uses.

29. Gutheim (1977) p. 256; the NCRPC was later replaced by the Metropolitan Washington Council of Governments founded in 1957 see http://www.mwcog.org.

30. See National Capital Planning Commission (1961).

31. On the 1961 Regional Plan, its European antecedents and lack of implementation, as well as on Reston and Columbia see Ward (2002), pp. 260–262.

32. Just as the New Deal city of Greenbelt before them, Columbia and Reston achieved many of their long-range planning and social goals. See Forsyth (2002); Bloom (2001). However, most of the area next to or outside the Beltway has suffered from the usual North American symptoms of rapid growth and witnessed the rise of edge cities like Tysons Corner on the Dulles Corridor. Since the mid-1990s, the Smart Growth movement, endorsed with particular enthusiasm by Maryland officials, has been set in motion to find alternatives to suburban sprawl at the doorstep of

the Nation's Capital. See Maryland Department of Planning (2001); Maryland National Capital Park and Planning Commission (2001).

33. Schrag (2001) sees Metro as the 'proof that bitter debate can lead to negotiated compromise' and as Washington's 'third grand plan . . . every bit as visionary' as the L'Enfant and McMillan Plans.

34. US (1962), p. vi.

35. Once construction of the Federal Triangle was underway in the 1920s, officials assumed (erroneously) that DC's impressive administrative building, erected in 1908, would eventually be demolished. Projects for a Municipal Center at the lower end of Pennsylvania Avenue near the Old City Hall were issued in 1927 and 1939. For more information on the Municipal Center and urban and architectural parallel between Erickson's Embassy and the U.S. embassy in Ottawa, see Gournay and Loeffler (2002).

36. Rowe was a lifelong DC resident who had first become concerned with city planning during her term of service on President Eisenhower's DC Auditorium Commission from 1955–1958. She was appointed to the NCPC in 1962. See Gutheim (1977).

37. National Capital Planning Commission (2002). The District has four representatives on NCPC: the Mayor, the Council Chair and two Mayoral appointees.

38. The District had a Zoning Commission since the 1920s.

39. http://planning.dc.gov.

40. Washington has perhaps the most difficult problems with planning for diplomatic missions, with 169 foreign missions attempting to locate chanceries (offices) and ambassadorial residences within a nineteenth century urban fabric. See 'Foreign Missions and International Organizations' in US NCPC (2004).

41. National Capital Planning Commission (2004).

42. National Capital Planning Commission (2004), pp. 1–10.

43. National Capital Planning Commission (2004), p. 11, note 4.

44. National Capital Planning Commission (2001).

45. http://www.savethemall.org.

46. Gallagher *et al.* (2003).

ACKNOWLEDGEMENTS

I would like to thank John Fondersmith, David Gordon, Jane Cantor Loeffler and Bill Webb for their suggestions.

Chapter 10

Canberra: Where Landscape is Pre-eminent

Christopher Vernon

The transformation of an obscure inland plateau into Australia's national capital began in 1913. Now, less than a century later, Canberra has grown to a population of over 300,000 thus becoming not only Australia's largest inland metropolis, but also its greatest achievement in landscape architecture and town planning. Encountering the city today, however, is an ethereal experience. In Canberra, the civic grandeur associated with national capitals emanates not from concentrations of architectural magnificence, but from the provocative omnipresence of the city's landscape – especially its luminous Lake Burley Griffin centrepiece (see figure 10.1). This distinctive landscape pre-eminence is not accidental. In fact, mediated by a nationalistic preoccupation with landscape – both native and recollected – design visions for the national capital were securely in place before the city had a site, a plan, or even a name.

On 1 January 1901, six of Great Britain's antipodean colonies federated to form the Commonwealth of Australia. Ambition to build a new national capital arose quickly from this ethos of political reconfiguration. Convened that May in the temporary national capital Melbourne (also Victoria's state capital), a 'Congress of Engineers, Architects, Surveyors and Others Interested in the Building of the Federal Capital of Australia' galvanized interest in the enterprise.[1] In harsh contrast to Great Britain's 'emerald green' and comparatively lush landscape, the Australian nation occupies a brown, arid continent. Unsurprisingly, the role water would play at the new capital pervaded the Congress's wide-ranging deliberations. Delegates resolved, for instance, that water and its supply should be considered not only for 'sanitary services' but also 'the creation of artificial lakes, maintenance of public gardens, [and] fountains'. This resolution had aesthetic implications. Use of the term 'lakes' – in lieu of 'reflecting pools' or 'basins' – suggests not only water bodies of considerable scale, but also ones of irregular outline and 'natural' appearance.

Congress delegate 'Mr A Evans' made this aesthetic dimension explicit in his paper 'A Waterside Federal Capital'. For him, 'the

Figure 10.1. View of Canberra from Mount Ainslie, overlooking the War Memorial, down the Land Axis, and across Lake Burley Griffin to the new Parliament House.

close proximity of a large sheet of water to palatial buildings enhances their appearance immeasurably' and affords a 'grand perspective to a noble city'.[2] Although thinly veiled by its title, Evans's essay was actually a propaganda piece; it advocated less the generic concept of a waterside capital and more the banks of Lake George in New South Wales (NSW) as its ideal location. Indeed, the Congress proceedings featured Sydney architect Robert Coulter's graphic representation of Evans's vision for the lakeside capital as its frontispiece (figure 10.2).[3] Evans observed: 'On the sloping hillsides and down to the water's edge are the palatial buildings of State and learning, whilst dotted amongst the foliage appear the villas of the residents and the spires of churches and public buildings'. '[P]icturesque boating sheds' and 'yachts on the Lake fill in the picture'.[4]

Coulter's rendering is not just a purposeful representation of a design proposition but a work of art. Considering this dualism, Evans's use of the term 'picture' is telling. Here, 'picture' refers not so much to the drawing as it does to the configuration and visual effect of the city it depicted. His description 'dotted amongst the foliage' also suggests that Evans visualized and, in turn, advocated the capital itself be like a 'picture' or 'picturesque'. Originating in Renaissance England, picturesque landscapes take the natural world as their model and rely upon irregular expanses of water and sylvan luxuriance for their effect. These considerable environmental requisites render the Australian application of picturesque technique problematic. Given this, Evans's view attests to the potency of nostalgia, if not imperialism. Although the city's architecture and symbolic content might be 'Australian' and its trees of local species, the twentieth-century nation's landscape taste – at least for Evans – remained colonial, rooted in eighteenth-century Britain.[5]

Figure 10.2. An Ideal Federal City, Lake George, NSW, 1901. Robert Coulter prepared this rendering for the Congress of Engineers, Architects, Surveyors and Others interested in the Building of the Federal Capital of Australia.

In Australia, the picturesque was introduced with the First Fleet in 1788. 'Almost Phillip's first act on land in a Sydney Cove', Paul Carter assessed, 'was to mark a line on the ground'. This line of enclosure also 'defined what lay outside it as no longer a continuum of gloomy woods but as a newly picturesque backdrop, a theatrical setting for the first act of the great colonial drama'.[6] The picturesque, however, was not simply a benign matter of optics. Along with bringing the unfamiliar Australian landscape into focus, colonists physically remoulded the terrain into aesthetic conformity to make 'landscapes that looked antique, wilderness-like, picturesque'.[7] Disguising the 'artificiality of their usurpation', the new 'owners' fashioned groves, 'intersecting slopes [and] glimpsed sheets of water' together 'like a jigsaw until it was hard to imagine it looking any other way'.[8] A picturesque Australian capital would similarly obscure the nation's youth and, through aesthetic and stylistic continuity, register its membership within the British Empire. If Sydney's founding was the colonial drama's first act, then the construction of a new national capital would be its last.

Evans's Lake George campaign furthered the 'battle of the sites' begun earlier in the debates surrounding Federation.[9] Adopting American precedent, Australia's new constitution required the national capital to be positioned within a larger federal territory. Seven contested years later, NSW's 'Yass-Canberra' district was selected in 1908.[10] Located inland from the eastern coast, the region's intermediate position between NSW capital Sydney and its Melbourne rival influenced the choice. Surveyor Charles Scrivener was next to identify the city's specific site. Scrivener's official instructions confirm that the national capital enterprise was as much a landscape design proposition as it was an engineering concern. As such, the selection criteria codified a picturesque approach from the outset. Potential locations were to be evaluated, for instance, from a 'scenic standpoint, with a view to securing picturesqueness, and with the object of beautification'.[11] In 1909, the surveyor selected the 'Limestone Plains' – a pastoral site in the broad valley of the Molonglo River – as fulfilling these considerable criteria. After debating the territory's final extent, NSW officially 'surrendered' it to the Commonwealth on 1 January 1911.[12] The national government would continue to own the land within the Australian Capital Territory (ACT) and control its development through leases until the advent of a local government in 1989.

Its federal territory delineated and the capital site fixed, the Commonwealth idealistically elected to launch an international competition to secure a city plan in 1911. The arid inland site, however, made the need for artificial water bodies acute. Consequently, amongst its myriad of requirements, the competition brief encouraged participants to consider damming the Molonglo to create 'ornamental waters'.[13] Not unlike the heated debate surrounding the site's selection, the competition also proved tumultuous. Professional dissatisfaction that a layman – the Minister for Home Affairs, King O'Malley – would serve as chief adjudicator led the Royal Institute of British Architects (RIBA) and other professional bodies to discourage their members' participation. Nonetheless, within weeks of the competition's 31 January 1912 close, one hundred and thirty-seven entries had been received from within and outside the Empire, including such geographically disparate places as North and Latin America, Europe, Scandinavia and South Africa.[14] That May, American architect and landscape architect Walter Burley Griffin's (1876–1937) submission was selected as the competition's winner. Although submitted in Walter's name, the plan was actually designed collaboratively with his architect wife and professional partner, Marion Mahony Griffin (1871–1961). Conceived at a distance in the United States and revised in Australia, the Griffins's design conceptually bridges the two nations.

Native Chicagoans, Walter and Marion Griffin's experiences in that city vitally informed their design approach. Although acclaimed for its progressive architecture, the burgeoning metropolis was also a locus of town planning innovation. In fact, Walter's visits to the city's 1893 World's Columbian Exposition were catalysts for his professional pursuits.[15] Curiously, the historical record suggests that his interests were further fuelled in the same decade when, as a university student, he first learnt of the Australian nation's progress toward federation and the prospect of a new capital.[16] Walter and Marion, practicing in close proximity in Chicago (first working together with Frank Lloyd Wright, next independently and then, from 1909, collaboratively), were inevitably familiar with the activities of Daniel Burnham, the city's leading urbanist. Propelled by the success of his contributions to the Columbian Exposition and then the City Beautiful transformation of Washington DC, Burnham's renown soon transcended the local. His (and Edward Bennett's) 1909 *Plan of Chicago* could hardly have escaped the Griffins's attention. Complementing their appreciation of American city planning ideals, the couple were also well-versed in British garden city design principles. Indeed, Walter's membership of Chicago's City Club provided him with opportunities to hear two of Britain's leading garden city protagonists first-hand. In 1911, the year the Canberra competition was announced, Raymond Unwin and Thomas Mawson travelled to Chicago and respectively lectured on 'Garden Cities in England' and 'Town Planning in England'.[17] By then, the Griffins's knowledge of town planning was considerable, albeit largely untested. If the couple did consult the RIBA's *1910 London Town Planning Conference* transactions (which included, amongst others, essays by Burnham and Unwin), as the competition brief advised, then its contents would not have been revelatory.[18]

But the Griffins injected a striking new dimension to large-scale city planning into their design for the Australian capital. Unlike most other submissions, the Griffins's plan was distinguished by its sensitive response to the site's physical features, especially its rugged landforms and watercourse (figure 10.3). This attribute proved paramount to their design's success. Composed on a cross-axial scheme, the plan fused geometric reason with picturesque naturalism.[19] Although indebted to City Beautiful ideals and more domesticated garden city planning principles, the Griffins's reliance upon geometry as an organizing device more profoundly registered their conviction that the natural world was the essential source for design. For these two designers, nature's primordial 'language' was a geometric one, as expressed for instance in botanical reproduction or crystal formation. At Canberra, then, the Griffins employed geometry to articulate strategically the site's otherwise latent geomorphic structure. When negotiating the fit of their geometric template to the actual site, the couple venerated existing landforms.[20] Hills and ridges, for instance, were not design impediments to be erased, but 'opportunities to be made the most of'. Discerning a linear correspondence between the inner hill summits and the more distant mountains, the couple accentuated the alignment with a 'Land Axis'. Anchored by Mount Ainslie at one end, the 'Land Axis' extends some 25 kilometres to its other terminus, Mount Bimberi. By using topographical features as axial determinants and visual foci, the Griffins 'monumentalized' the future city's physical site.

The Molonglo valley posed no less a design opportunity than did the site's landforms. Accordingly, the couple delineated a 'Water Axis' across its 'Land' counterpart at a right angle, aligning it with the river course. Answering the brief's call to establish 'ornamental waters', the Griffins reconfigured the river into a continuous chain of basins and lakes which stylistically reconciled 'formal' with 'natural'. As one moves out from its centre, the water body's outline and spatial character metamorphoses; the central basin's geometry gives way to the irregular margins of terminal 'East' and 'West' lakes. Here, the banks take on the character of a naturally-occurring wetland, a visual and spatial quality compatible with Australian anticipation of the picturesque.

Australia's fundamental allure was as an opportunity to perfect lessons learnt from America's shortcomings. Although it occupied an ancient continent, the new Australian nation lacked the cultural artefacts and other monuments typical of Old World and, by the opening of the twentieth century, even New World capitals. In compensation, the Griffins fashioned Australia's new national, cultural history from its ancient natural history – as illustrated by the design

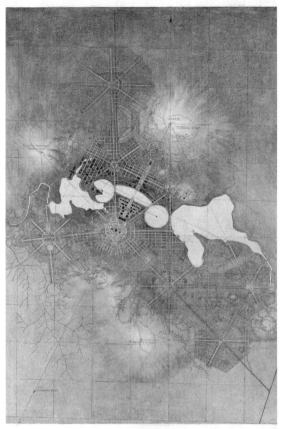

Figure 10.3. Commonwealth of Australia Federal Capital Competition, plan view of *City and Environs*, 1911. Walter Burley Griffin and Marion Mahony Griffin, landscape architects, Marion Mahony Griffin, delineator.

significance they awarded the Canberra site's physical features. This impulse to 'monumentalize' nature, most immediately, was a prescriptive reaction to the couple's Chicago experiences. That city was then in the midst of transformation by largely unregulated expansion; its remnant prairie and rural surrounds quickly subsumed by speculatively motivated city extensions and suburbs. Aiming to avoid this phenomenon, the Griffins envisaged Canberra as a designed alternative to urban indifference to the natural world.

Another seminal American source was the spatial and symbolic concerns of Pierre Charles L'Enfant's 1791 plan of Washington, DC. Of most import for Canberra, L'Enfant awarded landscape foci to both armatures of his cross-axial composition. An axial corridor projected west from the hill-top Capitol, the Mall took the view to the nation's vast interior as its focus. Although aesthetically indebted to the Picturesque, this use of landscape as an axial terminus was symbolically resonant. Then, the interior was perceived as an abundant 'wilderness' frontier, beckoning the fledgling democracy's westward expansion. The cross axis extended south from the elevated 'President's House', capturing the Potomac River convergence in its prospect.[21] L'Enfant's sophisticated landscape effects were later erased with the city's redesign by Burnham and the 1902 Senate Parks Commission. From then on monuments and other architectural objects incrementally usurped landscape as axial foci. Eschewing the City Beautiful's decorative aestheticism, the Griffins reinterpreted L'Enfant's aspirations and archaically revalued landscape as a spatial container and crucible of symbolic meaning. In their Australian transference of L'Enfant's technique of axial projections into a 'wilderness' (which the Griffins imagined Canberra's site to be), the couple similarly presented nature as symbolic of democracy.[22]

Elaborating their 'Land' and 'Water' cross-axial geometry, the Griffins organized the city centre's plan to form a triangle (known today as the Parliamentary Triangle), its points aligned with local summits (see figure 10.4). Concentrated within the triangle and its immediate environs, the capital's public edifices are arrayed in accordance with a systematic political symbolism.[23] Near the triangle's base, national cultural institutions line the central basin's northern margin. Here, set within a sweeping expanse of 'Public Gardens', one encounters the 'Zoological Garden' with its 'Aquarium' and 'Aviary', 'Natural History' and 'Archaeology' museums, galleries of 'Graphic' and 'Plastic' arts, 'Theatre', 'Opera', 'Stadium', 'Aquatic Garden' and 'Plant Conservatory', 'Gymnasia' and 'Baths'.[24] Prominently positioned on the 'Land Axis' at the foot of Mount Ainslie, a pleasure garden 'Casino' overlooks this cultural precinct.

Crowned with a landmark 'City Hall', the hill ('Vernon', now City Hill) at the triangle's north-west point became the nodal focus of a 'Municipal Centre' for the 'General Administration of Affairs'. Here, a 'Gaol', 'Criminal' and 'Civic' courts, 'Bank and Offices', 'Exchange and Offices' and 'Post Office' ring the city hall. A 'Mint' and 'Printery', although national in function, are also situated within this precinct. Just beyond the 'Municipal Centre', the Griffins located the 'University' (today the Australian National University) and a 'Hospital'. The triangle's north-east point is punctuated by another hill, providing the locus for the city's 'Merchandizing' or 'Market Centre'. Along with a 'Railroad Station', this centre also includes two 'Market' buildings and a 'Power Station'. Adjacent ridges provide platforms for the national 'Cathedral' and a 'Military College'. Parallel to the 'Water Axis', a secondary 'Municipal Axis' links the 'Market' and 'Municipal' centres to form the triangle's base. Collectively, the two Centres and the network of cultural institutions symbolize the 'People'.[25]

Across the ornamental waters at the basin's southern edge, the area gently rising to the triangle's apex (at the convergence of today's Commonwealth and King's Avenues) became the 'Government Centre'. Set within a topographically articulated hierarchy, a symmetrical ensemble of buildings accommodates the functions of government.[26] Beginning at the waterside 'Judiciary', one next ascends to the 'Legislative' precinct. Rising above these 'Departmental Buildings', the 'Houses of Parliament' rest atop 'Camp Hill'. Higher still, the triangle's apex culminates in Mount Kurrajong (now known as Capital Hill). Amidst this hill's lower slopes, the official residences of the 'Governor-General' and the 'Prime Minister' express the 'Executive'. Symbolic occupation of Kurrajong's summit, however, was awarded not to the government, but to the 'People'. Here at the highest elevation within the city's centre, the Griffins positioned a monolithic 'Capitol' (see figure 10.4). But unlike its American namesake, Australia's counterpart was envisaged as a ceremonial building to enshrine the achievements of its citizens.

At first, the Griffins's American-borne design for the Australian capital appeared to dovetail with local sensibilities, especially notions of landscape beauty. Unlike Chicago's increasingly urbanized hinterland, Australia remained the place where, as novelist D.H. Lawrence asserted, 'people mattered so little'.[27] Partly owing to the spatial insignificance of human occupation, the native landscape – known colloquially as the 'bush' – was pre-eminent. At the time of the 1912 competition, fuelled by domestic sources such as Heidelberg School landscape paintings, idealized images of the bush were gaining iconic status as symbols of an inextricably 'grounded' national identity. Marion Griffin's exquisite renderings emphasized their submission's landscape imagery and may have lured the adjudicators to see the design as a celebration of the bush (see figures 10.3 and 10.5).[28] The two American designers, however, were probably unaware of

Figure 10.4. Commonwealth of Australia Federal Capital Competition. The Griffins' city plan as drawn on the Map of Contour Survey of the Site for the Federal Capital of Australia provided to competitors, 1911.

Figure 10.5. Commonwealth of Australia Federal Capital Competition. *View from Summit of Mount Ainslie*, 1911. Walter Burley Griffin and Marion Mahony Griffin, landscape architects and architects, Marion Mahony Griffin, delineator.

the local landscape's increasingly nationalistic connotations.

The decision to award the Griffins's design first prize, however, was not unanimous. In fact, a dissenting judge had given first place to a design by Sydney architects Griffiths, Coulter and Caswell. This consortium's 'Coulter' was the same architect who made the 'Waterside Capital' rendering for the 1901 federal capital congress. Now, a decade later, his collaborative submission similarly featured a watercolour perspective view (figure 10.6).[29] In this scene, the ground's surface is awash in green and, unlike in the Griffins's scheme of geometric containment, the margins of these 'ornamental waters' meander. This was a city more at home in the Northern hemisphere; one evidently more compatible with local anticipation. Ultimately, however, King O'Malley endorsed the Griffins's victory.

Vital to their design's success, Marion Griffin's renderings contrasted dramatically with the Sydney group's submission. Infused with sepia, gold and other luminescent tonalities, her graphic ensemble evoked the site's more authentic colouration. An English critic rhapsodized that 'the buildings are spread so thinly on the ground, are so masked with trees, and are so small relative to the majestic roll of the terrain, that you see, not them, but Australia'.[30] Despite its laudatory intent, this assessment also reveals that the drawings could be deceptive in their persuasiveness. Marion's portrayals of the site's vegetation and landforms diminished, if not camouflaged, the visual impact of the proposed city's geometry and Chicago-like density (see figure 10.5).

Despite the competition's final outcome, however, the government controversially set aside the Griffins's design and appointed a 'Departmental Board' to derive a new plan from the various submissions. Profoundly disappointed, Walter Griffin resolved to have their design reinstated. In an impassioned January 1913 letter, he offered to explain it in person. Delayed by a change of government, the Commonwealth belatedly responded in July and invited him to visit. Construction of the now officially-named Canberra, however,

Figure 10.6. Commonwealth of Australia Federal Capital Competition. *View of the Lake at Sunset*, 1911. This rendering was included in a collaborative submission made by Sydney architects WS Griffiths, RCG Coulter and CH Caswell.

had already begun to the Departmental Board's hybrid plan. In August 1913, Walter at last toured the capital's site. He also consulted with the Departmental Board and even the new Prime Minister. Bolstered by local professional support, Walter's campaign to have his and Marion's design reinstated met with success. In October, the Board was disbanded and Walter accepted an official position as Federal Capital Director of Design and Construction.

The Griffins moved to Australia in May 1914. Working at a distance from Melbourne, Walter began Canberra's detailed design. In continuity with his American practice, he awarded priority to road layout and planting. Buildings were to be constructed afterwards, carefully inserted within this structural template. The circumstances of the future city's windswept site also mandated advance planting. This was no 'wilderness'; extensive grazing had left the once forested slopes largely denuded and the river banks eroded. This degradation made the need for rehabilitative planting urgent. Undertaken collaboratively with afforestation officer Thomas Weston, Walter's scheme to revegetate the city's summits was the most remarkable of his advance planting projects. Fusing utility with beauty, he initiated a plan to 'cover' the city's 'bare hills' with native trees, shrubs and 'carpet plants' in 1916. These plantings were to be differentiated according to colour; Mount Ainslie, for example, was to be planted with species of 'yellow flowers and foliage' and Black Mountain with 'white and pink flowers'. Through such expansive, colour coordinated plantings, Walter sought to transform the entire Molonglo valley into a cultivated, luminescent garden.

Walter Burley Griffin's Canberra tenure, however, proved short-lived. A series of political antagonisms led to a 1916 Royal Commission investigating the performance of his contract and the implementation of the city plan. This, along with the First World War's financial restraints, conspired against the complete realization of the couple's design. Walter's official association with the national capital ended controversially with the abolition of his position in 1920. Walter and Marion nonetheless elected to remain in Australia and devote themselves to private practice.

A succession of advisory bodies replaced Griffin's singular position as Federal Capital Director of Design and Construction.[31] The first was the Federal Capital Advisory Committee (FCAC), established in 1921. Chaired by nationally prominent architect and town planner John Sulman (1849–1934), the FCAC was charged with implementing the Griffins's design. Long a champion of a new capital (although the RIBA censure precluded his participation in its design competition), he had publicly advocated the Griffins's design. Now, however, his support was less fulsome. Despite the FCAC's mandate, Sulman and others proceeded, incrementally, to graft their own divergent visions onto Canberra's original frame. Soon the Chicago-like urbanity envisaged by the Griffins would be insidiously transformed into a disparate collection of garden suburbs.

The FCAC's selection of a 'provisional' (known today as 'Old') Parliament House site in 1923 was an early and prominent deviation. Departing from the original design, the FCAC located the edifice not on, but at the foot of the Griffins's symbolically-charged 'Camp Hill' site. With Parliament House underway, another important project gained momentum. This, too, would result in a significant departure from the plan. Australia's participation in the First World War profoundly affected its nascent national psyche and led to enthusiasm for establishing a War Memorial in Canberra. Taking up the popular initiative, in 1924 the FCAC elected to position the memorial and museum complex on the 'Land Axis' at the base of Mount Ainslie. Originally, however, the 'Casino' was to have occupied this dramatic site. Along with registering its disregard for the Griffins's disposition of buildings and land-use allocations, the FCAC's decision also demonstrated its apathy toward the couple's symbolic intent for the Capitol building and its surrounds. Compared to Ainslie itself, however, the War Memorial is diminutive and the summit prevails visually as the axial terminus. Nonetheless, the War Memorial's presence on this site layered a new commemorative patina onto an otherwise residential precinct. Amplifying this new overlay, the Griffins's 'Prospect Parkway' summit approach was renamed 'ANZAC Parkway' to honour the Australia and New Zealand Army Corps. Today known as ANZAC Parade, the axial thoroughfare is now lined with military memorials and effectively becomes an open-air extension to the War Memorial. Departures from, not adherence to, the original plan became the *status quo*.

The FCAC's activities culminated in 1924 with the preparation of a new city plan. After the establishment of a new Federal Capital Commission (FCC) in July 1925, the plan was officially gazetted (enshrined in Commonwealth law) that November. The gazetted plan, however, merely encapsulated a portion of the Griffins's street layout, omitting the design's land-use and other structural elements let alone its symbolic content. As the gazetted plan attests, the FCAC and FCC literally and metaphorically regarded the Griffins's design as little more than a road map. Nonetheless, with the compromised plan in place, Parliament opened in the new national capital on 9 May 1927. While its transfer from Melbourne began earlier, the government's presence in Canberra was more ceremonial than actual. Within a few years, the shift all but ceased. Although the looming economic depression contributed to the ebb, Parliament's own lingering antipathy towards the new capital was a factor no less potent. This antipathy climaxed with the FCC's abolition in 1930. For almost the next three decades, Canberra would be administered by the Department of Works and governed by the Department of the Interior, with little development progress.

Canberra languished into the 1950s. Known derogatorily as the 'Bush Capital', the city still lacked palpable urban fabric. The national capital, however, found a champion in Prime Minister Robert Menzies. With his support, a Senate Select Committee was appointed to investigate the city's development in 1954. The committee found, somewhat predictably, that administrators personally then saw Canberra not 'as a national capital', but 'as an expensive housing scheme for public servants'.[32] The inquiry also had a more unanticipated outcome. Sydney town planning academic Peter Harrison reported that realization of the Griffins's design was not contingent upon 'the construction of grand buildings'; arguing instead that buildings were 'made important' by 'their setting'.[33] Canberra, Harrison concluded, was 'not an architectural composition but a landscape composition'.[34] Whilst he accurately identified landscape's pre-eminence within the Griffins's scheme, Harrison's conception of it as simply architectural setting illustrates the contemporary power of the Modernist viewpoint. In this, architecture is seen in rational opposition to the chaotic natural world. Architecture, in turn, is held as the only means by which to structure that chaos. Landscape, instead of a formal entity in its own right, is regarded as merely setting or the space in between buildings. Nonetheless, Harrison's close study of the original plan amounted to a rediscovery and, at first, it appeared to resurrect the Griffins's vision.

Bolstered by the Senate Committee's bipartisan support, the Prime Minister acted decisively and sought expert advice to ignite the city's development. In 1957, the Commonwealth again looked overseas. This time, however, Menzies focused the gaze upon London, not Chicago, and Modernist town planning authority William (later Lord) Holford (1907–1975) was solicited for design recommendations. Having recently served as international adjudicator for Brazil's national capital design competition, Holford now accepted Australia's invitation. After touring Canberra that June, he completed his *Observations on the Future Development of Canberra, ACT* in December. Forsaking the Griffins's symbiotic pedestrian and tram city, Holford's Canberra – most fundamentally and like its Brazilian counterpart – was to be a 'City of the Automobile'. In 1958, Menzies established the National Capital Development Commission (NCDC) to implement the report's initiatives, including an extensive motorway network.[35] The capital's new Modernist landscape would now be increasingly experienced at high-speed through a windscreen frame.

Central to his vision for Australia's capital, Holford developed two inter-related urban design proposals which nested within his overall plan. Like Harrison and the Senate Committee, Holford also saw Canberra more as a landscape design than an architectural proposition. Indeed, the most dramatic outcome of his consultancy would not be a new government edifice, but the capital's much-anticipated lacustrine centrepiece. With Lake Burley Griffin's completion in 1964 (see figure 10.7), the city at last was unified in a manner compatible with the couple's vision. At the same time, however, the lake encapsulated prominent departures from its namesake's original design. Despite the significance the Senate Committee officially awarded the Griffins's plan, Holford was no less determined to evince his own hand in the city's design. British Modernism, paradoxically, was historicist in its landscape expression; drawing upon and re-vivifying the eighteenth-century picturesque. Believing it necessary 'to amend [its] formal symmetry', Holford revised the couple's waterbody design in its belated execution. Instead of the geometric clarity the Griffins envisaged for

Figure 10.7. View down the Water Axis across Lake Burley Griffin. Along with the lake and its parkland margins, the motorway visible in the lower left was a product of the Modernist visions of William Holford and the National Capital Development Commission.

the central basins, Holford now sought a 'frankly picturesque treatment'.³⁶ For him, this alternative approach ostensibly 'would be more in keeping' with the city's 'beautiful background of hill and valley'. As suggested by the new appellation 'lake', Holford's waterbody was executed with an irregular edge and its margins cloaked by picturesque parklands. To some degree, the lake's form was revised in economic concession to the steep topography at its edges. More emphatically, however, the new configuration and attendant parklands expressed Modernism's benign landscape imagery.

When re-conceptualizing Canberra's waterbody, Holford's design concern extended beyond its outline. In fact, he saw the lakeside as the ideal locus for Australia's still unrealized permanent Parliament buildings. This was a view informed not by the Griffins's design but by his own Brazilian experiences. Taking Brasília's 'Place of the Three Powers' as precedent, Holford now proposed to position the government at the lake's edge. Set amidst the lake's encircling parklands, when viewed at a distance Australia's Parliamentary buildings would resemble follies in an English landscape garden. Like the lake's new form, Holford's 'Lakeside Parliament' was also a dramatic departure from the Griffins's vision. Abandoning their original elevated site, Holford shifted the buildings further down the 'Land Axis' to the lakeshore. After more than half a century, Canberra was now poised to become a Modern variant of Evans's and Coulter's federation visions of a picturesque waterside capital. Implementation of Holford's scheme began in 1958. After a decade as the city plan's *status quo*, however, the 'Lakeside Parliament' was deleted in 1968. Nonetheless, the transformation of Canberra's landscape into a Modernist 'setting' had begun.

Holford's re-assertion of the picturesque was not without political dimension. By the 1960s, some came to see the Griffins's geometry as 'American' or 'un-Australian', if only by virtue of its authors' nationality. With the realization of Holford's ideals, the picturesque re-colonized the Australian capital and cast Canberra's die as British. Instead of the Griffins's populist ceremonial Capitol, for instance, Holford thought the inner city's highest summit ideal for a residential 'Royal Pavilion'. In his scheme, the British monarch would gaze down upon not only the lakeside Parliament, but also 'the people'. When introducing this proposal, Holford reminded the government that 'Her Majesty the Queen is also Queen of Australia'.³⁷ Menzies needed no reminder. Indeed, his commitment to the national capital was earlier galvanized by Queen Elizabeth II's impending visit; the Queen dedicated the Australian-American Memorial in 1954. Acutely aware of the episode's symbolism, the Anglophile Prime Minister went so far as to shift the memorial's original 'Land Axis' location to a less prominent position in advance of the Royal Visit. The pavilion, however, was never seriously pursued. If the Griffins sought to bring the Australian bush to the foreground, to give it primacy, then throughout the post-war era it was relegated to the background in lieu of deciduous trees, and memories of green, if not Empire. For Holford and the NCDC, 'Griffin was history'.³⁸

In the wake of Holford's proposals, the NCDC exercised its relative autonomy and considerable authority to accelerate Canberra's development. Between 1958 and 1988, the Commission orchestrated the construction of a number of landmark buildings within the Parliamentary Triangle. The National Library of Australia, opened in 1968, was an early showpiece.³⁹ By 1982, Australia's High Court and the National (art) Gallery, both designed by Edwards, Madigan, Torzillo and Briggs architects, were complete. Although the ensemble's brutalist concrete architecture proved controversial, this design was actually amongst the first to address Canberra on its own terms. Eschewing any pretence to a Griffins-like urbanity, these monolithic buildings instead respond to the capital's ethereal landscape actuality. The High Court and Gallery became a multi-faceted set piece within the open parkland expanse that now typified Canberra's ceremonial centre.[40]

Along with building projects and in response to projected population growth, the NCDC launched new planning initiatives. Most significant was its 'Y-Plan', a post-Griffin metropolitan growth strategy defined by the linear extension of satellite suburbs and sub-regional town centres (see figure 10.8). Framed by undeveloped hills and divided by open space, these new centres were linked by bushland motorways.[41] In parallel with new suburbs, the NCDC established an extensive open space network.[42] The city's early public gardens and parks, including some designed by Griffin and planted by Weston, were also now mature. As funding for significant building projects decreased, however, large-scale planting was employed to define and accentuate the city's axes and ceremonial spaces. Trees effectively became surrogates for buildings.

By 1988, the year marking the bicentennial of the British claim to Eastern Australia by Captain James Cook, the national capital had come into its own. Most of the inner-city infrastructure was established and new suburbs were expanding to fill the Y-Plan's outer reaches. The major construction event that year was the completion of a long-anticipated permanent Parliament House atop the triangle's Capital Hill apex (see figure 10.9). A separate development agency oversaw an international design competition for the building as well as its construction.[43] Not unlike the provisional building it replaced, however,

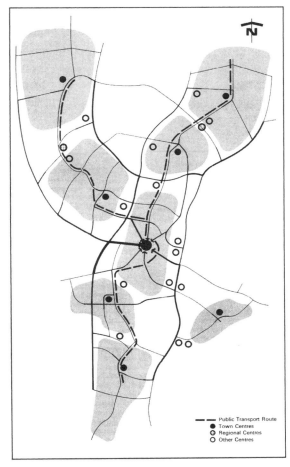

Figure 10.8. The National Capital Development Commission's *'Y'-Plan* for Canberra, 1970. This plan takes its name from the shape of its settlement and transportation corridors.

the siting of new Parliament House was also contested; for some, the government's relocation to the Griffins's site for the 'People' seemed to eliminate the possibility of ever realizing the original vision. The winning design by American architect Romaldo Giurgola, however, innovatively reconstructed the hill as architecture; earthen ramps enable citizens literally to walk atop Parliament House and their government (in the aftermath of 11 September 2001, however, this practice has been curtailed and the ramps barricaded). Encircled by gardens, Parliament House is at once landscape and architecture. A monumental flagpole structure, its pyramidal form evocative of the Griffins's unrealized 'Capitol', crowns the remarkable ensemble. This steel tower has now become a landscape icon within the national capital's skyline.

Also in 1988, the Commonwealth awarded, or rather imposed, self-government to the Australian Capital Territory (ACT). Under this new governance, the planning and development of the national capital now came under dual control. The ACT established its own metropolitan Planning Authority (ACTPA) and the Commonwealth replaced the NCDC with the National Capital Planning Authority (NCPA).[44] This dualistic planning arrangement – denoting the division between the day-to-day urban management of the city and the protection of its national capital functions – remains in place today.

Although the geographic area subject to national control, staff numbers, and funding were all diminished, the NCPA's central mission was to ensure new development was compatible with Canberra's national capital status. This was primarily achieved through its administration of what became the *National Capital Plan*. In 1997, its mandate was expanded. Enlarging its emphasis upon Canberra's national significance, the agency now proactively sought to 'build the National Capital in the hearts of all Australians'. This required new promotional initiatives to dispel negative perceptions of the city. In recognition

Figure 10.9. View of new Parliament House, completed in Australia's 1988 bicentennial year. Crowned by a monumental steel flagpole tower, Parliament House is at once landscape and architecture.

of its broader role, 'Planning' was omitted from the agency's title and it is known today as the National Capital Authority (NCA).

Throughout the 1990s, new suburbs burgeoned under the Territory government's guidance. At the same time greater development and infill pressures arose within more established suburbs, leading to numerous planning conflicts. After years of economically-driven policy planning, 'spatial' or physical planning now had to accommodate the new challenges of inner suburban redevelopment. In response, the ACT government restructured its planning and development activities. In 2003, the Department of Planning and Land Management was replaced by a new Planning and Land Authority (ACTPLA), along with a new independent Planning and Land Council to provide policy advice to the ACTPLA and the Minister for Planning. In parallel, a Land Development Agency was created to oversee development and sales of Territory lands.

Within the national context, the NCA promotes Canberra as 'a National Capital which symbolizes Australia's heritage, values and aspirations, is internationally recognized and of which all Australians are proud'. To this end, it has opened numerous memorials, facilitated many national institutions and events in the capital, and continues to manage a permanent exhibition on the city. In addition, the NCA oversaw the realization of a new National Museum of Australia (2001), one of the largest building projects since the NCDC era.

Since the NCA's establishment, interest in the national capital's symbolic content has surged. In its review of the Parliamentary Zone (1998–2000), the NCA identified that the precinct should become a true 'Place of the People', accessible to all so that they can more fully appreciate Australia's collective experience and rich diversity. Two recent projects are important benchmarks toward realizing this aim. Completed in 2002, Commonwealth Place and Reconciliation Place are positioned on the Griffins's 'Land Axis', near the edge of Lake Burley Griffin. Even before their completion, however, these works became fraught with controversy. This should come as no surprise. The occupation of land is always a politicized, if not contested, activity and this prominent site at the symbolic threshold of Parliament, cannot be anything but politically charged.

Commonwealth Place (see figure 10.10) is the outcome of an NCA design competition (2000).[45] The winning design, as described by the competition jury, 'cuts and forms the landscape rather than instigating a large building program' and, like Parliament House, is quite literally landscape architecture. The site has been re-fashioned into a cup-shaped amphitheatre, surfaced with highly-manicured turf. Architectural space is folded beneath the weft of the concave form, accommodating exhibition areas, a restaurant and offices. Sheathed with translucent glass, when illuminated at night, an ambient glow radiates from within the earth. Commonwealth Place itself is divided by a ramp. This incision resonates as a tectonic plate of sorts, metaphorically recalling the city's underlying geology. Through its provocative engagement with the ground itself, Commonwealth Place is one of the first designs to treat the 'Land Axis' as more than merely a tree-framed surface.

Also the result of an NCA competition (2001), Reconciliation Place is a cross-axial pedestrian way connecting the National Library of Australia with the National Gallery of Australia (figure 10.11).[46] A composition of sculptural 'slivers' of varying heights, each representing in word and image episodes in the reconciliation process, is positioned within this corridor. There are multiple possible routes through this grove of slivers,

Figure 10.10. View of Commonwealth Place, designed by Durbach Block and Sue Barnsley Design (2000). Positioned on the 'Land Axis' at the edge of Lake Burley Griffin, this commemorative project 'cuts and forms the landscape'. 'Speaker's Square', the pavement artwork visible in the left foreground, is Canada's gift to commemorate the centenary of Australia's 1901 Federation.

each offering a different historical reading. The sliver grove, like its vegetal counterparts, is to be dynamic. The promenade and initial slivers are already in-place as a skeletal framework; the ongoing unfolding of 'reconciliation' will be marked by the accretion of new slivers of varying heights. A dome of turf marks the cross-axial intersection between Commonwealth and Reconciliation Places. Partly inspired by Aboriginal middens, this earthwork functions as a viewing platform. Although unintended, the new landform can be read as a sort of spoils, evocative of the original hills and spurs levelled in Canberra's development. In the 1920s, however, the city's makers were confronted by the reality that antiquity on the 'Limestone Plains' was not limited to the geological. Excavations made in the construction of Old Parliament House not only altered topography but also unearthed Aboriginal artefacts.

Until Reconciliation Place, Indigenous presence in the Parliamentary Triangle was, at least overtly, concentrated at the Aboriginal Tent Embassy complex confrontationally located at the steps of Old Parliament House. Established more than thirty years ago, the Aboriginal Embassy and its dynamic, expanding surrounds is surely one of the most symbolically charged landscape spaces within the Parliamentary Triangle. Here, the landscape itself is employed as a medium of protest. Emanating from the embassy structure is an array of outlying '*gunyahs*' (timber and brush shelters), a perpetual (illegal) fire and clandestine plantings of eucalypts. Even the space within the flanking plantations has been colonized as a locus of habitation (also illegal). The controversial decision to distance Reconciliation Place from the Aboriginal Tent Embassy fuelled speculation that the new project actually aimed to relocate Indigenous presence and eliminate the embassy itself. As every (multi-cultured) generation is no less entitled to give built expression to its values and achievements, the erasure of the Aboriginal Embassy would be tragic. Ultimately, the national capital's commemorative landscape will always be necessarily incomplete.

Underpinning concern for Canberra's symbolic content, interest in the Griffins's ideas and their

Figure 10.11. View of Reconciliation Place, designed by a team led by Kringas Architects (2001). The turf dome marks the intersection of Reconciliation Place and the axial ramp entry to Commonwealth Place; two of the interpretative 'sliver' artworks are visible at the lower left foreground. The High Court of Australia appears at the upper centre.

formative impact upon the city has developed steadily, although even the *National Capital Plan* only obliquely acknowledges the inheritance of the original vision. Launched in 2002, the NCA's 'Griffin Legacy' project interrogated the relevance of the Griffins's design to the national capital in the twenty-first century. When reviewing Canberra's design evolution, it is vital to remember that the Griffins's enormous popularity is a relatively recent phenomenon. At various junctures in the past, alternative design visions for the national capital displaced the couple's ideals. This is not to suggest that all departures from their design are without merit. Griffin himself appreciated the organic nature of cities, modifying the Canberra plan as the need arose. Nonetheless, as detailed in its final report (2004), the project validated the original plan's cultural significance and conclusively established its contemporary relevance, even enlarging its legacy through a series of urban design initiatives.[47] As reasserted by this project, for the NCA, the Griffins's design is the most important reference in guiding future development.

As the national capital approaches its centenary, Australian mystique for the Bush is undiminished (with environmental concerns now providing additional impetus). In fact, despite its synthetic genesis as a designed city, Canberra, until recently, was promoted to tourists as the 'natural capital'. Many 'Canberrans', as the city's citizens identify themselves, have embraced the 'Bush Capital' moniker, appropriating the derogation as a term of endearment. No longer an allusion to geographical remoteness, the label is now taken to reference more literally the distinctive pervasiveness of actual bush within the city. The city's 'bush', however, is a cultivated mosaic of remnant indigenous vegetation, street trees and parklands of native and exotic species and of commercial timber plantations. Indeed, the capital's density owes more to vegetal architecture than to its built counterpart. Living in close proximity to this 'bush', however, is not without cost. In January 2003, out-lying bushfires quickly spread into a portion of the city, consuming nearly 500 houses and taking four lives. However, not all of the bushfire's consequences were tragic. Most prominently, the fires impacted the nation's perception of its capital. To its detractors, Canberra – like other *ex novo* capitals – is 'artificial'. However, as few Australian cities are immune from bushfires, the tragedy made Canberra 'real' in the national psyche – if only momentarily.

Today, the picturesque reigns triumphant at Canberra. To view the city from Mount Ainslie is to see it set against an 'emerald green' backdrop mosaic of parklands, timber plantations and phosphate-saturated hillside paddocks, occasionally tempered by vestigial bush. The sea of manicured turf emanating from Parliament House, cascading down its earthen ramps and pulsing throughout the city's ceremonial centre accentuates this effect. The twenty-first century city uncannily resembles more the picturesque imagery of Coulter's 1901 and 1911 'waterside capitals' than it does the Griffins's design. Yet, perhaps, it is equally through departures from the original vision, whether by intent or default, that the national capital becomes 'Australian'.

NOTES

1. *Proceedings at the Congress of Engineers, Surveyors and Others Interested in the Building of the Federal Capital of Australia, Held in Melbourne, in May 1901.*

2. *Ibid.*, p. 35.

3. Coulter (1901).

4. *Proceedings at the Congress of Engineers, Architects, Surveyors and Others Interested in the Building of the Federal Capital of Australia, Held in Melbourne, in May 1901*, p. 36.

5. Elsewhere in the *Proceedings*, for instance, architect G. Sydney Jones argued that the capital's architecture should 'be essentially Australian'. Horticulturist C. Bogue Luffmann advocated the planting of native species and urged that 'if we must have symbols, let us typify our own'.

6. Carter (1995): p. 6; also see his seminal text *The Road to Botany Bay* (Carter, 1987).

7. *Ibid.*, p. 4.

8. *Ibid.*, p. 6.

9. On the site selection process and related political battles, see an excellent study by Pegrum (1983).

10. Gibbney (1988), pp. 1–2.

11. Instructions from Minister for Home Affairs in 'Yass-Canberra Site for Federal Capital General (1908–09) Federal Capital Site – Surrender of Territory for Seat of Government of the Commonwealth', National Archives of Australia (NAA: A110, FC1911/738 Part 1).

12. Gibbney (1988), pp. 1–2.

13. Government of the Commonwealth of Australia (1911).

14. On the competition see Reps (1997). For the Australian planning history context see Freestone (1989) and Hamnett and Freestone (2000).

15. On Griffin as a landscape architect and town planner see Harrison (1995) and Vernon (1995). Other important references on the Griffins, for instance, include Turnbull and Navaretti (1998) and Watson (1998).

16. News of the Federation movement coalescing in the distant Antipodes captured Griffin's attention in 1896, whilst he was still a student at the University of Illinois. Convinced that a new capital city was an inevitable necessity, Griffin, his father recollected, 'then decided to build it'. See Vernon (1998).

17. The lectures appear in Unwin (1911) and Mawson (1911).

18. Royal Institute of British Architects (1911). The competition brief advised that the 'Conference held under the auspices of the Royal Institute of British Architects in October last, at which many authorities on the subject of town planning were present, must have a marked influence upon city Design from the utilitarian, the architectural, the scientific, and the artistic standpoints'. Government of the Commonwealth of Australia (1911), p. 9.

19. For Walter's own explanation of their design see

Griffin (1914), Griffin prepared this text during his first Australian visit.

20. Griffin (1912).

21. Thomas Jefferson was the next to further, in technique and symbolism, L'Enfant's conception of landscape as the primary, noble focus of axial compositions in his design of the University of Virginia. On Jefferson, see Creese (1985), pp. 9–44.

22. On L'Enfant's landscape vision see Scott (1991). Lantern-slides of a topographic map of Washington DC and a facsimile of L'Enfant's plan are amongst the surviving records of the Griffins's practice (private collection).

23. For an excellent exposition of the design's political symbolism, see Weirick (1988) and Sonne (2003), pp. 149–188.

24. These buildings are labelled and depicted in the Griffins's original competition plan, as drawn onto the contour base plan supplied to competitors.

25. Weirick (1988), p. 7.

26. Along with provision of buildings for 'Future' uses, the Griffins accommodated the following departments in their plan: 'Postmaster General', 'Trade and Customs', 'Home Affairs', 'Treasury and Commonwealth Bank', 'Attorney-General', 'Defense', 'External Affairs' and the offices of the 'Prime Minister'. The couple also included a 'Library' within the 'Houses of Parliament'.

27. Lawrence (1923, 1995).

28. The Griffins's submission included the following renderings: plan of 'City and Environs'; a triptych 'View from Summit of Mount Ainslie' and a series of cross-sectional drawings depicting the 'Northerly Side of Water Axis' (4 panels), 'Easterly Side of Land Axis' (4 panels) and a detail of the 'Government Group' on the 'Southerly Side of Water Axis'. These were accompanied by another plan drafted on the contour plan supplied to competitors and a typescript report.

29. 'Competitor number 10 WS Griffiths, RCG Coulter and CH Caswell. Perspective – view of the lake at sunset, also showing the continuation of avenue over the railway line with stairways etc', National Archives of Australia (Item no 4185410, Series A710, Series accession A710/1).

30. L. W., 'Canberra', unknown periodical [newspaper cutting] (London), p. 151. (Mitchell Library collections, State Library of New South Wales).

31. For a comprehensive overview of Canberra's administration and design evolution see Reid (2002) and Fischer (1984).

32. Senate Select Committee Appointed to Inquire into and Report upon the Development of Canberra (1955).

33. *Ibid*.

34. *Ibid*.

35. Appointed inaugural Commissioner, architect John Overall (1913–2001) soon infused the NCDC with his enthusiasm, technical and political skills and considerable managerial abilities. See Overall's (1995) own account of Canberra's development.

36. Holford (1957), p. 6.

37. Holford, (1957), p. 10.

38. Reid (2002), p. ix.

39. Designed by Sydney architect Walter Bunning, the National Library of Australia was positioned in relation to Holford's unrealized proposal for a lakeside Parliamentary complex. This building is stylistically resonant with Oscar Niemeyer's counterpart public edifices at Brasília.

40. The National Science and Technology Centre, Japan's gift to commemorate Australia's 1988 bicentennial, was the next major building to be erected within the Parliamentary Triangle. The building was designed by Sydney architect Lawrence Nield.

41. National Capital Development Commission (1970).

42. *Ibid*.

43. This agency was also headed by Overall.

44. On Canberra's metropolitan planning see, for example, Conner (1993).

45. Commonwealth Place was designed by Durbach Block and Sue Barnsley Design.

46. Reconciliation Place was designed by a team led by Kringas Architects.

47. See National Capital Authority (2004). The author participated in the project as Design Advisor to the NCA.

Chapter 11

Ottawa-Hull: Lumber Town to National Capital

David L.A. Gordon

Context from the Nineteenth Century

Ottawa[1] was not the first choice as the seat of government of the United Canadas. It was initially located at Kingston, then Montréal and after 1849 moved between Toronto and Québec City. Canadian politicians could not decide on a fixed site for the seat of government and finally asked the Crown in England to make the selection for them. Directed by her advisors, Queen Victoria chose Ottawa, a lumber town on the border between Québec and Ontario.[2]

Development of Canada's national capital did not start with a vacant site and a new plan, as was the case for Canberra or Brasília. When Queen Victoria made her choice in 1857, there were over 10,000 people living in the town. There was no immediate need for a plan for the new seat of government, since Barracks Hill was the obvious site for the parliament buildings, and there was the remainder of a 400 acre Crown land reserve available for future expansion. Perhaps another reason why no plan was prepared for the new capital was that few of the legislators cared for the place; while it may have been the second choice, it was certainly not the first choice of most politicians and officials either as a capital, or as a place to live.

Luckily for Ottawa, the huge public expenditure on the parliament buildings made it difficult to re-open the issue of the capital's location when Confederation with other British North American colonies was negotiated between 1864 and 1867. Further, there was equally little inclination to discuss the governance of the capital at this time. One of the negotiators proposed a federal district similar to Washington, but the idea was ignored and Ottawa was left as any other city in the new dominion under the direct control of the new province of Ontario.[3]

The antipathy of the legislators was not surprising, because at that time Ottawa was not an attractive place; it was a one-industry town, and that industry was lumber, not government. The politicians and 350 civil servants occupied only the picturesque trio of Gothic Revival buildings on Parliament Hill. The legislators typically boarded in hotels, and the civil servants barely made a dent in the society of 'one of the roughest, booziest least law-abiding towns in North America'.[4] In the late nineteenth century the capital had none of the utilities found in many cities of the day: no paved streets, no sewers, no gaslights and no piped water supply. The considerable natural beauty of the site was marred by timber-based industry, although the city was proud of its vigorous industrial image at the time.

The federal government of Canada made repeated attempts to plan and develop its seat of government during the twentieth century, but until the 1950s capital planning was primarily shaped by the design and siting of buildings. After World War II, urban planning accommodated dramatic growth and facilitated major improvements in Ottawa and Hull.

There were six phases in the twentieth-century planning of Canada's capital:

- the Ottawa Improvement Commission (OIC), 1899–1913;
- the Federal Plan Commission (FPC), 1913–1916;
- the Federal District Commission (FDC), 1927–1939;
- the immediate post-war period, 1945–1958;
- the National Capital Commission (NCC), 1959–1971;
- the transition to regional government, 1971–2001.

The Ottawa Improvement Commission:
The Washington of the North?
(1899–1913)

The official neglect of Canada's capital began to change in the 1890s, under Prime Minister Wilfrid Laurier. Laurier did not have a good early impression of the capital, but in 1893, he promised:

to make the city of Ottawa as attractive as possibly could be; to make it the centre of the intellectual development of this country and above all the Washington of the north.[5]

Unfortunately, 'Washington of the North' became the slogan for Ottawa's improvement as a national capital, establishing a precedent that was not always appropriate. Laurier established the Ottawa Improvement Commission (OIC) in 1899.[6] The Commission was granted $60,000 per year, partly to compensate the city for services by improving the capital's appearance. It reported directly to the Minister of Finance, but Laurier took a personal interest in its work.

At first, there was general acclaim for the OIC's work. It cleared some industries from the west bank of the Rideau Canal and built a parkway that was both popular and improved the view when entering the capital by train. In 1903, the OIC commissioned Montréal landscape architect Frederick Todd (1876–1948) to prepare a preliminary plan for Ottawa's parks and parkways. Todd had trained in the office of Frederick Law Olmsted and was one of Canada's first landscape architects and town planners.[7] He prepared a preliminary plan for the open space system of the national capital, including the first proposal to acquire Gatineau Park in Québec (see figure 11.1). Todd respected the unique natural setting of the city and its Gothic Revival parliament buildings, and recommended avoiding any literal planning of a 'Washington of the North'. His inter-connected parks system, regional approach and admiration for natural systems reflected the best of the Olmsted tradition and modern ecological planning principles. Unfortunately, the OIC chose to ignore the report and to proceed with incremental additions to the Ottawa parks, without the guidance of architects, planners or landscape architects. The city's Rockcliffe Park was enlarged along the Ottawa

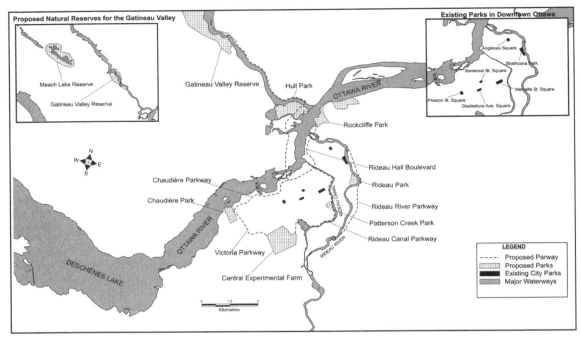

Figure 11.1. Parks and parkways proposed by F.G. Todd in 1903 for the Ottawa-Hull region.

River, another park was built on the Rideau River and several small squares in the city received their first landscaping.[8]

The new greenery sprouting throughout the city delighted its citizens and the Laurier government. Plan implementation for Canada's capital during this period showed an adequate financial strategy and a good political champion in Laurier, but the implementation agency suffered from a lack of design skills and administrative expertise.

The Federal Plan Commission: City Beautiful on the Ottawa River, 1913–1916

Although Prime Minister Laurier was satisfied with the work of the OIC, criticism of Ottawa planning gradually grew in the first decade of the century. Governor General Earl Grey, a patron of several English town planning movements, closely followed Ottawa planning issues.[9] He sponsored tours by British experts Raymond Unwin, Thomas Mawson and Henry Vivian, MP.[10] Ottawa architect Colborne Meredith started a well-coordinated lobby to take control of a new plan for the nation's capital, aided by Noulan Cauchon, a local railway surveyor. Meredith's objective was an elite commission of technical experts to supervise preparation of a comprehensive plan. This model was based upon Washington's successful experience with the 1902 Senate Parks Commission, which was well known at the time.[11]

The new Conservative Prime Minister, Sir Robert Borden, proceeded cautiously, quietly dropping the idea of directly commissioning Mawson. The government published a policy paper including critiques from the Royal Architectural Institute of Canada, Unwin,

Mawson and Meredith.[12] Borden wanted a process that was under his direct political control, rather than an independent panel of expert professionals. Senior staff discreetly assembled a group of prominent Conservative businessmen to act as a planning commission chaired by Herbert Holt, president of the Royal Bank. The federal government attempted to co-opt the local governments by appointing the mayors of both Ottawa and Hull as *ex-officio* members of the new Federal Plan Commission (FPC). Adding Hull to the FPC's mandate was an astute political move, since the Québec side of the Ottawa River had realized few benefits from Ottawa's designation as the seat of government, and received little attention from the OIC.

The FPC retained Edward H. Bennett of Chicago as their consulting architect and planner. Bennett (1874–1954) was born and raised in England, and educated at the prestigious École des Beaux Arts in Paris. He was responsible for several major plans, including the landmark 1909 *Plan of Chicago*, co-authored by Daniel Burnham. In the absence of capable Canadian planners, Bennett's English heritage, French education and American experience made him perhaps uniquely suited for the Ottawa-Hull commission.[13] He prepared a plan for the capital in the City Beautiful style (see figure 11.2), with comprehensive technical planning for infrastructure and zoning.[14] Although the Borden government tabled the FPC report in Parliament

Figure 11.2. Edward Bennett's 1916 plan for a municipal plaza astride the Rideau Canal in Ottawa in the City Beautiful style. The Parliament Buildings are on the bluff on the upper left. The consolidated train station and Château Laurier hotel are shown on the upper right. The site of the plaza and proposed City Hall on the mid-left are now occupied by the National Arts Center. Rendering by Jules Guérin.

in March 1916, it quickly disappeared from sight. The First World War was the preoccupation of the government, and, as the Centre Block of the Parliament Buildings had burned down a few weeks before, rebuilding would absorb any funds the government could devote to Ottawa outside the war effort.[15]

The political structure that Borden's office established for the FPC may also have hindered implementation of the plan. The Commission was disbanded and its staff dispersed after the report was printed. The FPC's Tory commissioners moved on to other concerns during the war, and they had no political access to the Prime Minister's office after the Liberal Party won the 1921 election. The mayors changed frequently in those days, so there were no powerful local advocates of the plan when it was released; it simply sat on the shelf.[16] This period is a classic 'good plan/poor implementation' scenario – the consultant team's technical expertise was wasted for want of political support, funding or administrative capacity. Its dramatic, large-scale proposals were inappropriate when the nation was focused on war, and its financial requirements were too large during the weak economic recovery from 1919 to 1929.

A King for a Town Planner: The Federal District Commission, 1927–1939

William Lyon Mackenzie King (1874–1950) was Canada's longest-serving prime minister, holding that office for most of the period from 1921 to 1948. Like his mentor Laurier, King was dismayed by Ottawa when he arrived as a civil servant in 1900. Unlike previous prime ministers, he had a strong personal interest in town planning. Although his main professional interest was labour relations, he regarded town planning as a key component of an overall programme for social reform.

Mackenzie King's interest in planning was complemented by a growing personal commitment to Ottawa's development as a capital worthy of the growing nation. He personally managed almost every planning and design proposal of the federal government over the next thirty years.[17] He took control of the Ottawa Improvement Commission during his first term of office (1921–1930), recruiting the energetic Ottawa utilities tycoon Thomas Ahern as the new chairman. Mackenzie King dissolved the agency in 1927 and established a Federal District Commission (FDC) with a wider mandate and larger budget. Originally he favoured the federal district concept, but the idea was unpopular in Québec due to concerns over language and culture. So the FDC became a parks agency operating on both sides of the river, but with no local government powers and little planning capacity.

The Prime Minister had ambitious plans during the improved economy of the late 1920s. He and Ahern planned an urban renewal scheme to create a major public plaza between Elgin Street, the Rideau Canal and Wellington Street. This scheme was loosely based upon Edward Bennett's 1915 proposal for a civic plaza in Ottawa (see figure 11.2). After a hotel on a key site was destroyed by fire, Mackenzie King pushed a bill through parliament to amend the FDC's act and provide a fund of $3 million to redevelop the core of the capital.[18] He also used federal investment to push Ottawa City Council into an agreement that they would relocate City Hall and widen Elgin Street. The federal government was determined to remake the historic core of the city in its own image.[19]

To give some political impetus to the Elgin

Street plaza, Mackenzie King named the project Confederation Square and proposed it as the site for the national memorial to those who gave their lives in the Great War. A memorial had been commissioned from an English sculptor, but its site had not been selected from among several locations on Parliament Hill and its surroundings.[20] Mackenzie King lost the 1930 election before he could start construction of the plaza, but he never gave up. When he returned to power in 1935, he vigorously pursued plans for the new square, perhaps embarrassed by the delays for the National War Memorial. However, despite his enthusiasm for the project, the complicated tangle of bridges, streetcars, streets and a canal resisted the efforts of a generation of planners to design an elegant solution. Canada simply did not have much urban design talent in the 1930s.

Mackenzie King found his planner during a tour of the 1937 Paris World's Fair, led by the Fair's chief architect, Jacques Gréber. The two men immediately established a good relationship. Gréber (1882–1962), who was near the peak of his

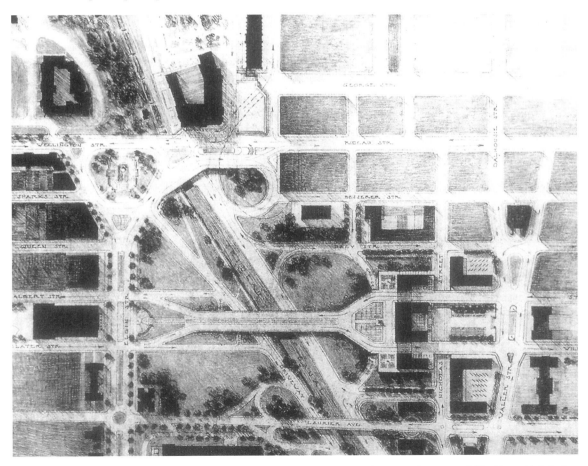

Figure 11.3. Jacques Gréber's plan for Confederation Square, *c.* 1938, showing elements of City Beautiful design and City Scientific traffic engineering. The National War Memorial is located in the centre of the triangular plaza at the request of Prime Minister Mackenzie King. Note the new bridge (later named for Mackenzie King) in the middle of the drawing.

career as a classically-trained architect, planner and professor, was invited to come to Ottawa as soon as possible to prepare plans for Ottawa's core.[21] He designed the new square (see figure 11.3) in time for the War Memorial to be unveiled during the 1939 Royal Visit. The rest of Gréber's plans for downtown Ottawa were put on hold during World War II.

These inter-war years saw limited implementation of the federal plans for Canada's capital, because its champion, Mackenzie King, lacked political support, and the responsible agency had poor funding and limited project management capacity. In addition, the economic and political environment for implementation of capital plans was weak in the 1920s and absolutely hostile in the 1930s. King had to wait.

Local Government Planning to 1945

The federal government's modest planning activity dwarfed most local efforts in the first half of the century. The Québec provincial government did not enact community planning legislation until after 1945 and local governments on the Québec side of the Ottawa River did not engage in formal community planning until the 1960s. The Ontario provincial government passed permissive town planning legislation in 1917 at the behest of federal advisor Thomas Adams. The City of Ottawa then established a town planning commission in 1921 chaired by local activist Noulan Cauchon. However, the OTPC was purely advisory, under-funded and had little impact. Cauchon and his aide John Kitchen prepared several schemes for traffic improvements in the City Scientific mode, and a zoning bylaw for part of the city. Cauchon's esoteric designs and penchant for publicity did not sit well with Mackenzie King, who ignored him and negotiated directly with the Mayor on Confederation Square and the replacement of Ottawa City Hall.[22]

Local governments focused on improving private property, which was the primary tax base in the early years of the century. While mayors on both sides of the river changed frequently, they rarely supported urban planning. Powerful national politicians such as Laurier and Mackenzie King had some money and staying power to pursue their own interests, often running over the local officials.[23] The federal government appropriated most of the planning initiatives in Ottawa, perhaps weakening local groups such as civic improvement leagues and town planning commissions, which were more active in other Canadian cities. Perhaps the most important local planning action in the first half of the century was a building height limit enacted by the city of Ottawa in 1914, at the suggestion of a reform mayor who was a member of the Federal Plan Commission. The 110-foot height limit was suggested by the FPC's consultant Edward Bennett, based on a US law relating to the width of Washington streets. It regulated building heights in Ottawa for a half century, protecting the primacy of the Parliament Buildings on the city's skyline.[24]

The Canadian community planning movement collapsed during the Depression and the Town Planning Institute suspended operations from 1932 to 1952. There was only a handful of planners operating in municipalities across the country and the federal government almost had the field to itself in the Ottawa-Hull region. Prime Minister Mackenzie King and the federal government made community planning a central element of its national post-war reconstruction programme.[25] Mackenzie King also launched a major planning initiative in the national capital that became a pilot project for the slow reconstruction of the profession.

The Immediate Post-war Years, 1945–1957

Mackenzie King intended that construction of a national capital for Canadians would be the principal memorial for those who fell during the Second World War. He established a National Capital Planning Committee (NCPC), independent of the FDC, with representatives from across the country. He chaired early meetings of the Committee, and frequent references in his personal diary shows that he followed its every move (see figure 11.4). Gréber was installed as head of the National Capital Planning Service, with an ample budget, numerous staff and a wide mandate. The mistakes of the Holt era were rarely repeated, since the NCPC consulted with local and provincial governments on both sides of the river. It built public support with newsreels, radio interviews, newspaper inserts and exhibitions of a large model of the future capital held in cities across Canada.

Mackenzie King's health was failing in 1948, but he hung on as Prime Minister until the draft plan was prepared and pushed an unusual commitment of $25 million to implement it through Cabinet as his final act.

After five years of research and consultation, Gréber's National Capital Plan was published in 1950 (see figure 11.5). It built upon previous plans including:

• relocation of the railway system and industries from the inner city to the suburbs;

• construction of new cross-town boulevards and bridges;

• decentralization of government offices to the suburbs;

• slum clearance and urban renewal of the LeBreton Flats district;

• expansion of the urban area from 250,000 to 500,000 in neighbourhood units;

• surrounding the future built-up area with a greenbelt;

Figure 11.4. Mackenzie King (*left*) and Jacques Gréber review the plan for Canada's capital, *c.* 1948.

• a wilderness park in the Gatineau hills and a parks system along the canal and rivers.

The railway relocation was the key element that unlocked the rest of the plan. Removing the east-west Canadian National line and its adjacent industry in the centre of Ottawa reconnected the road grid, separated noxious industries from residential areas, and provided rights of way for cross-town boulevards. Relocating the two railway stations to the suburbs permitted construction of a union station and freed up the yards in the heart of Ottawa for a convention centre, shopping and a hotel. The tracks leading to the station were replaced by a parkway along the east bank of the Rideau Canal. These proposals were an elaboration of the previous plans, except for the station relocation, which was not contemplated in the downtown plans prepared by Gréber in 1938–1939. The railways were replaced by new boulevards and an expressway.

Government departments and national institutions that were essential for diplomatic or parliamentary purposes were located in high-quality stone buildings close to Parliament Hill. Research laboratories, back office functions and administrative departments were decentralized to four suburban office parks in Ottawa and Hull. This decentralization enabled the many 'temporary' war-time buildings to be removed from the central city and freed sites for national institutions like a library, theatre and art gallery. It also allowed many civil servants to purchase inexpensive suburban houses with a short drive to work.[26]

The 1950 National Capital Plan became a landmark in Canadian planning history, setting the standard for the comprehensive plans which followed in the next decades.[27] Despite its status as a war memorial and Mackenzie King's legacy, the plan had a slow start. The FDC initiated some railway relocations, but the remaining elements stalled due to weak provincial planning legislation and lack of consensus among the local

Figure 11.5. Watercolour rendering by Jacques Gréber of the 1950 *National Capital Plan*, showing Gatineau Park extending from the north-west almost to the city core. Note the greenbelt surrounding Ottawa on the south side of the river (built), and the radiating avenues from the proposed new railway station (never built).

governments. Ottawa and Hull were given strong links to the plan by appointing both mayors to the FDC and councillors and senior staff to the NCPC. Ottawa's major complaint, the fiscal impact of tax-exempt federal property, was addressed by adjustments to the grants-in-lieu formula in 1944 and 1950.

Ottawa established its own planning department in 1951 but the City did not adopt an Official Plan until December 1967. Instead, the 1950 National Capital Plan served as a quasi-official regional plan: the city adopted its infrastructure maps and the local planning board tried, with little success, to defend Gréber's greenbelt in the surrounding townships. The problem was that the federal government had no legal jurisdiction for local land-use planning. In the end, the NCC had to expropriate the greenbelt on the Ontario side to prevent the suburban townships from filling it with unserviced residential subdivisions. The local property owners and suburban municipalities complained bitterly and appealed all the way to the Supreme Court, where they lost. However, the greenbelt was quietly dropped on the Québec side of the river.

The major political issue on the Québec side was the spectre of a federal district similar to Washington or Canberra, which would detach Ottawa, Hull and environs from their municipal and provincial governments. This idea was sometimes popular in Ottawa and it was the first recommendation of the 1915 Holt Commission.[28] However, a federal district was completely unacceptable to Québec politicians at all levels. They were not willing to give up the protection of their language, education, law and culture afforded by their local and provincial governments, so the 'federal district' issue poisoned all attempts at regional planning. Mackenzie King raised the issue again in Parliament in 1944, but did not take action. The Prime Minister finally reversed his position in 1946, but the issue continued to cloud politics in Québec. Even the name of the relatively toothless Federal District Commission was an affront, so it was re-christened as the National Capital Commission/La Commission de la capitale nationale in 1959.

Implementation of the 1950 plan was slow in its first decade due to suburban opposition, but the FDC built its capacity for project management. The FDC hired expert landscape architects, planners, engineers and project managers, developing a reputation for good fiscal management. As their organizational competence increased, they were given responsibility for landscaping federal buildings in the capital, project management of infrastructure and land-use planning approval for federal properties. However, responsibility for constructing public buildings remained with the Department of Public Works, setting the stage for conflict in the 1960s.[29]

The National Capital Commission, 1959-1970

A 1956 joint Senate-Parliamentary committee concluded that the federal government would have to act alone. The new National Capital Commission (NCC) absorbed the National Capital Planning Committee and Gréber's staff. It was given powers to expropriate land, build infrastructure and create parks. It used these powers to expropriate the land for the greenbelt.

The NCC's good managerial reputation in the 1950s and 1960s allowed them to move quickly to implement elements of the Gréber plan. By 1956 it was clear that the $25 million National Capital Fund would not be enough, due to inflation, expropriation of the greenbelt and better cost estimates for the infrastructure. The joint Senate-Commons committee recommended that the NCC's annual capital grant be at least doubled. By 1970, when most of the plan had been implemented, the NCC had spent $243 million.[30]

Surprisingly, much of the work was carried out after the Conservative party took power under John Diefenbaker from 1957 to 1963. Mackenzie King gave the project enough political momentum to last almost two decades. The FDC/NCC was virtually unstoppable as an implementation agency for twenty years after the war (see figure 11.6). All the elements were in place – political support, long-term finance, good economic conditions, skilled planning staff and strong project management.[31] Ottawa and Hull were transformed from dreary industrial towns into a green, spacious capital that was visited by millions of Canadian tourists.

Local and Regional Government Planning after 1945

The City of Ottawa supported the FDC by establishing the Ottawa Area Planning Board (OAPB) in 1946 to control unregulated suburban expansion. However, the suburban townships continued to approve low-density subdivisions without municipal services. The city reacted in 1948 by attempting to annex all the land inside the proposed boundary of the greenbelt. The rural townships fought the annexation, and lost. They also fought the greenbelt, refusing to incorporate it into their zoning bylaws and approving subdivisions. After six years of conflict, it became clear that Ontario and Québec planning legislation was not strong enough to establish a greenbelt by regulation, as in the London model.

However, planning remained co-operative in Hull, where NCC planning staff were consultants to the City of Hull, preparing an urban renewal plan in 1962 and expansion of Gatineau Park into a major regional wedge of open space. The NCC lost its first major battle with the City of Ottawa in 1965 over building height limits. A private developer with close connections to the Department of Public Works and the national

Figure 11.6. Gréber and staff locate the new National Library for FDC members, 1954. The professional planners were dominant in this era.

Liberal Party convinced Council to abandon the 110-foot height limit to create a high-rise central business district. Once again, the NCC lacked the legal jurisdiction at the federal or local level to prevent development that did not conform to its plan, and the view of Parliament Hill from the south was lost.[32]

The NCC's primacy in regional planning disappeared in the 1970s. The Ontario government established the Regional Municipality of Ottawa Carleton (RMOC) in 1968, and the Communité Regional de l'Outaouais (CRO) was set up by Québec in 1970. They both completed regional land-use plans in the mid-1970s which co-ordinated with suburban township plans prepared by local professional staff. The plans called for extensive low-density, automobile-serviced suburban development on land held by private developers at the periphery of the metropolitan area.

The NCC countered with an innovative plan for a higher density growth corridor served by public transit, and a new town on a site already assembled by the federal government south-east of the greenbelt. The 1974 *Tomorrow's Capital: An Invitation to Dialogue* plan was developed in secret, and it outraged local and regional governments and community groups when it was released at the end of a complex, five-year regional planning process. They refused to consider it, and lobbied the federal government to remove the NCC from all land-use planning activity. RMOC assigned the federal new town site its lowest priority for development and eventually dropped it from the regional plan altogether.[33]

Gréber's 1950 plan guided the growth of the Ottawa-Hull region from 250,000 to 500,000 people from 1946 to 1966, filling out the area inside the greenbelt. The local and regional plans guided growth as the region doubled its population again to 1.1 million by 2001. The regional governments on both sides of the river developed into sophisticated planning agencies and, during the 1990s, began to question the decentralized suburban model, with the RMOC developing an advanced bus transitway system co-ordinated with nodes of federal employment.

However, true regional planning was almost non-existent, since there was little co-ordination of transportation and land use across the border of the Ottawa River. The two provinces do not co-operate and the federal government has built all the bridges. As noted by John Taylor: 'either the major roads have no links to the bridges, or the bridges have no links to the major roads'.[34] Urban development of the region was becoming unmanageable with overlap and conflict between numerous local, regional, provincial and federal planning agencies. Both provincial governments took action early in the new century – Ontario dissolved all the local governments and the RMOC to create a new City of Ottawa which embraces almost all of the National Capital Region south the of the river. Québec consolidated most of its urban municipalities into a new City of Gatineau. Regional planning in the twenty-first century should have only three actors at the table – Ottawa, Gatineau and the NCC – with perhaps a better chance at reaching a working consensus.

Federal Planning in the Late Twentieth Century

Powerful independent implementation agencies like the NCC may have been needed in the early stages of planning and development of Ottawa but they became less relevant as the city became properly established. The benevolent dictatorship implied in the powers vested in these agencies is harder to justify once the principal activity shifts from rapid physical development to routine local governance, especially after the rise of citizen participation in the 1970s.

The National Capital Commission lost control of federal planning initiatives in the mid-1960s. Although all federally-built buildings were approved by the NCC, the Department of Public Works (DPW) began to lease space from private developers to accommodate the rapid growth of the civil service at lower cost.[35] The DPW never agreed to prohibit leases in private buildings which did not conform to the 1950 plan. Ironically, most of the mediocre high-rise office buildings which block the view of Parliament Hill are occupied by federal agencies.

After Pierre Elliot Trudeau was elected prime minister in 1968, other changes to the national capital occurred quickly in response to rising Québec nationalism. The National Capital Region of Ottawa-Hull was officially declared Canada's capital and 18,000 federal employees were moved to buildings on the north side of the Ottawa River to ensure that 25 per cent of the civil service were in the Québec portion of the region. Hull's urban renewal plans were implemented almost overnight as bulky privately-built office buildings sprouted in its downtown, and the DPW built a bridge to central Ottawa. The federal government also located major new public buildings for the national museum and archives in Québec.

The NCC was given added responsibility for programming the national capital, ensuring that its image was bilingual, while its planning capacity was reduced. The agency produces major public festivals which animate the capital and promote national unity, including Canada Day, Winterfest and a host of special events. Canada Day festivities are televised and other events are promoted nation-wide to encourage tourism. The NCC organizes and interprets the national capital for visitors.

The NCC did not withdraw completely from planning as its local critics demanded in the 1970s. It refocused upon the portfolio of federally-owned property in the region, preparing large-scale master plans based upon ecological principles for the greenbelt and Gatineau Park. The agency regained the initiative in the core of the capital through urban design projects which improved public spaces, most notably in the Parliamentary Precinct and a Confederation Boulevard creating a ceremonial route linking Ottawa and Hull. The continuing lack of co-ordination across the Ottawa River gave the NCC an opening as a facilitator in regional transportation and land-use issues. These themes were integrated in new plans for the core of Canada's capital and the national capital region prepared with much public consultation over the turn of the century.[36] These plans are controversial, but the NCC is still in the game, using land ownership, financial resources and professional talent to influence the development of Canada's capital. The game is quite different from the mid-twentieth century, when the federal government was the only player, but the recent changes in local government structures may allow the NCC to continue to play an important role in planning the future Canadian capital.[37]

NOTES

1. Ottawa was founded in 1826 and named Bytown in 1827. The Canadian seat of government was initially confined to this Ontario city, but the National Capital Region was expanded to Hull and the Outaouais region in the twentieth century. The metropolitan area was known as 'Ottawa-Hull' for most of the century. For the best general history of Ottawa see Taylor (1986); for a general history of the Outaouais see Gaffield (1997).

2. Knight (1991), especially pp. 67–70 and 90–91. For more on the political influence of the governors-general and prime ministers in Canadian capital planning, see Gordon (2001a).

3. Young (1995); Gray, quoted in Eggleston (1961), pp. 145–146.

4. Gwyn (1984), p. 40.

5. Laurier (1989), p. 84; *Ottawa Evening Journal*, 'The Washington of the North', 19 June, 1893, p. 3.

6. Aberdeen (1960) pp. 478–479, Nov. 19, 1898; Parliament of Canada (1899).

7. Jacobs (1983); Todd (1903).

8. Gordon (2002c); Ottawa Improvement Commission (1913).

9. On Grey and Letchworth, see Miller (1989); Miller and Gray (1992).

10. Gordon (1998).

11. For the Senate Parks Commission (chaired by James McMillan) see Moore (1902); Reps (1967); Peterson, (1985). On Daniel Burnham's role as principal consultant to the commission, see Hines (1974), chapter 7. The District of Columbia political model was also popular in Ottawa at the time. In May 1912, Ottawa electors decisively supported the creation of a Federal District in a non-binding referendum.

12. Parliament of Canada (1912).

13. For Bennett's background, see Draper (1982); Burnham and Bennett (1909); Gordon (1998).

14. Federal Plan Commission (1916); Gordon (1998).

15. FPC Commissioner Frank Darling's firm got the commission for the new building, in joint venture with Omer Marchand of Montréal. See Kalman (1994), pp. 712–772.

16. The plan was attacked by opponents of the City Beautiful approach. See Adams (1916). The federal district proposal was a complete non-starter in Québec, and Hull refused to pay its share of the costs. The court case dragged on for years, souring local opinion of the plan. See Gordon (1998).

17. See Gordon (2002b).

18. See Gordon and Osborne (2004).

19. Taylor (1989); Taylor (1986), chapters 4 and 5.

20. Gordon and Osborne (2004).

21. Delorme (1978), pp. 49–54; Lortie (1993), pp. 325–375; Lavedan (1963), pp. 1–14; Lortie (1997).

22. Gaffield (1997); Simpson (1985); for Ottawa planning see Taylor (1989); Hillis (1992); Ben-Joseph and Gordon (2000).

23. Taylor (1989).

24. Gordon (1998); Gutheim (1977).

25. Gordon (2002a); Wolfe (1994).

26. Office decentralization was also promoted for civil defence against bombing, although this argument disappeared after the power of the hydrogen bomb was understood.

27. Gordon (2001b).

28. See Federal Plan Commission (1916), p. 13; Cauchon (1922), pp. 3–6; Rowat (1966), pp. 216–281.

29. The Department of Public Works continued its role as the client and lesser of federal buildings, see Wright (1997).

30. Corrected for inflation this is C$1.5 billion in 1999. See Gordon (2001b).

31. The external environment was finally supportive in this period. The post-war economic boom meant there was money available, and the political culture had changed. Large-scale planning was now an accepted strategy and Ottawa-Hull changed from a blue-collar to white-collar population with the decline of the lumber industry and vast increase in civil servants during the wars. (Thanks to John Taylor for this observation.)

32. Collier (1974); Babad and Mulroney (1989).

33. Ottawa-Carleton, Regional Municipality (1976); Taylor (1996).

34. Taylor (1996), p. 792

35. Wright (1997).

36. National Capital Commission (1998).

37. The City of Hull was amalgamated in to the new City of Gatineau during the 2002 reorganization of local governments on the Quebec side of the National Capital Region. The metropolitan area is now referred to as Ottawa-Gatineau.

ACKNOWLEDGEMENTS

This paper is based upon archival research contributed by Aidan Carter, Emma Fletcher, Aurélie Fournier, Tiffany Gravina, Michael Millar, Jerry Schock, Inara Silkalns, Daniel Tovey and Miguel Tremblay. The research was funded by the Social Sciences and Humanities Research Council of Canada, a Fulbright Fellowship, the Advisory Research Committee of Queen's University and the Richardson Fund. Many librarians and archivists helped, but special thanks should go to staff at the National Archives of Canada, Rota Bouse at the National Capital Commission, and Mary Woolever at the Art Institute of Chicago. John Taylor provided an important and thorough critique of the draft chapter; and two anonymous referees improved an earlier version in a 2002 issue of *The Canadian Journal of Urban Research*.

Chapter 12

Brasília: A Capital in the Hinterland

*Geraldo Nogueira Batista, Sylvia Ficher,
Francisco Leitão and Dionísio Alves de França*

I do not refer to the width the land of Brazil has from the sea back, because, up until now, no one has walked its extent due to neglect of the Portuguese who, being great conquerors of lands, do not take advantage of them, but content themselves with scratching along the shore like crabs.

Frei Vicente do Salvador, *História do Brasil*, 1627

There is an *Atlantic vocation*, owing to the extensive continental coast, which obliges us to look toward the vast oceanic horizon, off to the other side of the sea. And there are the hills, the forest, the *sertão*, the immensity of horizons that are behind the coastal hills, and which very soon stirred up the curiosity and covetousness of adventurers . . .

Cruz Costa, *Contribuição à história das idéias no Brasil*, 1967

The idea of moving the capital of Brazil to its central plain dates back to the mid-seventeenth century, when the country was still a colony of Portugal and there were thoughts of transferring the Portuguese Court to the new continent. The aims of the most important Brazilian separatist movement, the Inconfidência Mineira (1789), included an interior capital seat, in the town of São João del Rei. With the rise of Napoleon – which would make the British Prime Minister William Pitt an advocate of a Brazilian capital in the hinterland[1] – the Portuguese Royal Family was forced to move to Rio de Janeiro in 1808.[2] From this time on, the proposition of an interior capital city, whether for political or strategic reasons, would always be present in discussions of the territorial and administrative organization of the country.

At the time of the first election of representatives of the Brazilian provinces to Lisbon in 1821, an important document by José Bonifácio de Andrade e Silva synthesized the issue:

It also seems to us advantageous that we raise a central city in the interior of Brazil to receive the Court or the Regency, which could be at a latitude of more or less 15 degrees, in a healthy, inviting place, with fertile land watered by some navigable river. In this way the Court or the seat of the Regency would be free from any external attack or surprise, and this would also attract the excess of idle population of the maritime and mercantile towns to the central provinces.[3]

After Independence in 1822, the debate continued, with greater or lesser intensity, but without practical results.[4] However, with the

Proclamation of the Republic in 1889 and the promulgation of the Constitution of 1891, the establishment of the capital on the central plain would become a constitutional precept.

Towards the High Central Plain

They want it, without wanting it.

Eliseu Guilherme,
Anais da Câmara dos Deputados, 1922

By 1892, an Exploratory Commission of the High Central Plain (*Comissão Exploradora do Planalto Central*) had already been nominated, headed by astronomer Luis Cruls and charged with choosing the location of the new capital. The extent and depth of its studies,[5] in the erudite and elegant tradition of nineteenth-century natural science, makes its *Relatório da Comissão Exploradora do Planalto Central* (or *Cruls Report*, 1894) the first technical document pertinent to the planning of Brasília.

The selected area – situated in the State of Goiás, which would become known as the 'Cruls Quadrilateral'[6] – met completely the suggestions of Andrade e Silva. Among its numerous advantages, the site – as if predestined, due to the 'great rivers that start in the region . . . and [that] by a singular caprice of nature, have their springs beginning as it were at a single point . . .'[7] – reinforced the symbolic dimension of national unity and integration ascribed to the capital transfer.

Despite the repercussions of the *Cruls Report*, the sparse measures taken can be summed up in a few railway connections with the region. Only in 1922, in the Nationalist context of the commemorations of the Independence Centennial, would Congress approve the establishment of the Federal capital in the 'Cruls Quadrilateral'.[8] Nevertheless, the Getulio Vargas dictatorship (1930–1946) had other priorities: the establishment of agricultural colonies,[9] and the betterment, though slowly, of the accessibility of the area, due to the improvement in navigability of some rivers and the construction of new railways. But legions of specialists – geographers, military and engineers – would not relinquish the issue, as several technical studies almost unanimously defended a hinterland capital.[10]

Choosing the Site

. . . a little further south, more to the north, farther east or west, it doesn't matter. But in the central high plain.

Everardo Backheuser (1947)

With the end of the Vargas dictatorship and the subsequent election of Marshal Eurico Dutra as President, the location of the capital became a point of controversy.[11] The Constitution of 1946, however, limited itself to reasserting the precept of an interior capital city.

To an even greater degree than before, the issue would be taken up by the military. And in that same year of 1946 the Commission of Studies for the Location of the New Capital (*Comissão de Estudos para Localização da Nova Capital*) was organized, under the presidency of General Polli Coelho. After some reconnaissance of the area chosen in 1894 and of some other alternatives, in 1948 the Commission presented important preliminary reports[12] and the final technical report (*Relatório Técnico*).[13] The more palpable outcome was a new proposal of delimitation, the so-called 'Polli Coelho Perimeter', an enlargement towards the north of the 'Cruls Quadrilateral'.

In 1947, before any decision was taken, the State of Goiás Legislative Assembly authorized the donation to the Federal government of 'all devoid lands within the area to be chosen for the site of the Future capital of the Republic ...'.[14] From that time on, the political commitment of the State of Goiás to the transfer of the capital would be unwavering. However, the first expropriation – of the Bananal Estate, where Brasília would actually be built – only took place in 1956.[15]

Eventually, in 1953 the Congress defined a third area, the 'Congressional Rectangle',[16] to be analysed by a new committee, the Commission for the Localization of the New Federal Capital (*Comissão de Localização da Nova Capital Federal*), headed by General Caiado de Castro. Among other measures, it contracted an American firm, Donald J. Belcher & Associates, to interpret aerial photos of the area and to indicate the five best sites for the undertaking.[17] These tasks were fulfilled in their valuable *Technical Report on the New Capital of the Republic* (*Relatório técnico sobre a Nova Capital da República*) (or *Belcher Report*, 1957).

In 1954 Marshall Cavalcante di Albuquerque replaced Caiado de Castro and, in the next year, the Commission was re-organized as the Commission for Planning, Construction and Transfer of the Federal Capital (*Comissão do Planejamento da Construção e da Mudança da Capital Federal*). Its comprehensive report, *New Metropolis for Brazil* (*Nova Metrópole do Brasil*),[18] was the last technical document on the location of the new capital city. Besides establishing the definitive site, it also presented an urban proposal for the new town of Vera Cruz, by Raul de Penna Firme, Roberto Lacombe and José de Oliveira Reis,[19] thereby creating further controversy.

Heroic Times

We need to build the superfluous ... because the essential will be done no matter what ...

Juscelino Kubitschek, quoted by Lúcio Costa (1995)

In an April 1955 speech in Jataí (GO), then presidential candidate Juscelino Kubitschek promised, if elected, to comply with the constitutional article in favour of transferring the capital to the hinterland.[20] Inaugurated in January 1956, on 19 September of the same year, he obtained the approval of Congress[21] for the necessary measures: the authorization to move the Federal capital from Rio de Janeiro to Brasília (which made its name official), the establishment of the boundaries of the Federal District (DF),[22] and the creation of the Company for Urbanization of the New Capital (*Companhia Urbanizadora da Nova Capital*) (NOVACAP).

A state company, reporting directly to the President and with headquarters in a city that did not yet exist, NOVACAP would be the main agent in the urbanization process. It had a wide range of powers, was the owner of almost all of the land in the DF, and the promoter of all kinds of construction. Financially, it had the authority to give guarantees of the National Treasury for the credit operations and could contract services without a call for bidding, in other words, independent of the usual official controls.[23] In practice, its institutional structure reduced the possibility of political interference in the enterprise and disassociated local decisions from the Federal administration.

From this moment on events would accelerate. On 24 September, Kubitschek appointed the NOVACAP board of directors, naming architect Oscar Niemeyer as technical director in charge of all architectural design. In October, Niemeyer designed the first government building of Brasília, the provisional Presidential Residence.[24] In turn, work began on the Paranoá river dam,[25] an airport, a hotel, and some Air Force barracks. By November Candangolândia, the first NOVACAP encampment, was opened; by December Niemeyer was concluding the design

Figure 12.1. From left to right: Oscar Niemeyer; Israel Pinheiro, Chairman of NOVACAP during the construction; Lúcio Costa; and President Juscelino Kubitschek examining a model of the Three Powers' Square.

for the definitive Presidential Residence, the Alvorada Palace,[26] perhaps his masterpiece in Brasília.

The output of Niemeyer and his team, ranging from apartment buildings, commercial centres, churches and hospitals, to government and monumental buildings, would be extraordinary. For the last he sought to emphasize their visual impact – sometimes with great success, as in the Cathedral (1958) or the Ministry of Foreign Affairs (Itamaraty Palace, 1962) – which would place him at the forefront of Formalist architecture typical of the period.

The exceptional trajectory of Oscar Niemeyer (b. 1907) started in 1936 when, as a member of Lúcio Costa's team, he worked directly under Le Corbusier in the design for the Ministry of Education in Rio de Janeiro. Again together with Costa, he designed the Brazil Pavilion for the 1939 New York World's Fair. In the early 1940s he met Juscelino Kubitschek, then Mayor of Belo Horizonte. From Kubitschek, he received a considerable commission, the Pampulha Park, that would launch his name internationally.

Thenceforth, Niemeyer would mature an architectural language of his own, far removed from functionalism and characterized by formal and structural invention. His prestige was such that in 1947 he collaborated in the design of the United Nations headquarters in New York. After Brasília, he consolidated a career of great productivity, making his architectural oeuvre one of the largest ever. And even today, Niemeyer exerts almost monopoly control on Brasília's 'Federal architecture'.[27]

The urban design selection would be more controversial. Besides the Penna Firme, Lacombe and Reis proposal, some professionals[28] were in favour of inviting Le Corbusier (then at the height of his fame, thanks to Chandigarh and, as was his routine, already offering his services to the Brazilian government). On the other hand, Kubitschek hinted at concentrating all urban decisions in the hands of Niemeyer. These manoeuvres were not well received by Brazilian architects, then celebrating the international repercussion of their Modernist achievements. The compromise solution, reached by the Brazilian Institute of Architects (IAB), was the promotion of a competition. Therefore, some ten days after NOVACAP was created, the 'Call for the National Competition for Brasília's Pilot Plan' was presented.[29] Twenty-six projects were submitted, all examples of functionalist urbanism, and the first prize was awarded to Lúcio Costa.[30]

Deeply influenced by Le Corbusier's ideals and designs, Lúcio Costa (1902–1998) was one of the leading proponents of *avant-garde* architecture in Rio de Janeiro in the 1930s. He had a key role in the conception of the Ministry of Education building (1936), for whose design he assembled a team of young architects[31] and persuaded Le Corbusier to come to Rio as a consultant. In 1937 he joined the National Heritage Service (*Serviço do Patrimônio Histórico e Artístico Nacional* – SPHAN), where he developed most of his professional work. In his later urban work, he would apply principles similar to those of the Pilot Plan in the Barra da Tijuca plan (1969), an area extending nearly 20 km along the south coast of Rio de Janeiro.

Brasília's Pilot Plan

Nothing is so dangerous as being too modern.

Oscar Wilde, *An Ideal Husband*, 1895

Costa's project was presented in a text of exceptional clarity – *Report of the Pilot Plan for Brasília* (*Relatório do Plano Piloto*),[32] a general plan for the town and a series of sketches. Taking the circulation system as his starting point, he proposed a road layout comprising parallel and slightly curved expressways in a north-south direction, the main one being the 'residential road axis'.[33] Perpendicular to this axis, and connected to it by a set of platforms which houses the bus station, the 'monumental axis'[34] gives access to the institutional areas: the Ministries Mall and the Three Powers Square to the east, and the Federal District Administration to the west.[35]

Urban activities were segregated in distinct sectors (banking, commercial, recreational, residential etc.) distributed along the residential axis in two symmetrical wings, north and south. Residential areas were organized into sequences of 'superblocks', 300 by 300 metres, reserved for apartment buildings (in general up on *pilotis*).[36] In spite of the stress on symbols of modernity, Costa favoured a town of low densities and heights, with a maximum of six floors for residential buildings and sixteen floors in other sectors.[37]

Brasília and Its Urban Design Paradigms

Even though an original indigenous Brazilian creation, Brasília – with its axis, and its perspectives, its *ordonnance* – is intellectually of French extraction.

Lúcio Costa, *Registro de uma vivência*, 1995, p. 282

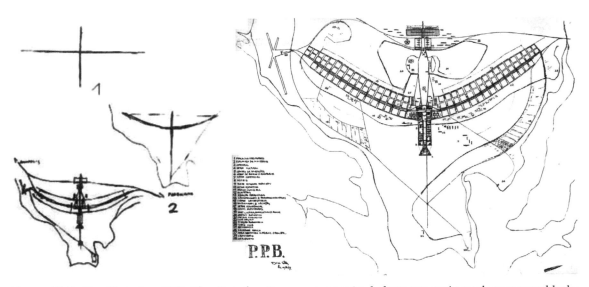

Figure 12.2. The Pilot plan, 1957. The city's functions were organized along two main road axes: superblocks extend along the 'residential road axis' (curve); the Ministries Mall and the Three Powers' Square are located at the east end of the 'monumental axis' (straight); the central bus depot is around their intersection.

From the end of the nineteenth century to the middle of the twentieth, in the West, urban theoretical speculation and intervention – often of an utopian bent – had as its main objective the mitigation of difficulties arising from exponential growth and the need to resolve certain issues then considered problematic: health and sunlight, traffic circulation and transport, spatial hierarchy, and control of the location of activities.

Some actions would become exemplary, such as the demolition and reconstruction of extensive urban areas, with a view to making them healthier and/or more attractive, and nearly always resulting in higher real estate values.[38] In the case of enlarging cities, besides the urbanization of adjacent areas,[39] another trend was the construction of suburban residential neighbourhoods.[40] An alternative would be the proposal, in the Renaissance tradition, of new urban forms, such as the *ciudad lineal* of Soria y Mata;[41] the garden city of Howard,[42] or the satellite-towns, defended by Hilberseimer.[43]

As for traffic circulation, worsened by the advent of the automobile, an effective measure would be the construction of metropolitan transportation systems, underground or elevated.[44] More as a theoretical approach, the specialization of roads with the concomitant separation of pedestrians and vehicles would be the object of several designs.[45] From the middle of the last century on, a commonplace solution would be the introduction of huge expressways into the urban tissue, breaking up its cohesion and continuity,[46] in an obsessive doctrine that could well be called 'roadway urbanism'. As for the distribution of activities in urban space, various zoning instruments should be mentioned, which would lead to a functionalist vision of the city, as upheld by the CIAM in the Charte d'Athènes (1943).

In the 1950s, this assortment of urban design paradigms gained currency in international and Brazilian professional *milieux*,[47] even to the point of being subjected to new critical scrutiny.[48] Nonetheless Brasília's Pilot Plan would be its most outstanding synthesis, obtaining such recognition that it would inspire projects of great visibility, like La Défense (Paris) and the Rockefeller Empire State Plaza (Albany, NY).

On the other hand, although Brasília's design has as its main influences Le Corbusier's prescriptions,[49] this linear city, an archetype of roadway urbanism[50] and of sectorization, to be expanded only by the addition of satellite-towns, is nonetheless a Beaux Arts interpretation of the functionalist vocabulary, as made evident by its symmetrical framework – chosen without a doubt to guarantee 'the desired monumental quality' of the capital city.[51]

In real life, the conflicting juxtaposition of expressways and urban tissue – with its viaducts, cloverleaf crossings and trenches – resulted in hollows, earth fills and retaining walls, all mistreating the ground, creating barriers and making even the circulation of automobiles difficult. The extreme sectoral division placed a rigidity on locations and imposed severe typological limitations. Finally, the closed symmetrical form would not turn out to be favourable to the articulation of the whole with its environment.

Lúcio Costa overcame such handicaps with the superblock, the most distinctive and inspired physical-spatial element of Brasília. A further example of Le Corbusier's influence,[52] its foremost precedent is found in Costa's design for the Guinle Park (Rio de Janeiro, 1948–1954).[53] In this set of three apartment buildings (originally six), he brilliantly explored the possibilities created by the *pilotis* to lay out the paths of pedestrians and vehicles at ground level and to adjust the buildings to the slope of the site.

Carrying on that experience, in Brasília's superblocks he opted for a much simpler type of traffic separation than elevated expressways – the cul de sac, similar to the access roads of the neighbourhood unit[54] and the Radburn superblock.[55] In this manner he was able to avoid those residual and unavailable spaces present in other sectors of town, which had a disagreeable effect on the urban tissue, and to obtain a more pleasing scale. However, even though a success as a design solution, the superblock continues to be an expensive and elitist answer. Accessible only to the few, it ended up not being widely applied in the remainder of the DF.

The Urbanization of the Federal District

I have solidarity with the aspirations of the people, but our relationship is ceremonious.

Lúcio Costa (1995), p. 276

A first wave of migration to the area of the future DF was unleashed with Kubitschek's promise to build Brasília. Encouraged by the approval for the capital city transference and the beginning of its construction, that migration intensified and in less than half a century led to the more than two million inhabitants of today. The chief indication of the success of the capital's transfer to the hinterland, this population increase induced an intense urbanization process that went far beyond the original expectations of the planners.

Building Brasília, 1956–1960

In those early years when Brasília was still just a huge construction site administered by NOVACAP, Kubitschek, assured by the way the project was evolving, abandoned the previous idea of a gradual move, to be done over a period of fifteen years, and set the inauguration date of

Figure 12.3. Brasília, 21 April 1960. The 'residential road axis' just after completion.

the capital for the 21 April 1960,[56] before the end of his term of office.

The demographic data for the area are impressive. In January 1957, there were close to 2,500 regularly contracted workers[57] and in July of the same year, 6,283 inhabitants (4,600 men and 1,683 women); a census of May 1959 indicated a total of 64,314 inhabitants.[58] With civil construction – which offered some 55 per cent of existing jobs – as its main economic means of support, the *'candango'*[59] population spread around in different places: NOVACAP civil servants lived in Candangolândia and Cruzeiro; the building workers proper lived in the encampments of their different companies, always close to the workplace;[60] and the migrants without regular jobs lived in spontaneous settlements or slums, here known as 'invasions'.[61] Given the lack of nearby cities, a camp that began to be settled in 1956, the Free Town, met their demands for provisions, services and entertainment, and functioned as the articulation centre of this improvised urban system.[62]

The construction of public buildings was brought about by an assortment of private enterprises, under contract to and supervision by NOVACAP. As for residential buildings in the Pilot Plan,[63] their demand would always be greater than the supply and it only got worse as the inauguration date approached. At first, NOVACAP engaged social security institutes of several workers categories;[64] but soon it was forced by the pressure of events to turn to state organs and even private companies.[65]

To meet the cost of the undertaking, an initial proposal was self-financing by means of the sale of close to 80,000 lots to raise an estimated 24 billion cruzeiros. The NOVACAP land sales, however, became a shoddy business liable to charges of corruption.[66] The highly inflationary but the ultimate solution was to request resources from the National Treasury, which financed the major part of the operation, consuming from 2 to 3 per cent of the GNP of the period,[67] something close to 250 to 300 billion cruzeiros or 400 to 600 million US dollars in values of the time.[68]

But figures would not dim the brilliance of the endeavour. The implantation of the Pilot Plan's entire urban framework and the completion of nearly all palaces dominated the stage, exciting the imagination and splitting national and international opinions. Brasília's saga had reached its summit; and even though many essential buildings were not finished and the powerful had nowhere to live, the inauguration occurred on the set date.[69]

A New Capital City, 1960–1976

A telephone isn't much for one who loves crazily and lives in the Pilot Plan . . .

If the girl he loves lives further back in Gama . . .

Song by Renato Matos

After inauguration, the high population growth rate continued, due to the gradual transfer of civil servants from the old capital and the continued migration. The DF had become an attraction for immigrants from every region of the country, from every social strata and from the most diversified branches of activity, in such a way that by the end of 1960 the population had reached 141,724 (68,665 in Brasília) and in 1970 surpassed the half million mark, with 546,015 inhabitants (149,982 in Brasília).[70]

The resolve, or rather, the tenacity of the migrants of all sorts to stay in Brasília (in flagrant conflict to the elitist urban programme, which established a ceiling of a half million inhabitants) would, in 1958, eventually cause NOVACAP, instead of expanding the Pilot

Plan area,[71] to adopt a policy of urbanization in dormitory suburbs for those of lower income.[72] These so-called 'satellite-towns' – resulting from the expansion of pre-existing villages, such as Planaltina (1859) and Brazlândia (1933); from the consolidation of encampments, such as the Free Town (1961, then named Núcleo Bandeirante); and the creation of new settlements such as Taguatinga (1958), Sobradinho (1959), Gama (1960), Guará (1968), Ceilândia[73] (1970) – were located according to a strategy that would favour the isolation of the Pilot Plan area, justified by a sanitation discourse.[74]

Given the Federal government's extensive land ownership, a consequence of the expropriations carried out from 1956, conflicts over land possession became frequent. The unappeasable housing demands, along with these disputes, led to the creation of the Economic Housing Society (*Sociedade de Habitações Econômicas de Brasília* – SHEB),[75] in 1962, which had as its agenda the creation of new satellite towns to settle invasion dwellers, a course that would become 'the housing policy of most if not all the DF governments . . .'.[76]

Therefore, besides the Pilot Plan's occupancy and the growth of the satellite towns, the urbanization process also included the proliferation of slums and the establishment of a *cordon sanitaire*[77] from 10 to 40 kilometres wide around the gentrified capital. A multi-centred or polynuclear pattern of urban development thus arose, characterized by scattered agglomerations, extremely low densities and strong spatial segregation. The 'posh' part – Brasília and its surroundings – had been appropriated by those involved in the state machine and by higher income groups, functioning as a focus for jobs and services around which gravitated a periphery lacking in equivalent blessings.

Ordering the Federal District, 1977–1987

As the population approached the one million mark,[78] the urban area was also stretching beyond the limits of the DF into the bordering territory of Goiás and Minas Gerais, in the so-called 'Entorno', always with the same pattern of extensive demographic voids. In the mid-1970s, some attempts at territorial planning were made in answer to a need made obvious with the institution of the DF government in 1969;[79] the focus was on sanitation and transport.[80]

The first official proposal was the Structural Plan of Territorial Organization (*Plano Estrutural de Organização Territorial*, PEOT, 1977).[81] Its analysis highlighted the dilemma of two contradictory objectives: the preservation of the Paranoá hydrographic basin for sanitation reasons, which implied the prohibition of new settlements in the area, versus the reduction of transportation costs and time, which would require a more continuous and compact urban structure than the polynuclear matrix.[82]

The focus on sanitation prevailed and PEOT recommended urban expansion should be diverted away from the Paranoá basin, in the DF south-eastern quadrant.[83] PEOT was complemented by two other studies: the Plan of Territorial Occupation (*Plano de Ocupação do Território*, POT, 1985)[84] and the Plan of Land Occupation and Use (*Plano de Ocupação e Uso do Solo*, POUSO, 1986).[85] These plans, essentially physical, established the DF's environmental zoning and maintained the interdiction of further occupation of the Paranoá basin. However, this taboo was broken by Lúcio Costa himself, with the plan *Brasília Revisited*,[86] in which he proposed the construction of 'proletariat blocks' along the DF main access roads and the legalization and/or creation of six new residential neighbourhoods, some of them in the Paranoá basin.[87]

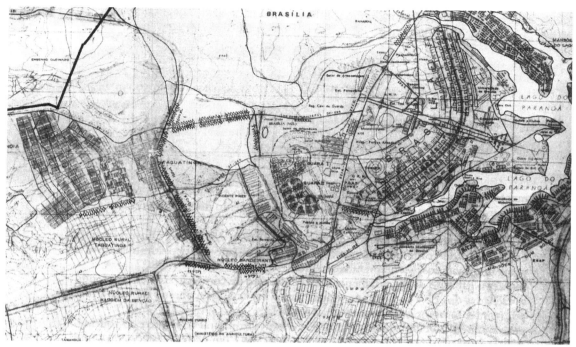

Figure 12.4. Brasília revisited. The plan, presented by Lúcio Costa in 1987, but not yet fully implemented, featured the construction of 'proletariat blocks' along the Federal District's main access roads and the creation of six residential neighbourhoods, some of them breaking the everlasting taboo of no further occupation in the surroundings of the Pilot Plan.

Meanwhile, another peculiarity of the DF urbanization became evident: its housing deficits had not remained restricted to the less fortunate classes. The swift increase in the cost of land – in the Pilot Plan[88] and in the South and North Lake neighbourhoods, as well as in some satellite-towns, such as Taguatinga, Guará and Núcleo Bandeirante[89] – also brought about a repressed demand for the middle and higher classes. As a consequence, illegal appropriations of public land by way of false property titles – often in environmental protection areas highly unsuitable for settlement – became quite frequent. In the hands of private developers, these subdivisions would result in the so-called 'clandestine' or 'irregular condominiums', small closed-off neighbourhoods of high standard single-family houses built outside routine urban regulations.[90] From this time forward, the condomini-ums multiplied rapidly, and today account for more than 40 per cent of the DF urban area.[91]

By the mid 1980s, the capital city was consolidated, and was the home to almost the entire Federal administrative machine. The main growth tendencies were established and the urbanized area had extended significantly to the Entorno. The DF urban complex, with an estimated population of 1,392,075 inhabitants (267,641 in Brasília) in 1986,[92] had become a typical Brazilian metropolis presenting location patterns of prestige, life quality and values decreasing from the centre to the periphery.

Urban Preservation and Political Autonomy, 1987 Onwards

The end of the military dictatorship in 1985[93] marked the start of a phase of profound institutional changes for Brazil. With the sanction of a new Federal Constitution, in 1988 the DF acquired political autonomy and so acquired an elected Governor and a Legislative Chamber of Representatives, responsible for land-use master plans.[94] About the same time, attaining an objective expressed since the time of the town's inauguration,[95] in 1987 Brasília – understood as solely the Pilot Plan – was included in the UNESCO's list of World Heritage Sites,[96] becoming the first twentieth-century city to receive such a distinction.

The issuing of specific preservation legislation[97] and the approval of a Constitution of the DF[98] made Brasília the object of actions of both Federal and District agencies, in a not entirely harmonious fashion. Divergences began with the Santiago Dantas Law,[99] which attributed the DF urban decisions to agents external to its administration.[100] The ambiguity and divergence of aims would only get worse in the 1990s.

The first Master Plan of Territorial Ordering (*Plano Diretor de Ordenamento Territorial*, 1st PDOT, 1992)[101] maintained the ever present orientation of occupying the south-western quadrant, now definitively polarized between the two largest urban centres – Brasília and Taguatinga.[102] To attenuate the growing discrepancies between satellite towns, it established the requirement for local directive plans for each administrative region of the DF, to identify specific attributes and to indicate measures for social and economic development.[103]

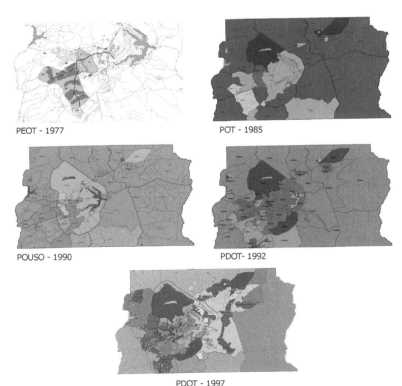

Figure 12.5. The Federal District master plans. The ordering plans induced urban dispersion over the entire territory and neither contained urban sprawl nor formed a cohesive urban tissue.

Conversely, the same administration that prepared the first PDOT, after 1989, adopted an aggressive policy of slum removal, the 'Low Income Population Settlement Programme'. Of a clearly populist orientation, this programme – carried out by the donation of lots served by minimal urban infrastructure and relegating the construction of houses or shacks to the dwellers themselves – would promote more urban sprawl.[104] In only four years it led to the institutionalization of six more satellite towns: Candangolândia, São Sebastião, Samambaia, Santa Maria, Recanto das Emas, and Riacho Fundo,[105] making the free distribution of land the main electoral currency of the DF.

With the second Master Plan for Territorial Ordering (*Segundo Plano Diretor de Ordenamento Territorial*, Second PDOT, 1997),[106] macro zoning was established that considered, at least formally, the Entorno as part of the DF's urban management. Besides a new metropolitan centre, consisting of Taguatinga, Ceilândia and Samambaia (conforming to the conventional policy of south-western occupation), the Second PDOT recognized the problems caused by the clandestine condominiums and introduced a polemic directive for their regularization: the extension of Sobradinho and Planaltina's urban perimeters, towns where they are more densely concentrated. As an outcome, a hasty urbanization of the eastern and western quadrants is already in progress.[107]

In 1996, the DF had 1,821,946 inhabitants (257,583 in Brasília)[108] and its urbanization process had a dismal tale to tell. The exclusionary Pilot Plan design induced urban dispersion over the entire territory; the ordering plans – accepting the Pilot Plan's cannons – neither contained urban sprawl nor formed a cohesive urban tissue, rather most of the time they served to legalize situations already in existence. Meanwhile, the DF administration itself had a leading role in disregarding its own directives, even with the promotion of programmes and projects that were not of first priority.[109]

Brasília Today

BSB sacked the Bauhaus . . .

Song by Renato Matos

Unlike some new capital cities, such as Canberra and Ottawa, Brasília became an outstanding metropolis on its own[110] and is now the heart of the Integrated Region of the Federal District and Entorno (RIDE).[111] With 2,948,421 inhabitants, of which 2,051,146 are within the DF (and, of these, 256,064 in Brasília), it is the ninth largest urban concentration in the country and the one with the highest rate of demographic growth (3.41 per cent per year).[112]

In fact, it encompasses many worlds and spaces. The capital city proper concentrates the political decisions and financial resources of the state and is a sophisticated place connected to local, national and international circuits of power. As such, it offers an exceptional quality of life while at the same time only housing a tenth of the metropolitan population, with numbers of residents falling each year. The World Heritage Brasília – the world's largest urban complex designed along rigorously functionalist lines – really only exists in the imagination of its champions, since its listing resulted more in the Pilot Plan's consecration than in consistent preservation measures.

The expectation that a planned core would induce an orderly occupation of the territory – an essential utopia of Modernism – did not come to pass. Brasília's paradigms were repeated in the metropolitan areas but were done on the cheap and in haste. Spread out over an area several

Figure 12.6. Contrary to expectations, the 'monumental axis' revealed itself as a propitious location for popular demonstrations. In the foreground, a farmer's protest; in the background, an upper-class district (right) and illegal condominiums (left).

times larger than that of a traditional city of equal population, the metropolis experiences tumultuous suburban growth with high degrees of migration.[113]

Nevertheless, Brasília is a remarkable accomplishment that launched the successful occupation of the Brazilian hinterland.[114] The site chosen by Cruls and Cavalcante de Albuquerque is of unequalled beauty, further enhanced by the Paranoá Lake and inspired landscaping. The relative youth of the city, together with its proximity to the seat of power and the concentration of resources,[115] offers opportunities that attract both rich and poor. Even the satellite towns, with all their hardships, offer incomparably better social services than those of other regions of the country. Although the less privileged live outside the Pilot Plan, they have already secured its surroundings as their own.

NOTES

1. In a speech given in Parliament in 1805, Pitt espoused the idea, suggesting some locations and even a name, Nova Lisboa. Brasil (1960), vol. 1, pp. 34–35.

2. When Brazil became a United Kingdom of Portugal.

3. Quoted in Brasil (1960), vol. 1, p. 41. An anonymous pamphlet published in 1822 suggests Brasília as the name for the future capital city; later on, Andrade e Silva would suggest both Petrópole and Brasília.

4. A noteworthy contribution was the long campaign of historian Francisco Adolpho de Varnhagen, the Viscount of Porto Seguro. In his pamphlet *Memorial*

orgânico (1850), Varnhagen lists twelve reasons for the construction of a new capital city, suggests a name, Imperatoria, and – agreeing with Andrade e Silva – proposes that it be located at latitude 15° or 16°; as for its altitude, he proposes that 'it be at least 3,000 feet above sea level', quoted in Brasil (1960), vol. 1, p. 139. In 1877 he made a journey to the then Province of Goiás and published his definitive study, *A questão da capital: marítima ou no interior?*.

5. In an expedition made from July 1892 to March 1893.

6. This area was first shown in the *Relatório Parcial* of 1894, published in 1896. It is a spheroid quadrilateral of 160 by 90 km, with an area of 14,400 km².

7. Cruls (1894), p. 18.

8. On the date of the Centennial, 7 September 1922, a cornerstone for the future city was laid a few kilometres away from where Brasília would in fact be built.

9. By the Central Brazil Foundation, in a policy that would become known as the 'March to the West'.

10. See, for instance, Castro (1946); Guimarães (1946); Backheuser (1947–48) or Demosthenes (1947).

11. During the Constitutional Convention in 1946, there were defenders of the 'Cruls Quadrilateral', of the cities of Goiânia (State of Goiás capital) and Belo Horizonte (State of Minas Gerais capital), and of the so-called 'Mineiro Triangle' (the westernmost salient of Minas Gerais), this latter one defended among others by Juscelino Kubitschek, then a Minas congressman. Brasil (1960) vol. 3, p. 12; Demosthenes (1947) pp. 13–19).

12. Brasil (1960), vol. 3, pp. 288–376 and 388–415.

13. *Ibid.*, vol. 3, pp. 415–36.

14. State of Goiás Law no. 41, 13 December 1947.

15. After the establishment of the Federal District definitive boundaries. The greater part of the DF lands was in fact acquired with Federal funds, even though done through the State of Goiás government. And since not all of the expropriations were properly registered, in the future the question of land ownership would be extremely complicated, leading to long drawn out legal disputes.

16. With 52,000 square kilometres.

17. With 1,000 square kilometres each.

18. Albuquerque (1958). Albeit published in 1958, it covers the works done until 1 September 1956.

19. *Ibid.*, pp. 190–193

20. Brasil (1960), vol. 3, p. 41.

21. Federal Law no. 2.874, 19 September 1956.

22. Inside the 'Cruls Quadrilateral', with some 5,800 square kilometres.

23. Moreira (1998).

24. A plain wooden structure, the so-called 'Catetinho' was built in only 10 days, being inaugurated on 10 November. Ficher and Batista (2000), p. 80

25. In order to form Lake Paranoá, a foremost element of the city's physiognomy.

26. Brasil (1960), vol. 4.

27. The list of examples is lengthy. Before the inauguration, he designed the Congress, the Presidential Palace (Planalto Palace), the Supreme Court, the Ministries and the National Theatre among others; in its first decades, the Central Institute of Sciences at the University of Brasília (1963), the General Headquarters of the Army (1977), the JK Memorial (1980) and all the annexes of the Congress and Ministries; in more recent years, the Superior Tribunal of Justice (1993), the new annex of the Supreme Court (1997) and the Public Ministry (2000).

28. Architect Affonso Eduardo Reidy and landscape architect Roberto Burle Marx.

29. This was a summary document, asking only for the basic scheme of the city and a justifying report, that leaves out the public buildings, implicitly to be designed by Niemeyer. GDF (1991), pp. 13–16. A little while later, NOVACAP stated that the city should have a political and administrative character, with limited industrial development and a maximum population of 500,000 (pp. 16–17).

30. The jury – composed of Oscar Niemeyer, Luiz Hildebrando Horta Barbosa, Paulo Antunes Ribeiro, William Holford (England), André Sive (France) and Stamo Papadaki (USA) – reached its decision in a little more than 10 days, announcing it on 23 March 1957.

31. Carlos Leão, Jorge Moreira, Affonso Eduardo Reidy and the latecomers, Ernani Vasconcellos and Oscar Niemeyer.

32. GDF (1991).

33. Composed of three parallel avenues, with a total of fourteen traffic lanes.

34. With eight lanes separated by a central 200-metre wide grass strip.

35. For many, this framework suggests an airplane.

36. In the working out of the details, additional residential sectors were created: a sequence of blocks for economical row houses and two neighbourhoods of detached houses on the other side of the lake (South and North Lake).

37. The one exemption being the Central Bank (1976–1981), with twenty floors. Today the tallest structure in Brasília is the Television Tower, 224 metres high; the tallest buildings are the twin towers of the Congress, with twenty-six floors each.

38. Along the lines of the Beaux Arts works of Haussmann for Paris (1854–1868), that would become an important trend after the City Beautiful Movement – or 'city of monuments', in Peter Hall's happy description (2002, p. 189), particularly appropriate to Brasília.

39. As the famous Barcelona *ensanche* (1859), by Cerdá, or the elegant extensions of Amsterdam (1913–1934).

40. Tried out initially in the USA, with Llewellyn Park (New Jersey, 1853), Chestnut Hills (Pennsylvania, 1854), Lake Forest (Illinois, 1856) and Riverside (Illinois, 1865), the latter designed by Olmsted.

41. Conceived for Madrid (1882) and taken up by Garnier in his *cité industrielle* (1901) and by Le Corbusier, from the studies for Rio de Janeiro (1929) until his *cité linéaire industrielle* (1944).

42. Presented in Ebenezer Howard's *To-Morrow: a peaceful path to real reform* (1898) and resulting in important designs by Parker and Unwin, in Letchworth (1904), Hampstead (1905–1909) and the garden neighbourhoods of São Paulo (1917–1919).

43. Leaving out the preoccupation with transport, Hilberseimer's exposition in *Großstadt Architektur* (1927) gives a prophetic image of the DF urbanization process: 'This big city separation or dissolution in work and residence zones leads, as a consequence, to the formation of the satellite system. Around the big city core, the central city, that in the future will be only a town of work, are situated, circularly and at sufficient distances, residential neighbourhoods closed in upon themselves, satellite-towns with a limited population, whose distance can be considerable, with all the modern transportation means and an adequately designed high speed train system. Even though they have local independence, such residential neighbourhoods are members of a common body, they stay closely united to the central core, constitute with it an economic and technical-administrative unity'.

44. Starting with the London underground in 1863.

45. Employed by Olmsted and Vaux in Central Park (New York, 1853), developed by Hénard in the *rue future* (1910) and obsessively defended by Le Corbusier, the separation of traffic types would be advanced by Stein and Wright, in a completely different frame of mind, from Radburn (Fairlawn, NJ, 1928–1933) on.

46. Along the lines of Sanders and Rabuck researches, described in *New City Patterns* (1946).

47. As attested, in the case of Brazil, by Szilard and Reis's book, *Urbanismo no Rio de Janeiro* (1950).

48. As in classic books such as Lynch's *The Image of the City* (1960) and Jacobs's *The Death and Life of Great American Cities* (1961).

49. In architecture, he recommended unifunctional detached buildings on *pilotis* (freeing the ground for pedestrians), with independent frames, glass façades and flat roofs; in urban design, strict activity separation, spatial class segregation, specialization of roads, and pedestrian and automobile separation through viaducts and overpasses, with the consequent dissolution of the traditional street. For the urban form, he proposed three kinds of agglomerations: units of agricultural exploitation, linear industrial cities, and radio-concentric cities of business, government, 'of thought and of art'. See 'Towards a Synthesis', 1945, in Le Corbusier (1946), pp. 69–71. Only one of his directions was not observed in the Pilot Plan design, that regarding the urban form, for which Costa adopted two branches of linear city, of industrial character in this line of divagation. However, the other two forms were represented in Brasília's competition: Rino Levi's design bears a clear influence of the Ville Radieuse, see Le Corbusier (1935), and the Roberto brother's recollects a cluster of Corbusian agrarian villages, see Le Corbusier (1959), p. 73.

50. This leaning, made clear early in the opening of the Pilot Plan Report ('. . . to apply to the technique of town planning the free principles of highway engineering, including the elimination of intersections . . .', GDF (1991), p. 78, complied with Kubitschek's express aspiration of creating a 'city for the automobile'.

51. GDF (1991), p. 78; see also 'Conceito de monumentalidade' (1957), in Costa (1967), p. 281.

52. As the *gratte-ciel cartésien* (1935), implanted in the centre of a block enclosed by expressways with cloverleaf intersections. Le Corbusier (1947), pp. 74–77.

53. Costa (1995), pp. 205–212.

54. A principle proposed by Clarence Perry in books such as *Wider Use of the School Plant* (1910), *Community Center Activities* (1916) and *Neighborhood and Community Planning* (1929). Among his prescriptions are new neighbourhoods designed with walking distances between housing and elementary schools (600 metres), and isolated from main through roads, to avoid 'the automobile menace'. See Perry (1929), p. 31.

55. Developed by Stein and Wright, it is characterized by complete separation of pedestrians and automobiles by means of over-passes and under-passes. Its morphology is composed of clusters of detached and semi-detached houses distributed around dead-ends in order to free the interior block for gardens, the so-called 'inner parks'. See Stein (1951), pp. 37–73.

56. Law no. 3,273, October 1st 1957. The chosen day coincides with the commemoration of Tiradentes, the Martyr of Independence.

57. By NOVACAP and by private builders.

58. Brasil (1960), vol. 4, pp. 54 and 243, and GDF (1984), vol. 1, p. 10. No data were found about the previous population within the limits of the DF. In the present chapter, the figures for Brasília always refer to the Pilot Plan and South and North Lake populations together. As for the origins of the new population, whereas civil servants and technical cadres came mainly from Rio de Janeiro, the labourer majority was from the north-east. Such a miscegenation would become an essential trait of the Brasília identity.

59. The popular appellation given to those who came as workmen and, by extension, to those born in Brasília, it is a word of African origin that means 'inferior' or 'vulgar'.

60. Such as Vila Planalto, for works in the Three Powers Square and the Ministry Mall, or Vila Paranoá, for the Paranoá River dam.

61. Such as Vila Amauri, Vila Sarah Kubitschek or Lonalândia (Quinto Júnior and Iwakami, 'O canteiro de obras da cidade planejada e o fator de aglomeração', in Paviani (1991)).

62. 'The Free Town was born . . . hitherto the largest agglomeration, with the basic function to provide services for the rest of the population: shops, free fairs, bars, restaurants, builders' supply shops, and whatever was then necessary. In order to "motivate" those who where arriving, apart from being tax exempt, they got lots of land, with the provision that these would have to be given back at the Pilot Plan inauguration. All buildings were necessarily made of wood, since the settlement was not to be a permanent one . . .', Ribeiro (1982), p. 116. The Free Town would not lose its importance after the inauguration, since even the Pilot Plan would go on depending on its commerce for several years, Pescatori (2002), p. 1.

63. The offer of housing and the social gradations among the Pilot Plan inhabitants had been taken into account by Costa; in his Report it is recommended that 'the mushrooming of hovels either in the urban or the rural areas' should be avoided; 'it is up to the Urbanization Company, within the proposed plan, to provide decent and economic accommodation for the entire population', GDF (1991), p. 83.

64. The only institutions then in the country with experience in large-scale housing, their engagement allowed the Federal government to secure the payment of their public debt, Tamanini, (1994) p. 197 and França (2001), p. 5.

65. França (2001).

66. Moreira (1998).

67. Lafer (1970), p. 210.

68. Given the strong devaluation of the cruzeiro in relation to the dollar for the whole period, there have been controversies concerning these figures. See Mindlin (1961) and Vaitsman (1968).

69. Astonishing the most incredulous, like Norma Evenson. Accepting simple-minded prejudices about Brazil (possible only in someone unaware of the degree of administrative competence required to organize a single Carnival pageant, deploying precisely from 3,000 to 6,000 people in only 70 rigorously timed minutes, at what is regarded as the largest popular spectacle in the world), overlooking the hard working routine of its people, and disregarding considerable accomplishments such as Belo Horizonte, built between 1894 to 1897, and Goiânia, built between 1933 to 1942, this author felt justified in stating that 'The creation of Brasília represented a triumph of administration in a country never noted for efficient administration; it represented adherence to a time schedule in a society where schedules are seldom met; and it represented continuous hard work from a people reputedly reluctant to work either hard or continuously', (1973, p. 155).

70. GDF (1984), vol.1, p. 10.

71. Which would not be conceivable, given the understanding of the Pilot Plan as a 'complete city', with the shape of a closed figure.

72. The decision was later criticized by Costa: 'The growth of the City was anomalous. There was the inversion known by everyone, for the Plan was supposed to be such that Brasília could be within the limits of 500 to 700 thousand inhabitants. When it approached these limits, then the satellite-towns would be rationally projected and architecturally defined, so that they could orderly expand', quoted in Tamanini (1994), p. 440.

73. Implemented by the Campaign for Eradication

of Invasions, whose acronym – CEI – explains its appellation.

74. This bias would never be cast aside, as shown by the Directive Plan of Water, Sewage, and Pollution Control (Planidro), which recommended a population ceiling for the Lake Paranoá basin. GDF (1970).

75. Subsequently, Social Housing Society, SHIS (1966), Housing Development Institute, IDHAB (1989), and since 1999, Secretariat of Urban Development and Housing, SDUH Vieira (2002).

76. Pescatori (2002), p. 3. An attempt to reconsider such a philosophy was the organization, in 1983, of the Executive Group for the Settlement of Slums and Invasions, GEPAFI; though more concerned with community priorities, it was extinct in 1985.

77. Or, in Le Corbusier's words, 'a protection zone without constructions' (1925, p. 181).

78. 937,600 inhabitants (228,141 in Brasília) in 1976; 1,002,988 (228,386 in Brasília) in 1978; and 1,176,748 (275,087 in Brasília) in 1980. GDF (1984), p. 10.

79. No longer under a Mayor, but a Governor designated by the Presidency. At that time, NOVACAP was absorbed by the GDF executive structure; of its previous powers, it only retained a prestigious name and today it is the agency in charge of parks and gardens.

80. Batista, 'The view from Brazil', in Galantay (1987), pp. 355–364.

81. Brasil, 1977. PEOT was based in previous studies; see GDF (1976).

82. Batista, 'Problemas e respostas de uma metrópole emergente', in Paviani (1987), pp. 208–220.

83. However, it allocated some areas for non-residential uses in the Paranoá basin and recommended the implementation of a mass transport system.

84. GDF (1985).

85. GDF (1986).

86. Costa (1995).

87. Some were implanted, such as the Vila Planalto, at last legalizing the old and picturesque encampment, and the Southwest Sector, located above the South Wing; of the others, only the Northwest Sector, symmetrically located in the North Wing, has been the object of successive designs.

88. One of the factors of this extreme value was the retention of plots for apartment buildings in the North Wing by their major landowner, the University of Brasília, delaying for decades the occupation of almost a fifth of the total area available for residential use in superblocks and introducing a strong asymmetry in the Pilot Plan.

89. Given the lesser zoning restrictions, these towns soon acquired weight in the DF economic dynamics and started to lose their characteristics as low-income ghettos.

90. The first was the Quintas do Alvorada Condominium (1977), localized in the São Bartolomeu River basin, in the DF north-eastern quadrant. Malagutti (1996), p. 74.

91. According to geographer Rafael Sanzio, the urbanized area of the DF went from 40,000 to 72,000 hectares in the 1990s, due mainly to condominium proliferation (Nossa, 2002, p. C3). Today, in Governor Joaquim Roriz's administration (1999–2002 and 2003–2006), they are the DF's most serious political and legal issue, already reaching the scandal pages of national newspapers.

92. GDF (1986), p. 121.

93. Started with the coup of 1964, this military administration was essential for the irreversibility of the Capital's transfer, in an obvious 'Versailles' effect'.

94. Even before, Niemeyer had already indicated the necessity of preservation statutes for the Pilot Plan. See Niemeyer (1960), p. 518.

95. Thanks to the lobby by some sectors of the Brazilian intelligentsia – under the leadership of then Governor José Aparecido (1985–1988) – afraid that the country's redemocratization would result in alterations in Lúcio Costa's design.

96. The District government has both state and local responsibilities. Before 1969, it had a Mayor; since 1969, it has a Governor.

97. In answer to UNESCO's requirement for protection measures, in 1990 Brasília was listed by the Federal agency in charge of historic preservation.

98. DF Organic Law, 8 July 1993, which made mandatory the periodic elaboration of Directive Plans of Territorial Ordering.

99. The Federal Law no. 3,751, 13 April 1960, that established the rules for the DF administration, introduced a static vision of the Pilot Plan, as something to be kept up without alterations, a posture that would be maintained by the authoritarian regime.

100. First by a specific Senate committee and, after 1969, by the Architecture and Urbanism Council (CAU), later Architecture, Urbanism and Environment Council CAUMA), whose composition has always been far from representative of the community.

101. Law no. 353, 18 November 1992, and GDF, 1992.

102. A mass transport system, with 40 kilometres of service and linking Brasília to the main satellite towns of the south-eastern quadrant, is now in experimental operation and certainly will bring immeasurable changes to the urban context.

103. So far, of the twenty-six Administrative Regions, only Sobradinho, Candangolândia, Taguatinga, Samambaia and Ceilândia have had their respective Local Directive Plans approved.

104. Even though socially relevant, such a policy has systematically disrespected consistent directives for environmental preservation and priority expansion zones. The localization of the new settlements it promotes and their low densities make them dormitory suburbs whose characteristics do not encourage the development of economic activities that could engender a significant number of local jobs, and only reinforce the DF's spatial segregation.

105. Candangolândia corresponds to the expansion of the old NOVACAP's encampment; Samambaia's construction started earlier, in 1983; São Sebastião was an agricultural colony; the others are brand new towns.

106. Complementary Law no. 17, 28 January 1997, and GDF (1997).

107. The induction of this new vector goes against all standing directives, particularly the environmental ones.

108. GDF (2001), p. 7.

109. As happened with the recently inaugurated bridge over the Paranoá Lake, fruit of substantial expenditure in a roadway system that will lead inescapably to an intense urbanization of a controlled growth zone.

110. One hypothesis that could help to explain this fact concerns the site's choice. While the location of those towns was decided by a tug-of-war dispute among pre-existing important metropolises – in the Australian case, Sidney and Melbourne; in the Canadian case, Toronto and Montreal – Brasília's central-west location was the result of a long-term decision that drove away most of the regional disputes. As a consequence, the town was built in a region quite far from the influence of the major Brazilian metropolises, São Paulo and Rio de Janeiro.

111. Instituted by the Complementary Law no. 94, 19 February 1998, the RIDE is comprised of the DF, nineteen Goiás boroughs and three Minas Gerais boroughs.

112. IBGE, 2000 Demographic Census.

113. Since the end of the 1980s, the DF is the region with the highest relative rate of migrants in the country.

114. The capital city transfer was its main incentive, impelling an economic development that spreads to the entire centre-west region and starts to reach the north region.

115. According to the 2000 Census, the DF income per capita is the country's highest (R$ 14,405) and the DF participation in national GDP went from 1.37 per cent to 2.69 per cent between 1985 and 2000. But it must be remembered that such a performance includes the Federal budget, whose expenses represent almost 60 per cent of the DF GDP.

Chapter 13

New Delhi: Imperial Capital to Capital of the World's Largest Democracy

Souro D. Joardar

Urban historians may debate over the number, but the city of Delhi has seen the emergence of several capitals – capitals of different dynasties and political rules. In the twentieth century, however, the city was projected onto the global map of modern urban planning and development with the establishment of a majestic edifice of colonial power in the Orient – New Delhi, the new Imperial capital of British India. True to typical colonial urban morphology[1] the new capital city was juxtaposed yet distinctly set apart from its immediate predecessor, the Mughal capital of India – the walled city of Shahjahanabad – that, together with its surrounding outgrowth, became by default the native 'Old Delhi'. The history of the planning and development of New Delhi is quite distinct, but spans over only the first three decades of the twentieth century. For the rest of the century, it would be worthwhile exploring the identities and transformations of this capital city as it became more and more, physically and administratively, an integral part of the exploding and impersonal metropolitan Delhi and its region, especially after India's Independence.

The following distinct pre- and post-colonial phases of planning and development of the capital city may be identified:

(*a*) Decision-making on shifting Imperial capital functions from Calcutta to Delhi (1857–1911);

(*b*) Development of New Delhi as the Imperial capital of British India (1911–1932);

(*c*) Development of New Delhi during the last phase of colonial rule and early post-Independence plans and programmes (1932–1970);

(*d*) New Delhi in the context of an expanding metropolis and the National Capital Regional Plan (1970–2002).

Preamble to New Delhi: Events Leading to Building the New Imperial Capital

Ironically, New Delhi was the Imperial capital for less than forty years in the two hundred year history of colonial rule over the sub-continent, but the political and administrative rumblings over shifting of capital functions from Calcutta and building a new capital went on for nearly half a century preceding the new capital's establishment. The infamous Sepoy Mutiny[2] of 1857 had jolted the British into realizing the problem of ruling the vast sub-continent from its eastern corner. Despite its strategic advantage for a quick naval retreat through the Bay of Bengal, Calcutta, with estuarine marshes all around, was often criticized for its unhealthy climate. Following an earlier historical initiative by Warren Hastings,[3] a post-mutiny committee formed by Sir Stafford Northcote[4] favoured decisions to shift the colonial capital along with providing full governorship to Bengal.[5] The matter was revived again by Lord Lytton[6] in 1877 and in the Royal Durbar[7] of 1903 presided over by Lord Curzon,[8] although without any positive end.[9]

The nationalist movements that swept through Bengal, especially Calcutta, following Curzon's partitioning of this large province (78 million in 1905) was a catalyst in the colonial politico-administrative decision to shift the capital. As the administrative division split the linguistically and culturally homogeneous region, creating an ethno-geographical divide between Hindus and Muslims, the consequent irate Bengali reaction even amongst its elite class touched a raw nerve of the highest echelon of British Royalty. King George V deplored the unrest in the Empire and actively engaged himself in restoring normality through his emissaries, especially the new Secretary of State for India, the Earl of Crewe. Specifically, a scheme was devised that linked the proposal to undo the partitioning of Bengal and to provide the province with autonomy and full governorship, thus mollifying the aggrieved, with the long standing proposal of building a new capital away from the troubled place, thus killing two birds with one stone.[10]

In the past, many locations in northern India, including Delhi, had been contemplated as a potential new capital. Besides its centrality and connectivity within the Great Indian Empire, Delhi carried in the minds of the colonial rulers a symbolic value – as the age old saying goes: 'he who rules Delhi rules India' – a realization of the Indian ethos, especially across northern and central India, enhanced during royal contact with the innumerable minor and major princes.[11] The possibility of gratification of the Muslims by moving to a seat of historical Pathan and Mughal dynasties, especially in the wake of the Bengal turmoil, also carried political weight.[12] Furthermore, various establishments and infrastructure were already in place in Delhi by the turn of the century – British military and civil lines, and railway links to Calcutta, Bombay, Agra, Punjab, Rajputana and to the regular summer capital, Shimla.

The idea of shifting the Imperial capital from Calcutta to Delhi was obviously not without opposition and criticism. Besides the severe opposition from various stakeholders in Bengal – both native and European settlers – Lord Curzon tried to influence decision-makers in London against it. But Viceroy Hardinge[13] doggedly pursued the long standing idea until it became a concrete administrative and political decision that culminated in the surprising proclamation by the King at the much hyped Coronation Durbar of December 1911 that a new Imperial capital was to be built in Delhi.[14]

Development of New Delhi as the Imperial Capital

Hardinge possibly had relatively less say than the India Office in London in the constitution of the Delhi Town Planning Committee,[15] with its mandate to provide temporary government accommodation and the siting, design and development of the new capital, together with the appointment of Edwin Lutyens as the chief planner of the proposed capital. However, Hardinge had strong influence over its functions in India in the matters of site selection for the new capital and the locations and designs of its key buildings.[16] Further, he played a significant role *in absentia*[17] in the mobilization of finance and the implementation of the project.

Within the Delhi area, the main contenders for the site of the new capital were the northern and the southern sides of the historic Shahjahanabad. The other two sites considered had severe constraints – the eastern trans-Yamuna site was flood prone and the one west of the Ridge[18] had no visual link with historic Delhi (see figure 13.1). A site selection tussle among the planning committee, the Viceroy and politicians and administrators at large (including key persons in London) went on for a year. The northern site, which was already developed with the cantonment and the civil lines with its many beautiful bungalows, had the symbolic significance of hosting three previous Imperial Durbars and was imbued with scenic value because of its closeness to the river and the commanding heights of the Ridge. But the committee, especially Lutyens, stood firmly behind the southern site[19] as it provided more flexibility in designing a worthy capital city with room for future expansion and would prove more cost-effective than the former which would entail greater land acquisition and construction costs. It was also regarded as healthier. Ironically, the northern site at Delhi with its many historic bungalows[20] was used as a 'temporary capital'[21] for several years while New Delhi was on Lutyens's drafting board and then under construction.

A basic premise of New Delhi's planning and design was connectivity between the major capital elements and surrounding elements of historic Delhi,[22] and alignment of sweeping vistas. While Hardinge was very sensitive to the historical connectivity, the idea fitted Lutyens's concept of a 'Baroque' scheme structured with grand avenues and axes. Here, the location and orientation of the most symbolically significant building of the imperial capital – Government House (and its associated premises and offices) – and its physical and visual links with the others became a crucial issue and one open to varying opinions.[23] Firstly, overruling Lutyens and others, Hardinge pressed for Raisina Hills[24] as the location because the hilltop provided a sweeping panorama and opportunities for connecting vistas to important historical landmarks to the south, the river to the east and the

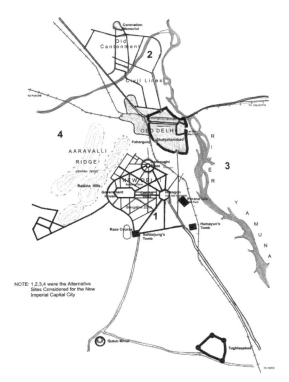

Figure 13.1. New Delhi and its surrounding context.

NEW DELHI: IMPERIAL CAPITAL TO CAPITAL OF THE WORLD'S LARGEST DEMOCRACY

Figure 13.2. New Delhi in the context of Metropolitan Delhi (1990).

walled old Delhi to the north (see figure 13.1). Secondly, a major scheme by Lanchestor to align the principal axis and vista of the capital city towards the north connecting the Raisina Hill complex with Shahjahanabad, especially the tall minarets of its Grand Mosque[25] was abandoned because of the anticipated cost of land acquisition in Paharganj,[26] which lay between them.[27] Thus, a historic opportunity to integrate the 'New' and the 'Old' Delhi physically was lost.[28] In the end, the axis was aligned towards the east (to the river) and became the 'Central Vista' with the capital city extending almost equally on either side (see figure 13.2).

Both Lutyens and Herbert Baker, his chief architect, conceived the Raisina complex as their 'acropolis' at the terminus of this grand vista. They fell out over spacing between Government House (designed by Lutyens as the focal element) and the two Secretariat buildings (designed by Baker – see figure 13.3) which flanked it vis-à-vis their relative heights and thus their relative visual dominance.[29] This dispute led not only to their acrimonious personal relationship but also to a general controversy in the progress of the capital city project.[30] Furthermore, Lutyens's determination to adopt a pure Western classical architectural style in the design of Government House gave way to Hardinge's insistence on incorporating traditional Indian features, while Baker with his international experience seemed more receptive to local issues.[31] But the appropriateness of Baker's circular design for the Council House[32] as well as its location in relation to the bold Central Vista became an issue.[33] However, the Council House was an obvious after-thought emerging from the British democratic reforms, especially the Montagu-Chelmsford reform of 1919[34] providing for self-governance.

Connaught Place, the elite shopping centre of the new capital is another majestic circular building complex. Its two-storey high buildings with colonnaded shopping arcades are set around a grand circus. The shopping centre was conceived by architect W.H. Nicholls and built later by R.T. Russell.[35] Lutyens located this commercial hub at the northern most point, closest to Old Delhi, but prominently connected with the Central Vista (through the Queensway) and the Council House. Over time, its location and connectivity would transform this quiet elite European node into the giant central business district of a sprawling metropolis (see figure 13.5).

Lutyens's New Delhi plan (figure 13.4) spread over a vast rolling plain of about 8,600 hectares. The plan included sweeping grand vistas, vast open spaces, gardens and street landscaping, monumental arches, sculptures and fountains, and majestic public buildings. A spreading network of diagonal avenues with circular intersections

Figure 13.3. The south Secretariat Building designed by Herbert Baker.

resembled the other Baroque influenced modern capital city, Washington DC.[36] New Delhi was conceived also as an English 'garden city' in terms of its very sparse built form, luxurious open spaces and street landscaping. When the capital city was inaugurated in 1931, its density was ludicrously low (less than 8 persons per hectare) compared to the nearly 200 persons per hectare in neighbouring Old Delhi.[37] Lutyens's vast bungalow area alone was twice the total area of the native city with only 640 bungalows in 1940 on plots ranging from 1.5 to 3.25 hectares each. The rigid zoning for social stratification and the hierarchical order of residential space matching the official ranks of the residents were the colonial dictates guiding New Delhi's plan, especially its bungalow area.[38]

Hardinge had put Chief Commissioner Malcolm Hailey in charge of the Imperial Delhi Committee (later changed to New Capital Committee) for execution of the capital project almost immediately after the Planning Committee submitted its final report. The Raisina Municipal Committee was initially established in 1916 to provide municipal services to the large number of construction workers. It eventually became the prime local body – New Delhi Municipal Committee[39] – of the capital city. Hardinge, however, had remained the prime engine of development of New Delhi. He shunted energetically between supervision of nuances of design and planning issues, scheduling construction and work progress, and tackling political opponents[40] and the uncertainties, during the Great War, of finance for the prestigious and extravagant project.[41]

The New Delhi capital was officially inaugurated in February 1931, nearly twenty years after the royal declaration at the 1911 Durbar and about fourteen years after official approval of the Imperial capital project in 1917. Ironically, by that time, Lutyens and Hardinge's 'Asian Rome'[42] conceived to strengthen the foundation of the Imperial colonial rule in the sub-continent, was sensing its twilight and the dawn of a new role. Within sixteen years of its inauguration it became the capital of a free country and most populous democracy of the world.

How has New Delhi adapted to this new role as a democratic political capital? And, what impacts have the contextual metropolitan growth have on perceptions of its future?

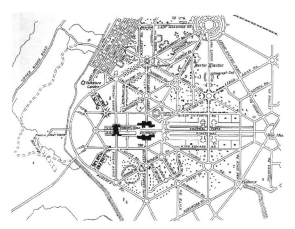

Figure 13.4. Lutyens and Baker's 1912 plan for Imperial Delhi, later re-named New Delhi.

New Delhi in the Context of Emerging Metropolitan Growth and Early Post-Independence Planning

Lutyens's New Delhi had 'tunnel vision' in its narrow mandate to incorporate only the highest establishments of the colonial administration and associated residential and social facilities, especially for the elite Europeans.[43] Its deliberate dissociation from the growing commercial-industrial activities and population of the pre-existing Delhi was a myopic vision of the future growth potential of a capital of a large, populous country.

A new cantonment to the south-west of the planned Imperial capital was an afterthought for which the British annexed a freshly acquired 10,000 acres (4,000 hectares) to Lutyens's New Delhi.[44] The population of New Delhi was less than 15 per cent of the total urban population of Delhi Territory in 1931.[45] The latter grew by 55.5 per cent over the next decade[46] as industries increased in the area surrounding New Delhi and the outbreak of World War II attracted large-scale labour migration from the surrounding region to Delhi. The War simultaneously increased both government and military activities with demands for civil and military accommodation. Their immediate impacts on the newly built capital were further housing development on the southern fringe of Lutyens's New Delhi,[47] development of the new cantonment, and the addition of hundreds of temporary hutment barracks around the Secretariat and on vacant sites around the Hexagon allotted originally to the princely states.[48]

Even before the outbreak of the war, the colonial government became aware of the metropolitan growth pressure in Delhi and the uncoordinated works of different urban local bodies in and around the new capital. While much of the construction and maintenance of government buildings, staff quarters and roads in New Delhi was entrusted to the Central Public Works Department (CPWD), the New Delhi Municipal Committee (NDMC) provided the urban services of water supply, sanitation, power supply, street sweeping, etc. Development and maintenance of the vast cantonment area was done exclusively by a Cantonment Board. The Delhi Municipal Committee (MCD)[49] had been active for many decades in development, regulation and urban services provision across Old Delhi including the walled city, Sadr Bazar, Sabzi Mandi and other suburbs, but the recent growth had generated several suburban Notified Area Committees.[50] Even though the government's Delhi Development Committee had recommended in 1939 the formation of a technical body to co-ordinate the programmes and activities of different agencies, there was little headway in further institutional reforms for urban planning and development in and around the capital city because the government was preoccupied with the war and India's Independence issues.

The growth pressure increased many fold immediately after Independence, with the influx of refugees to Delhi and the need for new capital functions in the new democracy. Delhi's urban population grew at the phenomenal rate of 107 per cent between 1941 and 1951.[51] The effects were different across New Delhi in contrast to the rest of the metropolis. Delhi had to accept about 500,000 refugees[52] and quickly provide camps[53] as well as hastily developed rehabilitation colonies[54] through piecemeal land acquisitions around the fringes of the metropolis. But in the heart of the capital city – New Delhi – the leafy green was yet unruffled, except for the construction of several new government buildings for the new democracy[55] on the vacant lots, and the change in names of the Imperial buildings and other capital elements.[56] Removal of many icons of the Imperial capital, notably, the statue of George V[57] and the Britannic lions, symbolized the national-political ethos immediately after the freedom from two hundred years of colonial rule. However, increasing need for accommodation for the rising numbers of government employees[58] necessitated development of vacant land on the south-east of Lutyens's bungalow zone for low-density residential areas and also some residential expansion on the southern fringe.[59] New Delhi's (correspondingly NDMC's) jurisdiction also expanded to incorporate a large new diplomatic enclave, named 'Chanakyapuri',[60] to its south-west, beyond the race course (see figure 13.2).

In 1957, the Delhi Development Authority (DDA) was established through a Parliamentary Act for integrated physical planning and land development across the entire Delhi State[61] cutting across jurisdictions of the MCD, the NDMC, the Cantonment Board and Notified Committees, and the villages in the surrounding fringe areas. Its first Master Plan (with a horizon of twenty years) came out in 1962[62] although a quick Interim General Plan (IGP)[63] came as its precursor in 1956. Both these plans lamented the historic failure of Lutyens's New Delhi to bind the capital city with the old city through Paharganj and the lost opportunity for their planned integration.[64]

While the land acquisition price and problems in Paharganj were given as the reason for ignoring an earlier proposed northern orientation of New Delhi, these problems increased many fold over time across the uncontrolled growth in the area. Consequently, urban renewal of the old city and its surroundings (which became increasingly congested with growth of small industries and wholesale trade) was almost doomed.

At the same time, metropolitan growth demanded a centrally-located, highly-accessible central business district and Lutyens's elite shopping area – Connaught Place – perfectly fitted the role with its sparse surroundings being soft targets for intense commercial growth. The DDA dramatically opened up the land for private sector development with high floor area ratios.[65] The bungalow area immediately surrounding the historic two-storey colonnaded structures encircling the Connaught Circus witnessed rapid transformation of the pristine skyline, rattling the very character of the garden city.[66] However, the 1962 Master Plan for Delhi (MPD-62) on the whole, had conceived New Delhi (Planning District-D of Delhi) as a 'conservation district' for which the statutory plan envisaged limited redevelopment, stress being on conservation of the garden city character. Much of the redevelopment was again confined to the north of the Central Vista and was in the form of government buildings and public institutions along the arterials and major roads. Lutyens's bungalow zone was assigned to housing VIPs[67] and a few diplomatic offices, especially south of the Vista[68] (except for a few patchy clusters where private sector[69] conversion has taken place from bungalows to flats). Further, the new southern residential areas of the extended NDMC were also planned for low-density housing. Remarkably, therefore, the early plans strove to fit the post-Independence capital functions into the sprawling colonial physical form of New Delhi.

But how does this physical form fit the potential role of the core of a growing metropolis? To this extent, the poly-nuclear concept of the MPD-62[70] and the emphasis of both the IGP and the MPD-62 on decentralization through development of Delhi's surrounding region should be helpful in sustaining the character of New Delhi. In particular, the fast growing ring towns around Delhi, which together formed MPD-62's concept of a 'Delhi Metropolitan Area (DMA)', had been envisaged as absorbing much of Delhi's growing migratory population and decentralizing its industrial and commercial growth. The MPD-62 even proposed the decentralization of central government offices into four ring towns by 1981.[71] Indeed, Delhi's adjacent urban centres, located mostly along the highways radiating from Delhi in the neighbouring states of Uttar Pradesh and Haryana,[72] have been exploiting the accessibility and proximity of the centre of national political power, the large metropolitan market with its low tax regime and the planned policies of Delhi to decentralize industries and government functions (see figure 13.5). New Delhi still remained almost a sleepy suburb right at the core of a large and growing metropolitan area bustling at its fringes. But its high accessibility, very low land utilization and excellent infrastructure relative to the rest of its surroundings, gradually created a debate as to its conservation versus its growth.

Late Twentieth-Century Planning and Development Context for New Delhi

Following the IGP and the MPD-62, policy on metropolitan growth management became even more 'outward looking' with the establishment by Parliamentary Act in 1985 of the National Capital Region Planning Board (NCRPB). The NCRPB prepared planning guidelines[73] for a large (30,242 square kilometres) National Capital Region (NCR) of Delhi comprising, the National Capital Territory (NCT) of Delhi[74] and the surrounding districts falling across three neighbouring states[75] (see figure 13.5). The first NCR plan for 2001, brought out in 1988, advocated the development of remote regional urban nodes further away from the DMA to reduce the pace of Delhi's growth. Delhi had been growing at a high steady rate of around 52 per cent each decade through the 1950s, 1960s and 1970s and a significant proportion of the growth had been through migration, especially from the neighbouring states. But, in spite of the regional plan, the turn of the century saw sustained growth of NCT Delhi itself and much faster growth of the adjoining DMA cities than those of the remote 'priority' towns of the NCR.[76] The neighbouring states enacted laws[77] to encourage development of these adjoining cities to take advantage of Delhi's market and, by

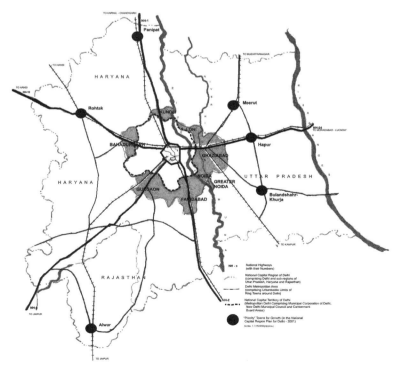

Figure 13.5. Connaught Place – the metropolitan hub – and the surrounding New Delhi in the context of Delhi's exploding metropolitan area and the National Capital Region.

the turn of the century a host of new towns had grown around Delhi. Consequently, a massive conurbation has emerged where, in reality, the NCT Delhi has appropriated its neighbours' jurisdictions. Further, urban land consumption within the Territory itself has been increasing from plan to plan creating the very real possibility that over three-quarters of the NCT will have been consumed by 2021.[78]

The revised master plan for Delhi brought out in 1990 for 2001 (Perspective Plan 2001) has been unique in terms of its 'inward looking' growth management policy. For the first time, a plan proposed densification of the existing urbanized area to accommodate the additional population, along with some urban extension. Among the eight Planning Divisions across Delhi, the plan suggested high population 'holding capacity' for five, including New Delhi (Division–D). It recommended a capacity for 754,685 persons in New Delhi – more than double the existing population. But even in 1991, the NDMC area recorded a resident population of only around 0.3 million, with a density of 71 persons per hectare compared with urban population densities of 124 persons per hectare for the entire NCT and 167 for the MCD area.[79] In contrast, the 'floating' population of New Delhi was more than a million. These figures indicate on one hand the attractiveness of New Delhi as a work centre as well as its capacity to house more population and, on the other, its resistance to change, even in the wake of much denser metropolitan growth around it.

Accommodating a mere 3 per cent of the total population and comprising less than 3 per cent of the total land of NCT Delhi, the future of New Delhi could be considered a non-issue in the development context of an exploding metropolis (see figure 13.6). Yet, across the wider metropolitan landscape, conservation versus redevelopment of the first capital city of the twentieth century, with its unique sprawling, largely suburban form, wide avenues, lush green open spaces and memorable vistas and architecture, has been an issue of continuous debate. The future, especially of Lutyens's bungalow zone, has been a contentious issue. To the elite professionals – architects, urban designers and conservationists – it is the very icon of Lutyens's garden city. The high profile politicians and officials living there also have a stake in its quality of life in the heart of a populous metropolis. At the same time this sparsely used prime real estate senses growth

Figure 13.6. Physical expansion of Delhi through development of planned residential suburbs.

pressures from its surroundings. A few small patches of privately leased land had already tasted the fruits of high return from conversion of bungalows to flats.

Conscious of the conservational value of the capital's urban landscape, the government had formed, as early as in 1960, a Central Vista Committee and later the New Delhi Redevelopment Committee with a technical Design Group in 1971 – all culminating finally in establishing the Delhi Urban Arts Commission.[80] Unfortunately, however, there have been limited studies of alternative possibilities of striking a balance between conservation and more intensive use of the land. On one hand, the Buch Committee set up by Prime Minister Vajpayee to look into the prospects of Lutyens's bungalow zone in light of the growing housing need in the city has come out against any change.[81] On the other, recently there have been strong overtures towards much more intensive land use in Delhi, such as the recommendations of the Malhotra Committee[82] to increase floor area ratios and reduce lot subdivision sizes across Delhi, or the recently announced Guidelines for the upcoming Master Plan 2021[83] favouring private sector land development and high-rise housing.

At the turn of the twenty-first century, the first capital city of the twentieth century, especially its bungalow zone, is in crossfire. In the spirit of the World Monuments Fund, we may say, New Delhi is an 'endangered site'. We need to search for a way to sustain New Delhi as a 'living monument' amidst the values and aspirations of a new society and the changing landscape of the dynamic metropolis that surround it.

NOTES

1. Gupta (1988), pp. 1–36; Joardar (2002); Rapoport (1972).

2. Insurgency by the natives employed in the British army, especially in the lower ranks, against the British rule.

3. Warren Hastings was Governor General of Bengal with supervisory powers over Bombay and Madras. In 1782 he wrote a memo citing the defects of Calcutta in functioning as the seat of British rule in India; Irving (1981), p. 16.

4. Sir Stafford Northcote was the British Secretary of State for India, 1867–1868. Irving (1981), p. 17.

5. Irving (1981), p. 17; Thakore (1962) as quoted by King (1976), p. 231.

6. Lord Lytton was the then Viceroy of India.

7. A Durbar was a traditional Mughal Royal court where the *Badshah* (Emperor) used to meet royal officials, local princes and other dignitaries (and occasionally also the common man in 'Aam', i.e. common Durbar) for royal decrees, announcements, hearings, etc. After taking over from the Mughals, the colonial government also held several Durbars in Delhi on very special occasions, especially related to a British King, Queen or Prince.

8. The charismatic Lord Curzon of Kedleston was the Viceroy of British India from 1899 to1905 preceding which he was the Governor General of Bengal who brought about Partition of Bengal, leading to intense native opposition.

9. Irving (1981), p. 17.

10. Chakravarty (1986); Irving (1981), pp. 18–19.

11. Irving (1981), p. 18.

12. Irving (1981), p. 27.

13. Charles Hardinge of Penshurst was the Viceroy of India from 1910 to 1916. He was instrumental in finally formalizing the longstanding proposal to shift the capital, first through official declaration at his Council in June 1911 and placing it as the royal declaration agenda at the Coronation Durbar in December 1911.

14. King George V made the royal public announcement at the Coronation Durbar (which until then was held a guarded secret by Hardinge and others) of shifting the Imperial capital from Calcutta to the historic Delhi. He subsequently laid the foundation stone of the new capital.

15. The three full members of the committee – Edwin Landseer Lutyens, John A. Brodie and Captain George Swinton – were decided by the London office of the Secretary of State, The Earl of Crewe, while Hardinge's initial choice, H.V. Lanchestor, was inducted later as a

consultant. Irving (1981), pp. 39–42. Herbert Baker was inducted as the choice of Lutyens as the chief architect in his team.

16. Especially, the design of the Government House (later named Viceroy's Palace) where he prevailed upon Lutyens to incorporate traditional Indian features – such as the dome resembling a Buddhist *Stupa* or the elephant sculpture – while Lutyens from the beginning had strong opposition towards traditional Hindu and Moslem architecture as well as the colonial style incorporating vernacular features then prevailing in India. Irving (1981), pp. 164–274.

17. Hardinge's term was completed in March 1916 and Viceroy Chelmsford succeeded him. But in London he influenced the Secretary of State and Buckingham Palace against the opposition of Curzon and Chamberlain to the New Delhi project on financial grounds. Irving (1981), pp.117–120.

18. The Aaravalli Hill range, popularly known as the Ridge, traverses the Delhi region in a north–south direction.

19. Two key reports of the Committee of 13 June 1912 and 11 March 1913 recommended strongly the vast plain lying south of the existing city – one could visually link the *Purana Qila* (old Fort) to its south-eastern fringe and the *Qutub Minar* further down south on the west – although opposition grew against this site.

20. Such as the famous Metcalf House.

21. The colonial government started shifting capital functions from Calcutta to Delhi right after the 1911 Durbar.

22. That is *Jumma Masjid*, Red Fort and *Chandni Chowk* to the north, Yamuna River and Humayun's Tomb to the east and *Purana Qila and Qutub Minar* to the south.

23. Irving (1981), pp. 55–90.

24. A foothill of the central Aaaravalli Ridge.

25. *Jumma Masjid*.

26. A mixed residential and commercial locality growing on the southern fringe of the walled city of Shahjahanabad.

27. Irving (1981), pp. 56–63.

28. King (1976), p. 235, quoting Nilsson (1973), p. 45.

29. Ultimately, the slope of the ramp connecting the Viceroy's Palace from the Central Vista became a contentious issue because it obscured the view of the Lutyens palace when travelling between Baker's two secretariat buildings – see Irving (1981), pp. 142–163.

30. Irving (1981), pp. 142–165.

31. Irving (1981), pp. 275–280.

32. Presently, the Indian Parliament House.

33. Irving (1981), pp. 295–310.

34. Which provided for self-governance for Indians and formed a legislative assembly.

35. W.H. Nicholls was the Architect Member of Imperial Delhi Committee from 1913 to 1917; R.T. Russells was the Chief Architect who took over the work of Connaught Circus after Nicholls left. Irving (1981), p. 314.

36. Irving (1981), p. 83.

37. King (1976), pp. 267–268.

38. King (1976), pp. 248–253.

39. In 1932; in 1925, it was called the Imperial Delhi Municipal Committee.

40. Lord Curzon in London and local politicians in India.

41. Irving (1981), pp. 109–116.

42. Hindu (2003).

43. Notable among them were the Irwin Stadium (now, National Stadium) the Race Course, Gymkhana and Golf Course, the conserved Ridge Forest, Welligdon (now, Safdarjung) and Irwin Hospitals.

44. Delhi Development Authority (1962), p. 6.

45. King (1976), pp. 267–268; Government of India (1951).

46. Government of India (1951).

47. Lodi Colony housing and Lodi Estate bungalows.

48. Gwalior, Jodhpur, Bundi, Bikaner. Delhi Development Authority (1962), p. 6.

49. Constituted first in 1863 under Act XXVI of 1850, the body was politically reformed with greater autonomy under the Municipal Act of 1884.

50. The first was Mehrauli formed in 1901 followed by Civil Station in 1913, Shahdara in 1916 (later became Shahdara Municipal Committee in 1943) Red Fort in 1924 and West Delhi in 1943.

51. Government of India (1971).

52. Mostly from West Punjab, Baluchistan, Sind and North Western Frontier Province of Pakistan.

53. The largest was the Kingsway on the north of Old Delhi near the historic Coronation Durbar ground; other camps were at Karolbagh on the west and Shadara on the east of Old Delhi.

54. Thirty-six colonies were built to accommodate 47,000 refugees: Nizamuddin was adjacent to the south-east corner of Lutyens's Delhi with Defence colony and Lajpat Nagar on its south and on further south were Kalkaji and Malviya Nagar. Several were built on the west and north-west of Old Delhi beyond Karolbagh.

55. Especially, 'Rail Bhavan' (Ministry of Railways Building) 'Krishi Bhavan' (Ministry of Agriculture Building) 'Shastri Bhavan' (Ministry of Education and Culture Building) on the north side and 'Udyog Bhavan' (Ministry of Industry Building) 'Nirman Bhavan' (Ministry of Works & Housing/Urban Development Building) on the south side of the Central Vista.

56. Notably, the Viceroy's House/Government House to 'Rashtrapati Bhavan' (the President's House) with its large President's Estate (the President of India is also the Commander-in-Chief); Council House – to 'Sansad Bhavan' (Parliament); Secretariat Buildings – North Block and South Blocks form today's 'Central Secretariat'; the Central Vista/Kings Way to 'Rajpath' (same meaning) the Queensway to 'Janpath' (People's Way); the Hexagon with the Memorial Arch and the cenotaph (with George V's statue) to India Gate.

57. A proposal to institute the statue of Mahatma Gandhi – the Father of the Nation – under the same canopy where King George's statue stood faced political critics; the canopy is still empty.

58. Government employment doubled between 1931 and 1941 when most offices had shifted to Delhi from Calcutta; but it jumped about 250 per cent over the next decade. See Government of India (1951).

59. Notably Pandara Road area, Kaka Nagar, Vinay Nagar, Sarojini Nagar, etc.

60. Named after a historical figure, Chanakya – the Prime Minister of the Hindu King Chandragupta Maurya – whose diplomatic skills were said to be legendary.

61. The previous Delhi District or Territory was ruled by the Chief Commissioner. The Government of India declared it a Part 'C' State under the Constitution of India on 17 March 1952, providing limited autonomy for its 48 member Legislative Assembly. Especially, law and order and all operations related to land and buildings, particularly those of the federal government, would remain the responsibility of the Union Government. Thus the DDA has been an agency of the Central Government, rather than the Government of Delhi State.

62. Delhi Development Authority (1962).

63. Prepared with the help of the central government agency – Town Planning Organisation (TPO).

64. Delhi Improvement Trust (1956), p. 5; Delhi Development Authority (1962), p. 5.

65. The initial floor area ratio was 400, which was reduced later to 250.

66. Much of the concentration of commercial development has occurred along Barakhamba Road and Curzon Road which has also the American Library and the British Council; the Parliament Street is a mix of office and bank buildings and public institutions (for example, All India Radio) along with the historic Jantar Mantar (Sun Path Observatory); Janpath (previously, Queensway) in the middle retains a low-rise shopping street character interspersed with the multi-storey Cottage Emporium building and the old Imperial Hotel; along Baba Kharag Sing Marg (previously Irwin Road) – radial at the south-western end – has developed the various state emporia and their tourist office and landmark Hindu temple and Gurdwara (Sikh temples).

67. Ministers, members of parliament, high profile bureaucrats, military personnel and diplomats, etc.

68. Who are yet to move to Chanakyapuri.

69. Under private lease.

70. It proposed a hierarchical order of commercial nodes, especially several district centres across the metropolis

71. Delhi Development Authority (1962), p. 106.

72. India's Punjab State was divided into Punjab and Haryana states with Chandigarh as their common capital.

73. There is no statutory enforcement as the plan traverses four different states having their own constitutional roles for development.

74. Delhi State was renamed National Capital Territory of Delhi under a Parliamentary Act in 1992 without any significant change in power, but with an enlarged Legislative Assembly and a Cabinet of Ministers.

75. The States of Uttar Pradesh, Haryana and Rahasthan.

76. The NCR plan put greater emphasis on the development of the relatively remote towns than the Ring Towns in the DMA. See National Capital Region Planning Board (1999).

77. Such as the Haryana Development and Regulation of Urban Areas Act, 1975, which promoted massive private sector, land development in Gurgaon and the

Uttar Pradesh Industrial Area Development Act, 1976 under which two industrial new towns, NOIDA and Greater NOIDA have been established.

78. Association of Urban Management and Development Authorities (2003), pp. 1–25; Joardar (2003).

79. Government of India (1991).

80. Although its jurisdiction covers the entire Delhi, it was formed primarily for New Delhi. Ribeiro (1983).

81. Government of India (1998) as quoted by Kumar (2000).

82. Government of National Capital Territory of Delhi (1997).

83. Delhi Development Authority (2003).

Chapter 14

Berlin: Capital under Changing Political Systems

Wolfgang Sonne

No other capital city, in the twentieth century, experienced political changes as extreme as Berlin: from the German Empire to the Weimar Republic to the National Socialist dictatorship, followed by democracy in the West and socialism in the East, before finally sailing into the safe harbour of democratic Europe as the reunited Federal Republic. Berlin served as a bridge from which the erratic course of this state ship was commanded, albeit answering to a wide range of demands and assuming many roles. Thus the principal political turning points (1918, 1933, 1945–1949 and 1990) are key markers in the timeline of Berlin's history as a capital city in the twentieth century.

Despite these political ruptures, and the contrasting urban motifs they sometimes inspired, Berlin displays a remarkable urbanistic continuity. To begin with, there are the surprisingly constant population figures: the estimated 4 million, which formed the basis for the Hobrecht Plan of 1862 (daringly forecast at a time when the city boasted no more than roughly half a million inhabitants), proved realistic for the entire century to come.

The continuity in the fundamental image of the city is equally remarkable. Despite the disparate demands for political representation and the resulting divergent models, Berlin has maintained the principal characteristics of its urban plan: in the long term, established patterns of ownership and the existing infrastructure of roads and sewers outlasted any revolutionary models for reconstruction. Even the typology of the urban block with a height limit continues to dominate the urban image – despite intermittent diatribes against Berlin's image as a 'city of stone'. Nor did political changes necessarily precipitate changes with regard to state representation through urban design: monumental axes defined the urban planning in the imperial era and the Weimar Republic as much as they did during National Socialism and socialism. A similar case can be made for the public forum, a typology that has survived in urban plans over the course of the most diverse political environments to this day.

In what follows, the focus will be on those elements that distinguish a capital city from any other metropolis: the urban plan for government buildings and the concomitant ideas on state representation. With its exposure to different political systems, Berlin offers a fertile ground for exploring the link between such systems and the urban image.

'The Evolution of Greater Berlin': Berlin in the Imperial Age, 1900–1918

The lack of state-sponsored initiatives to develop a comprehensive plan for Berlin as a capital city at the beginning of the twentieth century was partially due to the fact that its important institutions were already housed in monumental splendour in the *Stadtschloß* (Imperial Palace), slightly expanded under Wilhelm II, and the *Reichstag* (Parliament), erected by Paul Wallot between 1882 and 1894.[1] The political antagonism generated by the tension between the dynastic legitimacy of the imperial house and the democratic foundation of the parliament also contributed to a stalemate in urban development. The chancellor sat unobtrusively in the Baroque Radziwill palace on Wilhelmstrasse, where the principal ministries had also been housed in Baroque city residences of the nobility. The deliberate isolation of the judiciary, on the other hand, was clearly programmatic: the new *Reichsgericht* (Supreme Court) designed by Ludwig Hoffmann was built in Leipzig between 1884 and 1895, and not in Berlin – a statement on federalism and the independence of the judiciary.[2]

The greatest obstacle to developing an effective master plan for the capital, however, was the contrast between the Empire under a conservative Prussian influence and the increasingly social democratic city of Berlin. The Kaiser and the ministries of the Reich denied the city permission to incorporate the suburbs into the city proper in order to prevent the formation of an even more powerful social democratic municipality.[3] They casually accepted the fact that this also prevented the capital of the Reich from addressing urgent planning problems such as housing and transportation, the provision of green spaces, and last, but not least, the quest for a uniform image. In the end, it was left to artists and architects to put forward a comprehensive plan. The initiative by the two architects' associations of Berlin in 1906 brought about the first success.[4] In 1908 the city of Berlin and the associated cities of Charlottenburg, Schöneberg, Rixdorf, Wilmersdorf, Lichtenberg, Spandau and Potsdam, joined by the districts of Teltow and Nieder-Barnim with over 200 communities, launched a competition for the entire settlement area of Greater Berlin.[5]

True to its mission, the competition did not request a specific scenario for the capital of the Reich. Some of the twenty-seven participating architects nevertheless paid particular attention to the issue of appropriate representation of the state in public urban monuments. Thus the winners of the third prize – a team composed of architect Bruno Möhring, economist Rudolf Eberstadt and transportation engineer Richard Petersen – designed a *Forum des Reiches* (Forum of the Empire) for the area of the Spreebogen, bluntly stating its imperialistic and militaristic intentions (see figure 14.1). The authors placed the *Kriegsministerium* (War Ministry) directly opposite the Reichstag, accompanied by the *Reichsmarineamt* (Imperial Naval Office), the *Reichskolonialamt* (Imperial Colonial Office) and the *Generalstab* (Military Headquarters), and offered an emphatic interpretation of this arrangement: 'The army and the people, the corner stones of Germany's greatness and power, unified in architectural monuments . . . Surely an ensemble of this kind would speak powerfully to every German and unequivocally demonstrate the foundations of the Reich to every foreign visitor!'⁶ The team's nationalistic image for this urban square was by no means founded in national design roots alone. It was equally indebted to the French tradition of monumental urban squares, which, for example, had been promoted by the École des Beaux-Arts in Paris for the Prix de Rome in 1903 and analysed by art historian Albert Erich Brinckmann in his book *Platz und Monument* in 1908.

None of the competition's proposals was ever realized. Although the *Zweckverband Gross-Berlin* (Greater Berlin Association) was founded in 1912, it lacked the necessary planning authority: the association's principal function was to propose traffic routes, to purchase and preserve forests, and to define new building lines. It had no authority to influence existing development plans or to draft building codes. An independent housing policy was beyond its reach as was a social and education policy for the community.⁷

'Metropolis and Capital City': Berlin in the Weimar Republic, 1918–1933

The drastic political changes after the First World War, the deposition of the Kaiser and the overthrow of Prussia's conservative state authorities finally made possible the long

Figure 14.1. Bruno Möhring, *Forum des Reiches* (Forum of the Empire), competition Greater Berlin, 1910. A large urban square is supposed to announce the imperial ambitions of the Reich by assembling military monuments and buildings for the military administration.

overdue integration of the political community and the entire settled area of the city. The law on the creation of the *Stadtgemeinde Berlin* (Berlin Munipality) took effect on 1 October 1920; the new municipality comprised eight former cities, fifty-nine rural communities and twenty-seven farm districts. The distribution of government functions in the city remained virtually unchanged: the parliament continued to sit in the Reichstag, the ministries were still clustered around Wilhelmstrasse. The only building to undergo a fundamental functional change was the most prestigious structure: the Emperor's palace was converted into a museum.

The economic situation was difficult and several years passed before a new urban plan with a focus on housing and transportation was drafted under the newly appointed city architect Martin Wagner in 1925.[8] Still, the scattered distribution of the national government – a product of the ultimately anti-constitutional policy pursued by the Imperial House – continued to pose a challenge to architects. Thus, in 1920, Martin Mächler published his ideas (first conceived in 1908) on creating a north-south axis, whose principal aim lay in gathering the various ministries of the Reich in the centre of the Spreebogen.[9] The architect Otto Kohtz also pursued the idea of a centralized cluster for the Reich ministries in his project for a *Reichshaus* (Imperial House) on Königsplatz, which he proposed for the first time in 1920. However, this proposal replaced the conventional urban space with a pyramidal high-rise that would dominate the urban landscape in the form of a 200 m high Expressionist *Stadtkrone* (City Crown).

Even the most realistic project of the Weimar Republic was also initiated by architects. Several members of the *avant-garde* architects' association *Der Ring* presented new solutions at the Great Art Exhibition of Berlin in 1927 for transforming the Königsplatz into a *Platz der Republik* (Square of the Republic). Based on Mächler's plan, which was also shown at the exhibition, Hugo Häring envisioned the government district as a modernist administration complex with high-rise slabs, reminiscent of Le Corbusier's administration centre for the *Ville contemporaine* of 1922, although Häring's concept still emphasized the importance of axial links. Thus he interpreted his north-south axis crossing the street of Unter den Linden as 'a distinct and clear line through this axis of the rulers'.[10] And the square in front of the Reichstag was interpreted as a 'Forum of the German Republic' surrounded by grandstands for public events.[11]

The architects' hopes for an urban plan for a democratic government district remained unanswered, however. After the 1927 competition for the Reichstag expansion which failed to include the urban context of the building, a second competition was launched in 1929, which gave the architects the freedom to include the Platz der Republik in their design. Berlin's city planner Martin Wagner expressed the ambitions for a politically appropriate design as follows: 'What will happen to the Platz der Republik? A designer would have been found long ago for a "Square of the Monarchy". The new democracy still needs to develop its own design consciousness'.[12] The most significant design from a political perspective was the work of a non-participant: Hugo Häring. He modified his earlier design, the most important addition being a theatrical grandstand facing the Reichstag (see figure 14.2). Häring understood his design as a fundamental contribution to democratic building:

My studies aim first and foremost to outline and depict the changes to the urban plan which are due and clearly prescribed by the political changes, that is to explore today's task from the perspective of today's client, the now sovereign people, its contents and essential foundations.[13]

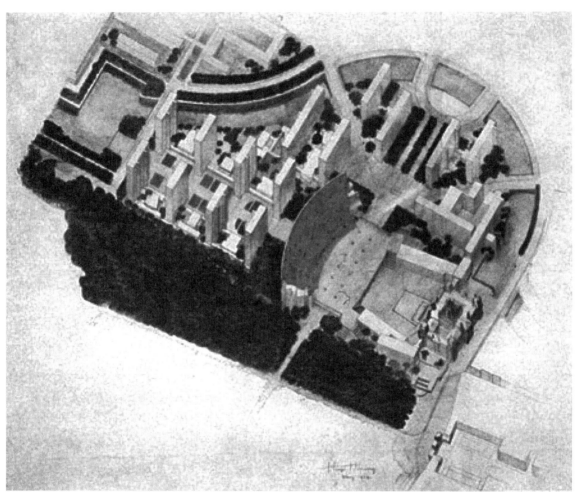

Figure 14.2. Hugo Häring, *Platz der Republik* (Square of the Republic), 1929. A large grandstand for public gatherings is designed to symbolize the new Republican state.

Monumental urban planning was arrested chiefly because of the economic situation in the Weimar Republic. During the challenging early years, new government buildings were simply unthinkable. And once plans were finally on the table, the global economic crisis of 1929 quickly put a stop to any further endeavours. The only remaining architectural mark from the Republic was the modest expansion to the *Reichskanzlei* (Imperial Chancellery) in Wilhelmstrasse by Eduard Jobst Siedler, 1927–1930.

'Reshaping the Capital of the Reich': Berlin in the Third Reich, 1933–1945

Once the National Socialists 'seized power' the era when the state displayed complete disinterest in devising an urban plan for Berlin as a capital city came to an abrupt end. Adolf Hitler took a personal interest in giving the city the representational character he deemed necessary for his political aims. First plans for a monumental north-south axis began as early as

1933. Hitler's view of urban planning in the Reich capital as a tool in his war policy is evident in the document he composed immediately following France's capitulation in 1940:

Within the shortest possible time, Berlin's urban renewal must invest with the expression of a capital of the new, strong empire that the magnitude of our victory deserves. In my view, the realization of this prime building task of the Reich is the most important contribution toward securing our victory for the long term.[14]

An autonomous municipal administration could only be a hindrance to these political goals. And – like many other institutions – it was promptly eliminated: the last session of the *Stadtverordnetenversammlung* (City Council) took place in November 1933. While a city council was at first maintained in name only, true authority lay with *Gauleiter* (District Leader) Joseph Goebbels, who also acted as Stadtpräsident (President of the City) from 1940 onwards. Since the local planning authorities were too slow in implementing Hitler's wishes, he appointed the young and malleable architect Albert Speer on 30 January 1937 as *Generalbauinspektor der Reichshauptstadt Berlin – GBI* (General Building Inspector of the Imperial Capital Berlin). Speer answered directly to Hitler and was given far-reaching planning authority, thereby removing power from all existing municipal and state planning authorities. The planning activities of the GBI were awarded an annual budget of roughly 60 million Reich marks.[15] As Speer put it, Hitler was dedicated to 'developing Berlin into a real and true capital city of the German Reich'.[16]

Speer's comprehensive plan proposed restructuring road traffic into a large cross of principal arteries complemented by numerous ring roads reaching all the way to the outer urban ring road and integrated rail traffic with the help of two central transit stations. This was further augmented by plans for new housing developments and a linked park system. But the core of the plan was the central section of the north-south axis between the two railway stations, for which Speer envisioned a series of classicist solitary buildings bordering the 7 km long and 120 m wide boulevard (see figure 14.3). The axis was interrupted on the one hand by a massive triumph arch and on the other hand by the 290 m high *Grosse Halle des deutschen Volkes* (Great Hall of the German People) which would hold up to 180,000 visitors. The traffic-free square in front of the hall and directly across from the Reichstag was dedicated for the new *Führer-Palais* (Leader's Palace), whose monumental scale would dwarf the Reichstag. This central section was set aside exclusively for ceremonial propaganda purposes dedicated to the 'Thousand Year Reich'. In his memoirs, Speer summarized the intent as follows: 'The idea was that [visitors] would be overwhelmed, or rather stunned, by the urban scene and thus the power of the Reich'.[17]

On the one hand, the comprehensive plan developed under Speer's authority was strongly influenced by ideas which Hitler had formulated as early as the 1920s: with his designs for a cupola-topped hall and a triumph arch, Speer remained faithful to his 'leader'. On the other hand, the essence of the plan was very much in the spirit of earlier planning proposals: similar concepts with regard to restructuring the rail system had already been developed during the competition for Greater Berlin in 1910 and Mächler's idea of a north-south axis with public buildings from 1920 remained an enduring favourite throughout that decade. Nor was the plan by any means original within the context of contemporary international capital city planning: the creation of a ring system of roads and monumental housing schemes is in tune with the planning approach for Moscow in 1935; the scale corresponds to the ensemble inaugurated

Figure 14.3. Alber Speer, north-south axis and *Grosse Halle des deutschen Volkes* (Great Hall of the German People), 1942. The overwhelming scale of traditional urban elements should demonstrate the overwhelming power of the National Socialist State.

in 1931 in the British colonial capital city of New Delhi; and the classicist solitary structures are reminiscent in their arrangement of the Mall in Washington, which was being realized in the 1930s based on the plan from 1902. While capital city planning was indeed a genuine concern for National Socialist politics, they resorted to few specifically National Socialist urban design means to realize their goal.

The plans were halted in 1943 and all means diverted to support the war effort, with Speer himself being appointed *Rüstungsminister* (Minister of Armaments). The components that had been executed by then remained no more than an ephemeral fragment, notwithstanding Speer's vision that his buildings should stand for a thousand years and should be beautiful ruins too. Some of this was quite deliberate: the new ministries erected on Wilhelmstrasse, such as the *Reichsluftfahrtministerium* (Imperial Air Force Ministry), 1935–1936, by Ernst Sagebiel would in fact have been rendered obsolete by the axis concept. The same is true for the most important public building to be realized in the National Socialist era, the *Neue Reichskanzlei* (New Imperial Chancellery), 1937–1939, in the Voßstrasse by Speer which would have been replaced by that newly planned for the square in front of the Great Hall. Hitler felt that the construction of this massive 'temporary' structure was necessary, however, in order to create an appropriate backdrop for the New Year's reception for diplomats after the annexation of Austria. At the inauguration ceremony on 9 January 1939, Hitler outlined the domestic political goals he aimed to achieve with architectonic monumentality:

Why choose the biggest scale? My German compatriots, I do this in order to give individual Germans their self-confidence back. To demonstrate to the individual in a hundred different ways: We are not at all defeated, on the contrary, we are absolutely equal to every other people.[18]

During the Third Reich, urban planning for Berlin was basically a matter of capital city planning: the new government buildings were to foster identity on the domestic front and create an impressive image to the world of the nation's foreign policy. All that remained of this new representation in urban design, however, was a massive pile of rubble.

'Capital City and Metropolis': Berlin in the Post-war Era and the Early Years of the Federal Republic of Germany, 1945–1990

After the war, Berlin's status as a capital city remained uncertain at first: since there was no central government in Germany, there was no need for a central capital city. Moreover, the four allied victors each claimed a section of the former capital of the Reich, and thus in July 1945 Berlin was divided into four independent sectors, the entire city being granted special status. In August 1946, the Allies ratified a preliminary constitution for the region of the former *Stadtgemeinde Berlin* (Municipality of Berlin) and in October 1946 the new *Stadtverordnetenversammlung* (City Council) held its first official meeting. It was only in 1950, that a new constitution for Berlin was ratified, giving the city the status of a *Bundesland* (one of the federal states within the newly founded Federal Republic of Germany) – although this status was only valid in West Berlin in view of the new political situation.

Urban renewal seemed a far more pressing issue than the political constitution as 30 per cent of the buildings were more or less destroyed. As early as May 1945 – a few days after the capitulation – Hans Scharoun was chosen as the new chief planner of Berlin by the Soviet occupying forces. The *Kollektivplan* (Collective Plan), for which Scharoun was largely responsible and which was presented at the 1946 exhibition *Berlin plant* (Berlin is planning) in the partially restored *Stadtschloß* (Imperial Palace), proposed a complete break with the traditional image of the city, dividing the city according to natural landscape patterns with a grid of highways.[19] None of the first plans proposed ideas for a government district: on the one hand, because there seemed to be no immediate need to create one, and on the other, because it seemed inconceivable even to broach the subject in the wake of the representational planning associated with the National Socialists.

However, any implementation of comprehensive plans was hindered by the growing political tension: Berlin became the city at the front of the Cold War between the Communist East and the Capitalist West, with the most obvious act being the construction of the Berlin Wall on 13 August 1961. The emergence of two

German states in 1949 – the Federal Republic of Germany and the German Democratic Republic (GDR) – obviated the need to establish one central capital city. While both German states declared Berlin as their capital city, the Federal Republic of Germany chose Bonn as its temporary seat of government in response to Berlin's precarious situation. In federalist tradition, the *Bundesverfassungsgericht* (Federal Constitutional Court) moved to Karlsruhe. Thus the pronounced modesty of the government buildings in Bonn designed without any attempt at endowing them with a representational character had yet another underlying cause:[20] in addition to the ostentatious rejection of the National Socialist building propaganda it was important to express visibly the transient nature of the government district in Bonn, to ensure that its temporariness would be remembered.

The state therefore had to strengthen Berlin's capital city status at the urban planning level. In 1957 – at a time, when the realization of such a plan was politically unattainable – the Federal Republic of Germany and the *Bundesland* (State) of Berlin launched the competition *Hauptstadt Berlin* (Capital City Berlin), whose boundaries included both the western and eastern sectors. The hope was that the competition would yield a new urban image for the capital city in the spirit expressed by Chancellor Konrad Adenauer in the foreword: 'The structural and urban renewal is intended to express Berlin's spiritual and intellectual role as Germany's capital city and as a modern metropolis'.[21] The symbolism of a capital city played an essential role and the new values of democracy and international associations were emphasized:

The complexion of the new capital will at any rate not be that of nationalist governmental power. Rather will it be dominated by the ideas of democracy and international collaboration on equal footing.[22]

The competition did indeed yield the first post-war proposals for significant government districts, manifesting on the one hand the rejection of National Socialism and on the other hand the confrontation with Socialism. The

Figure 14.4. Friedrich Spengelin, Fritz Eggeling and Gerd Pempelfort, Government Centre, competition *Hauptstadt Berlin* (Capital Berlin), 1958. A free arrangement of solitary structures within a landscape was intended to mark a break with the past and to symbolize a new free society.

moderately Modernistic jury thus opposed any proposals based on axes and symmetry and awarded distinction above all to designs that featured a liberal play of cubes in flowing space. Most participants located the government district on the Spreebogen, as the competition brief had in fact suggested. The winning design by Friedrich Spengelin, Fritz Eggeling and Gerd Pempelfort (see figure 14.4) showed the typical solution. Their description spoke of a 'government district with a verdant centre, arranged in a free albeit orderly group, a visible expression of the spirit of the democratic state'.[23] Thus the liberal arrangement of cubes and planes – adhering to Le Corbusier's Capitol in Chandigarh and to the canon of the International Modern of CIAM VIII – had received its political blessing. The escalating East-West confrontation banished any thought of a new government district for a democratic unified Germany. The only significant building after the erection of the Wall was the restoration of the Reichstag as the future parliamentary seat by Paul Baumgarten, 1961–1972.

'Capital of the GDR': Berlin in the German Democratic Republic, 1949–1990

After the foundation of the GDR, plans for the capital of the new republic were at first pursued according to Scharoun's planning ideas. While the western section of the city was characterized by a rather conservative traffic plan based on the existing urban pattern, the East revelled in a radically *avant-garde* approach to urban planning. This context would soon undergo a complete about face, however. Under Moscow's influence, the new order of the day was to design a monumental image for the capital that would demonstrate the victorious achievement of Socialism, modelled on the 1935 general plan for Moscow designed under Josef Stalin. The new uncompromising architectonic lines emerged over the course of the ensuing years: the East presented a conservative national image symbolized in Stalinallee, while the West responded with a modernist International Style epitomized in the Hansaviertel.[24]

At the *Sozialistische Einheitspartei Deutschlands – SED* (United Socialist Party of Germany) party conference in July 1950, state and party leader Walter Ulbricht outlined the political demands that the capital must answer to: 'The centre of the city should take on a characteristic image by means of monumental buildings and an architectonic composition, which does justice to the significance of the capital'.[25] In order to achieve this goal, a delegation of architects had been dispatched to Moscow in April of the same year to familiarize themselves with the new Moscow urban planning in the style of Socialist Realism. On 27 July 1950, the GDR government ratified the *Sechzehn Grundsätze des Städtebaus* (Sixteen Principles of Urban Design), which stipulated among other points that:

The centre of the city is the political focal point for the life of its population . . . The political demonstrations, parades and festivities marking public holidays take place on the squares in the city centre.[26]

In keeping with these guidelines, the planning commission, which was by then coordinated on the state level and administered by the *Ministerium für Aufbau* (Ministry for Development), envisioned a *Zentrale Achse* (Central Axis) from Stalinallee to the Brandenburg Gate, a *Zentraler Platz* (Central Square) that would replace the palace and a high-rise *Zentrales Gebäude* (Central Building) that would house government functions. The plan moved to implementation without delay: the palace was demolished as early as September 1950 to create space for the desired mass demonstrations. In 1951, the Stalinallee competition was held to accelerate the development of the central axis.[27] Egon Hartmann was awarded first prize for his proposal, which was subsequently reworked by several collectives prior to execution.

As the development of Stalinallee progressed, however, the young state lacked any means of realizing the central government building. Moreover, the GDR also saw a change in building policy in the wake of Stalin's death in 1953: economic constraints, especially, made an industrialized approach to building virtually inevitable. In direct response to the western *Hauptstadt Berlin* (Capital City Berlin) competition, the government of the GDR and the municipality of East Berlin organized the competition *zur sozialistischen Umgestaltung des Zentrums der Hauptstadt der DDR, Berlin* (for the Socialist Redesign of the Centre of the Capital of the GDR, Berlin) in 1958. The brief constituted a manifesto of the political demands: 'The economic, political and cultural life that arises from Socialism must be expressed in the urban plan for the centre of the capital'.[28] Yet the task of creating just such an expression of socialist life beyond Stalinist historicism and Western internationalism proved almost beyond solution for the architects. The result was thus rather unsatisfactory and no first prize was awarded. Second prize was granted to the collective led by Gerhard Kröber who had designed a high-rise which could also be understood as just another capitalistic office tower.[29] Beyond the scope of the competition, two other proposals were important. The first was put forth by functionary Gerhard Kosel, who suggested that half of old Berlin should be flooded to emphasize the impact of his monumental high-rise (see figure 14.5).[30] The other was a design by architect Hermann Henselmann, whose concept of a *Turm der Signale* (Tower of Signals) did not comply with the competition guidelines but ultimately provided the decisive impulse for the realized project.[31]

The central government high-rise, which had been planned since the foundation of the state, was in the end replaced by three buildings: in 1962–1964 the *Staatsratsgebäude* (Council of State Building) on Marx-Engels-Platz by Roland Korn and Hans-Erich Bogatzky as seat of the government, incorporating the remnants of the Eosander portal from the demolished palace; in 1965–1969 the *Fernsehturm* (Broadcast Tower) by Werner Ahrendt, Fritz Dieter, Günter Franke and Hermann Henselmann as a socialist claim to the tallest structure; and finally in 1973–1976 the *Palast der Republik* (Palace of the Republic) by Heinz Graffunder as a multifunctional building, which housed, among others, the parliament of the GDR, the *Volkskammer* (People's Chamber). This ensemble fitted perfectly into the framework of international utilitarian architecture: it could also be read as a collection of a congress centre, a corporation's headquarters and a broadcast tower from any Western city. Although the GDR was finally able to realize its socialist centre over the course of thirty years, despite crippling financial difficulties, it failed to develop a specific representational form that was significantly distinct from that of its class enemy.

'Capital of the Federal Republic of Germany': Berlin in the Federal Republic of Germany after Reunification, 1990–2000

The opportunity for a comprehensive urban plan arose only after the fall of the Berlin Wall on 9 November 1989 and the accession of the GDR into the Federal Republic of Germany on 3 October 1990. Following a lengthy debate,

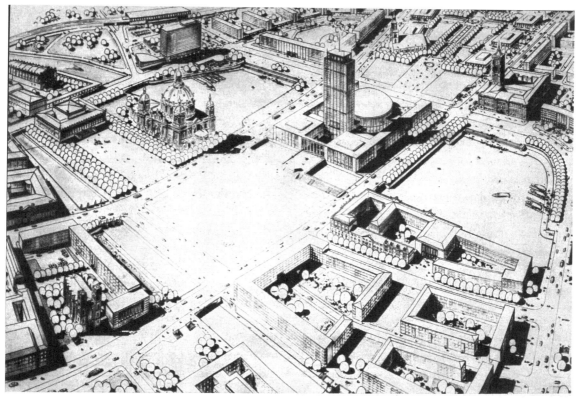

Figure 14.5. Gerhard Kosel, *Zentrales Regierungsgebäude* (Central Government Building), 1960. A large public square serves for state demonstrations while a skyscraper claims the victory of Socialism.

the German Bundestag finally agreed on 20 June 1991 to ratify the non-party motion for the *Vollendung der Einheit Deutschlands* (Completion of the Unity of Germany) and to declare Berlin as the country's future capital.[32] The necessity for a renewed capital city plan for Berlin was thus given. At the communal level, too, Berlin once again became a unified political body, forming an enlarged Bundesland Berlin with a *Regierender Bürgermeister* (Governing Mayor) and Senate as government.

The urban planning stages that followed reunification were accompanied by a broad public debate. The Berlin Senate entered into the debate primarily through a series of events called *Stadtforum* (City Forum) chaired by *Senatsverwaltung für Stadtentwicklung und Umweltschutz* (Senate Department of Urban Planning and Environmental Protection) and a series of reports entitled *Städtebau und Architektur* (Urban Design and Architecture) published by *Senatsverwaltung für Bau- und Wohnungswesen* (Senate Department for Building and Housing). In addition to seeking widespread public support, the goal was also to achieve a balance of federal and municipal interests for the capital city plan. The two governments (Federal and city) signed the contract *zum Ausbau Berlins als Hauptstadt der Bundesrepublik Deutschland* (for the development of Berlin as capital city of the Federal Republic of Germany) in 1992.[33]

In 1992 the Federal government held international competitions for the reconstruction of the Reichstag and the development of the government quarter at the Spreebogen, which had been discussed as a potential site for governmental purposes since the beginning of the twentieth century. Norman Foster finally renovated the Reichstag with a cupola in glass following the client's wishes – ecologically and politically correct by using daylight and offering access for visitors.[34] Thus the new Reichstag corresponds to the consensual idea of the Federal Republic on democratic building: transparent, open to citizens and casting technical innovation into a conventional formal mould.

Similar ideas also informed the *Internationaler Städtebaulicher Ideenwettbewerb Spreebogen* (International Competition on Ideas for Urban Design at the Spreebogen), which called for buildings for the *Deutscher Bundestag* (Federal Parliament offices), *Bundeskanzleramt* (Federal Chancellery), *Bundespressekonferenz* (Federal Press Conference), *Deutscher Presseclub* (German Press Club), *Deutsche Parlamentarische Gesellschaft* (German Parliamentary Association) and *Bundesrat* (Federal Council). Rita Süssmuth, the president of the Federal Parliament, had outlined the political expectations in the competition brief as follows:

The German Bundestag wants to meet the demand for a transparent and efficient parliament which is close to its citizenry. It is open to the outside and is conceived as a place of integration and as a centre and workshop of our democracy.[35]

Figure 14.6. Axel Schultes, *Band des Bundes* (Federal Ribbon), competition Spreebogen, 1993. The legislative and the executive branches are integrated into an east-west oriented structure which aims to symbolize the reunification.

Axel Schultes and Charlotte Frank's winning design found widespread public approval with the easily understandable figure of a *Band des Bundes* (Federal Ribbon) (see figure 14.6). The architects had developed a spatial arrangement which reflected the separation of powers by arranging the executive and the legislative around a 'Federal Forum'.[36]

The media were virtually unanimous in interpreting this linear city with an east-west orientation as a unifying tie between the two state and city halves. Historically, it is in the tradition of 1960s megastructures – for example, the fantastic-sculptural expression of the *Monumento continuo* by Superstudio from 1969, an abstract linear structure running through existing cities. Also the new building for the *Bundeskanzleramt* (Federal Chancellery) by Axel Schultes – chosen in 1995 by Chancellor Helmut Kohl after a competition – follows a monumentality inspired by 1960s Modernism, especially Louis Kahn.[37] From this perspective, the new government district of Berlin can also be understood as a long overdue victory of an *avant-garde* urban planning movement from the past, which was otherwise vehemently rejected in the critical reconstruction of Berlin's historic building blocks for the residential and commercial portions of the city.

Also for the ministries, the Federal government had first planned sweeping demolitions and new construction projects. One of several efforts in this direction was the *Internationaler Städtebaulicher Ideenwettbewerb Spreeinsel* (International Urban Planning Ideas Competition for the Spree Island), held in 1993–1994, which envisioned the creation of a second, central site for state representation in the city that would contain the *Auswärtiges Amt* (Foreign Affairs Office), the *Bundesministerium des Innern* (Ministry of the Interior) and the *Bundesministerium der Justiz* (Ministry of Justice).[38] From 1994 onwards, however, the new Federal Minister for Planning, Building and Urban Design, Klaus Töpfer, began to house ministries in existing buildings for financial and ecological reasons.[39] While this perpetuated the historically evolved disorganized distribution of state buildings across the city, it also anchored Berlin's varied and contradictory architectural history firmly in the present time: the *Bundestag* (Federal Parliament) holds its sessions in the former Reichstag, the *Bundesrat* (Federal Council) meets in the former *Herrenhaus des Preussischen Landtags* (Lord's Chamber of the Prussian Parliament), and some ministries are housed in former National Socialist buildings used also by the GDR for government purposes.

However one may dispute the political significance of Berlin's new government architecture, the capital city planning after 1990 has been an extraordinary success both in terms of organization and in terms of finances. Given a total cost for new buildings and infrastructure during the first post-reunification decade in Berlin of roughly 200 billion DM,[40] the government buildings stayed within 6 billion DM of the estimate.[41] This was achieved thanks to effective organization in addition to the cost-efficient renovation policy. The new capital was politically organized and realized with some Federal Commissions, the *Bundesbaugesellschaft Berlin mbH* (Federal Building Society for Berlin Ltd.), and finally a *Beauftragter der Bundesregierung für den Berlin-Umzug und den Bonn-Ausgleich* (Commissioner of the Federal Government for the Move to Berlin and the Bonn Compensation) within the *Bundesministium für Raumordnung, Bauwesen und Städtebau* (Federal Ministry for Planning, Building and Urban Design). Thanks to secured state financing, solid organization, non-party politics and an international level of urban planning, the seat of government was officially moved from Bonn to Berlin on 1 September 1999. A century of highly controversial capital city planning for Berlin as capital city had thus reached what one can safely assume to be a long-term conclusion.

Conclusion

Berlin's chequered planning history offers numerous insights into the factors that lead to the success or failure of capital city planning. The lack of political continuity was perhaps the most decisive factor contributing to the failure of countless plans: during the first half of the century, changes to the political system roughly every fifteen years made the implementation of far-reaching urban plans impossible. It was not until a certain continuity manifested itself during the second half of the century – the existence of the GDR for forty years and the stability of the Federal Republic of Germany since 1949 – that successful capital city planning came within reach. Broad political consensus was a decisive factor for state continuity: in the Kaiser era and in the Weimar Republic, political antagonism had stifled any realistic planning; dictatorial homogeneity in the Third Reich and in the GDR ultimately led to the downfall of both systems. Conversely, the democratic and federalist Federal Republic finally enabled the erection of a significant government district.

Whether the plans were implemented with dictatorial authority or by means of democratic processes of consensus building, turns out to be of secondary importance: Speer's plans would undoubtedly have been carried out if the Third Reich had continued to exist; yet the consensual approach adopted by the Federal Republic after 1990 was at the very least equally successful. Strong administrative authority in combination with the political stability of a democracy seems to be the most effective means. Nor does the process require the figure of a single political champion. Although Hitler invested tremendous personal effort in promoting the capital city planning, it was ultimately unsuccessful, while the successful planning after 1990 was executed without a single political leadership figure.

Political stability alone, however, will not achieve the creation of a representative capital city if financing has not been secured. Thus the failure of the central urban plan for the capital of the GDR, that is, the mediocrity of its execution, was less a question of a lack of political will or its being the victim of a political change, than the result of a lack of economic resources. By the same token, abundant financial means alone are equally insufficient when they are not supported by political stability, as demonstrated by the National Socialist planning debacles. Only political continuity and economic prosperity together allow for the implementation of comprehensive capital city plans, such as that realized by the Federal Republic after 1990. Successful implementation of meaningful capital city planning is impossible unless the state takes direct responsibility for funding its government buildings: precisely this level of financial support from the state was lacking in the Kaiser Era and in the Weimar Republic.

State organization has always been a part of realizing capital city plans. Initiatives by the city alone – usually inspired by architects, such as the Greater Berlin competition in 1910, the urban plans from the 1920s, the post-war plans or the *Internationale Bauausstellung – IBA* (International Building Exhibition) 1984–1987 – were able to set important urban development efforts in motion, but neither could nor wanted to promote plans specifically targeted at developing a capital city. A consensus founded in an exchange of interests between state and city has proved to be especially effective, the individual planning issues clearly divided among state and municipal institutions – accompanied by a marked will to co-operate – as was the case after reunification in 1990. However, a dictatorial imposition of state will on the city, such as it was practised during National Socialism or in the GDR, may well have been equally successful, if it had been matched by the relevant political continuity and financial means.

All urban designs for Berlin in the twentieth century were on a distinguished international level, even those that were marked by nationalistic and separatist overtones. The urbanistic quality of the designs for the 1910 competition and from the 1920s is uncontested both by their contemporaries and by historians. And even the plans developed during the Third Reich and in the GDR are by no means out of step on the international stage; their partial urban qualities cannot be eliminated by references to the political errors or crimes of their initiators. The plans pursued by the Federal Republic since 1949, conversely, deliberately sought international comparison, no matter how different the urban philosophies from 1958 and 1990 may have been. However, planning alone, no matter how good, is not a sufficient condition for the successful creation of a capital city, as is demonstrated only too strikingly in Berlin's history throughout the twentieth century.

There are further conclusions one can draw from the history of capital city planning for Berlin. On the one hand, there is no genuine and homogenous Berlin tradition. Nearly every design has its contemporary international counterparts. On the other hand, there are no urban forms of representation that are specific to a particular political system: nearly all systems resorted to the use of axes to render their central political institutions experiential; nearly all systems made use of the forum typology to set the stage for a real or imaginary communal spirit. Scale, height and centrality also invariably played a decisive role in the design of state buildings, regardless of the political values they were intended to represent. This is even true for the asymmetrical and anti-axial projects from the 1950s, whose partial rejection of state representation is an historically understandable exception. The frequently controversial political messages were often expressed in stylistic differences in the architecture rather than different approaches to urban planning.

The key factors to successful capital city planning are all the more evident in Berlin's planning history for the many failures they contrast. Effective implementation requires political continuity with a consensual will to political co-operation, secure state financing for government buildings and the infrastructure they require, state organization of the implementation phases that include municipal authorities and finally an urban planning philosophy that aspires to an international level and that is willing and able to render the importance of state institutions visible and experiential.

NOTES

1. Herzfeld (1952); Taylor (1985); Cullen (1995); Hoffmann (2000).

2. Cf. Stadtgeschichtliches Museum Leipzig (1995).

3. Nowack (1953). For the general planning history of Berlin cf. Hegemann (1930); Schinz (1964); Werner (1976); Lampugnani (1986); Kleihues, (1987); Brunn and Reulecke (1992); Schneider and Wang (1998); Kahlfeldt, Kleihues, and Scheer (2000); Jager (2005).

4. Vereinigung Berliner Architekten and Architektenverein zu Berlin (1907).

5. Anon. (1908); Cf. Posener (1979); Konter (1995); Bernhardt (1998); Sonne (2000); Sonne (2003).

6. Anon. (1911), p. 25; Cf. Anon. (1910); Hofmann (1910).

7. Cf. Nowack (1953); Escher (1985).

8. Cf. Scarpa (1986); Hüter (1988).

9. Mächler (1920); Kohtz (1920).

10. Häring (1926); also in: Häring and Mendelsohn (1929). Cf. Nerdinger (1998), pp. 87–99; Schirren (2001).

11. Berg (1927), p. 47.

12. Häring and Wagner (1929), p. 69. Cf. Wagner (1929); Anon. (1930).

13. Häring and Wagner (1929), p. 72.

14. Letter dated June 25, 1940; in: Reichhardt and Schäche (1985), p. 32; Cf. Miller Lane (1968); Larsson (1978); Schäche (1991).

15. Reichhardt and Schäche (1985), p. 37.

16. Speer (1939), p. 3. Cf. Stephan (1939).

17. Speer (1970), pp. 134–35; quoted from: Helmer (1985), p. 39.

18. Schönberger (1981), p. 184.

19. Anon. (1946), pp. 3–6. Other post-war plans are: Moest (1947); Bonatz (1947). Cf. Kleihues (1987); Durth (1989); Kahlfeldt, Kleihues and Scheer (2000).

20. Schwippert (1951); Arndt (1961). Cf. Bundesministerium für Raumordnung (1989); Flagge, Ingeborg and Stock (1992).

21. Berlinische Galerie (1990).

22. Bundesminister für Wohnungsbau Bonn und Senator für Bau- und Wohnungswesen Berlin (1957), p. 8.

23. Bundesminister für Wohnungsbau Bonn und Senator für Bau- und Wohnungswesen Berlin (1960), p. 29.

24. Nicolaus and Obeth (1997); Dolff-Bonekämper (1999).

25. Ulbricht (1950), p. 125.

26. Düwel (1998), pp. 163–187.

27. Magistrat von Berlin (1951).

28. Anon. (1958).

29. Anon. (1960)

30. Kosel (1958); Cf. Verner (1960).

31. Cf. Düwel (1998); Hain (1998); Müller (1999); Müller (2005).

32. Deutscher Bundestag (1991).

33. Senatsverwaltung für Bau- und Wohnungswesen, Berlin (1996), pp. 90–91; Cf. Senatsverwaltung für Bau- und Wohnungswesen Berlin (1992); Senatsverwaltung für Bau- und Wohnungswesen Berlin (1993); Wise (1998); Welch Guerra (1999).

34. Cf. Wefing (1999).

35. Zwoch (1993), p. 6.

36. *Ibid*, p. 46.

37. Burg and Redecke (1995); Cf. Wefing (2001).

38. Zwoch (1994).

39. Cf. Bundesministerium für Verkehr, Bau- und Wohnungswesen (2000); Wagner (2001).

40. Stimmann (1999), Berlin nach der Wende. Experimente mit der Tradition des europäischen Städtebaus, in Süß and Rytlewski (1999), p. 558.

41. Bundesministerium für Verkehr, Bau- und Wohnungswesen (2000), p. 6.

ACKNOWLEDGEMENT

I would like to thank the Society of Architectural Historians of Great Britain for awarding a Dorothy Stroud Bursary, supporting the publication of this chapter.

Chapter 15

Rome: Where Great Events Not Regular Planning Bring Development

Giorgio Piccinato

In 1870, after a bloody battle against the Pope's troops, Rome became the effective capital of the Kingdom of Italy, a State constituted in opposition to the temporal power of the Church.[1] With a population of 200,000 (less than Milan, Genoa, Palermo or Naples), the city was a small, compact nucleus, surrounded by fields, pastures, gardens, and interspersed with villas, convents and churches, all contained by the Aurelian Wall. The debate over the city's future began immediately. The intertwinement of powers and responsibilities between the central government and the local administration produced procrastination about important decisions on the urban strategies to be adopted.[2] While the Roman citizens, the aristocracy and the clergy banded together in a hostile stance against the new masters, the city was populated by public servants and employees of the new government, immigrants from the countryside in search of labour, professionals and entrepreneurs attracted by the prospect of a new development. During Mayor Pianciani's mandate,[3] the first regulatory plan (*piano regolatore* or official plan) for the capital was designed by the Chief Engineer Alessandro Viviani.[4] It was presented to council in 1873. However, it was the central government that managed the city, dealing directly with the private sector on the location of the new seat of government, buildings and offices, while the municipal council remained divided on the fate of the city and its political economy. At that time everyone – aristocrats, merchants, and dignitaries from the Roman Church – was engaged in land speculation. Even the Mayor lamented the fact that buy-and-sell activities were more widespread than building activities, exacerbating the housing shortage.[5]

The 1873 plan limited itself to the ratification of ongoing real estate operations,[6] such as the new Via Nazionale,[7] linking the railway station to the city centre, Piazza Indipendenza and Piazza Vittorio Emanuele. The plan appeared to follow the ideas of Quintino Sella,[8] a proponent of new expansion in the eastern sector. Several ministries settled in this area following the lead of the Ministry of Finance. A sector in the south was devoted to warehouses and industries, which had already starting settling there during the last years of the pontifical reign. The text accompanying the plan mostly dwells upon the street network and public hygiene, to justify the expected interventions in the historic centre. During the debate at city council, opposition to significant industrial development and working-class neighbourhoods became clear, for fear of recreating the social conflicts already experienced in other European capitals.[9] The plan, foreseen to last twenty-five years, was never officially adopted.

At the beginning of the 1880s, the national government considered contributing a long-term loan to build the public works infrastructure necessary for a functioning capital city, should there be an approved regulatory plan.[10] For this reason, in 1883, the so-called second Viviani plan was finally adopted. It was not very different from the first, but it did provide for a system of important public facilities and thoroughfares to connect the centre of the historic town to the new urban expansions (see figure 15.1).[11] Suddenly the construction of the capital appeared as a grand enterprise; money and businesses from northern Italy and Europe (France, Germany, England) poured into the city in a quantity that the small-scale Roman developers would not otherwise have been capable of producing. During this time, the government built seven bridges, the Polyclinic, the Central Court and the grand monument to Vittorio Emanuele II, the first king of Italy. However, the municipality was always on the verge of financial disaster for it was not capable of collecting tax on buildable land (with which the plan was to be financed) and was therefore forced to sign *convenzioni* (covenants) with the private sector for the necessary developments.

The 1880s, thanks to the intervention of the central government, will be remembered as the years of 'building fever'. Construction happened everywhere, both within and outside the plan; the grand patrician villas disappeared, hastily converted into rental housing by the nobility. The fever did not last long, but it was quite

Figure 15.1. The first Master Plan of Rome, 1873, which was revised and approved only in 1883.

intense. In 1886–1887, 12,000 rooms were built, in 1888–1889 only 800.[12] With no substantial economic and demographic growth, apartments remained vacant, investors were discouraged, projects started on the basis of bad loans were abandoned, numerous banks closed their doors, and tens of thousands of construction workers returned to their region of origin.

Modern City Management: The Nathan Administration

In the complex context of the first decades after Unification (1870), Rome's urban development was a result of conflicting forces from all quarters, often acting at different levels. The idea that the regulatory plan should synthesize collective needs and objectives seemed fainter than ever. There was a dearth of good ideas, yet there was no agreement on the functions to be assumed by the city – other than the representative role of a capital. The city also suffered from being subject to a double administration, the municipality and the national government, autonomously pursuing their own objectives. This resulted in short-sighted politics, where decisions were commonly inconsistent and disjointed. For instance, the national government had begun the construction of a new Rome, separate from the old, in the east, with the creation of an 'axis of the ministries' (the old Strada Pia renamed Via XX Settembre), while *extra muros* expansion gained approval. Grand villas were destroyed in the inner city. An annular (ring) road network was called for, but only radial roads were created. Land was reserved for low-density development but higher densities sprang up as soon as building permits were issued.

The recovery came only at the start of the twentieth century, in a different political context. The Prime Minister was Giovanni Giolitti, who until the start of World War I led the transformation of Italy from a conservative, agricultural country to an industrial society, through a social pact between the working class and the urban middle class. It was in this climate that the *Blocco popolare* (Popular Coalition) was formed, uniting radicals, republicans and socialists, and which in 1907 gained control of the municipality. The new mayor was Ernesto Nathan, a Jewish, Mazzinist freemason – the only remarkable figure in the history of modern Roman administration.[13] He stayed in power for seven years. Nathan tackled the areas of elementary education, public hygiene, democratization and the fight against land speculation. He municipalized the transit system and the power grid. He assigned to Milan's Chief Civil Engineer, Edmondo Sanjust di Teulada,[14] the task of developing a new regulatory plan. In 1909, Rome finally had an accurate regulatory plan, based on updated cartographic information (see figure 15.2). Sanjust presented his plan as an urbanistic instrument resting on a set of legal and financial provisions. The plan also defined a building typology – multi-storey buildings, houses, 'gardens',[15] in order of decreasing density – which provoked the landowners of the two latter categories, who joined the ranks opposing the *Blocco Nathan*. In 1914, allied to the new nationalist right, the old aristocracy regained power in the municipality. During Nathan's administration and after, the Roman Institute for Public Housing (IACP: *Istituto Autonomo Case Popolari*) flourished and kept growing. In 1930, its achievements came to represent more than 10 per cent of the total housing stock in the city. Thanks to the work of excellent architects, the Institute remained a locus of Roman architecture for a long time. The IACP strove to create buildings that blended social housing into the existing urban fabric; and on its initiative two garden cities, Garbatella and Aniene, were created in the 1920s.[16]

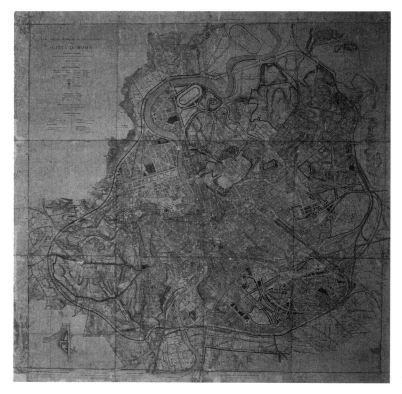

Figure 15.2. Master Plan of 1908–1909, by Edmondo Sanjust di Teulada.

The history of Roman urbanism is not just that of its official plans. It is also that of spatial transformations brought about by important events, and of ideas which arose independently from the plans. The 1911 Great National Exhibition is a good example. Organized to celebrate fifty years of Italian unity, the exhibition was an opportunity to affirm the role of Rome as capital of a great nation: a city where scientific congresses and art exhibitions would take place, where the idiosyncrasies of the regions would join together and where visitors from all parts of the country would meet. Sumptuous pavilions and model dwellings were built over a vast area adjacent to the Villa Borghese (recently acquired by the municipality to be turned into a public park) and crossing the Tiber on a new iron bridge. Buildings, the bridge and the new National Gallery of Modern Art were to remain after the exhibition ended. The area surrounding the Gallery was destined for the Foreign Academies, while the other side of the Tiber would host a new neighbourhood with projects by Edmondo Sanjust and Josef Stübben, among others.[17]

Meanwhile, a grand project was under way, across all the official regulatory plans: the archaeological research in the area extending south from the Campidoglio. That area was little urbanized in mediaeval and papal times, and was systematically excavated after Unification: the Republican Forum, the Palatine Hill, the Circus Maximus, the Imperial Forums, the Basilica of Massenzio, the Temple of Venus, the Markets of Trajan, and the Baths of Caracalla. The project continued during the Fascist era, in the shape of a green wedge, a triangle with Piazza Venezia opening along Via Appia towards the Colli Albani. In the 1970s, the Via Appia Antica

National Park was created, extending from Saint Sebastian Gate to the external ring road. The result is an archaeological precinct that traverses the entire southern quadrant of the city to end up at what is still considered the historic centre.

Reviving the Imperial City: Fascist Rome, 1922–1943

Two themes characterize Fascist planning: rural policy and the construction of Rome. The Mussolini decades were not a period of major economic growth, nevertheless, the country was largely spared the dramatic 1929 Depression. Italy reinforced its industrial structure thanks to significant public interventions in several basic economic sectors, and instituted a system of social services. Cities continued to grow, worsening the problems of local administrations and encouraging ways of life little appreciated by the regime. The Fascist campaign against urbanization started in 1928 with an article signed by Mussolini himself entitled 'Evacuating Cities'.[18] The campaign reached its peak in 1939 with a law forbidding people to migrate to towns of 25,000 or more, under the threat of losing their work permit.

Rome was the sole exception to this law. In Rome growth was accepted, encouraged and promoted. The city's grandeur had to reflect the 'greatness' of Fascism. This did not only refer to architectural splendour, but also to the size of the population, which had to be comparable to other European capitals. 'Romanity' and 'Latinity', already a part of Italian rhetoric, became the myth around which all the Fascist propaganda developed. In this respect, reorganizing urban spaces and building a new grandiose capital became the dream that only Fascism was capable of creating: the 'return' of Rome, with visible continuity between the past and the modern city. Also, the architects attempted to harmonize the new with the ancient, and looked to the stylistic paradigms of ancient Rome. Archaeologists, for the first time in Roman history, became advisers, designers and executors of urban programmes.[19] Even the most extreme options – fortunately realized only in part – of demolition and remodelling were put into practice, wherever one could 'liberate' or emphasize the signs of Roman tradition. The work of archaeologists was so appreciated by the regime that a new grand avenue was constructed in the Forum area, linking the Colosseum to Piazza Venezia. The inhabitants of the demolished central quarters were displaced to new towns outside the city, ironically designed along Modernist principles.[20]

Rome changed rapidly during Fascism (1922–1943). The city's influence on the country strengthened, with a strongly hierarchical and centralized government, and with increased emphasis on the economy and on social services. Industries also grew: construction, engineering, chemicals production, and telecommunications. The population doubled in twenty years. In an attempt to solve traditional conflicts between the local authority and central government, the position of Governor, nominated by the head of the national government and reporting directly to him, was created. Several organizations for social housing and co-operatives joined the IACP in building the city for both public servants and rulers. Bridges, hospitals and ministries were built, parks were opened to the public, and the banks of the Tiber were developed. Everything happened without an effective plan, but still within a coherent political design. In this respect, architecture and urbanism played an important role. The search for a Fascist 'style' saw a permanent squabble between the innovators

– following the European *avant-garde* – and the traditionalists, who were commonly inspired by classicist eclecticism. An already established architect, Marcello Piacentini,[21] emerged as one of the most faithful agents of the new ideology. In those years at least three outstanding projects were realized: Mussolini's forum, a great sports complex, notable for its landscaping and architecture in the northern sector; the new university campus; and the E42 world exhibition district.

A new Master Plan in 1931 (see figure 15.3) was not particularly innovative: the historic centre was subject to demolition largely already under way, while the expansion zones were characterized by high building densities, thus abandoning the garden city experiments of the 1920s. Obviously the 'new Rome' would be built over the old, demolishing, isolating and slashing into the fabric and the monuments.[22] For the first time, however, the plan proposed an innovative mode of execution, so much so that it served as a model, ten years later, for the drafting of the new planning legislation.

The 1931 plan was contradicted a few years later, when the old idea of expanding the city towards the sea resurfaced during discussions on the location of the upcoming Esposizione Universale 1942 (or E42, and later EUR).[23] An area of 436 hectares was expropriated a few kilometres south of the Aurelian Wall for the exhibition. It was intended that the site would house a true permanent town and would be a focus for the capital's future growth. Every new public facility was commissioned under competition, thus obtaining projects of great interest after the exhibition ended. The new quarter was designed according to a system of orthogonal axes, dotted with marble public buildings arranged symmetrically (see figure 15.4). The precisely drawn streets were to be wide and tree-lined, villas and gardens were to be developed around the nucleus (see figure 15.5), while an artificial lake would increase its attractiveness. The start of World War II interrupted the works leaving many buildings incomplete. Ten years passed before development of the area would continue.[24]

Growth and Land Speculation, 1945–1962

In 1945, Rome had one and a half million

Figure 15.3. Piacentini's Master Plan, 1931.

inhabitants. A large number of refugees arrived from war-ravaged areas all over the country. The city became the focus of various migrations affecting Italy: south to north, countryside to city, and from small towns to bigger ones. Each year added 50,000 to 60,000 immigrants, who required lodging, and were forced to adapt to cohabitation, and sometimes living in barracks, caves, or public monuments. The housing problem was dire – and stayed so for at least 20 years. The lack of services such as schools, hospital beds and public transport was also a problem. The largely uncontrolled expansion in all sectors, sustained by constant demand, brought buildable land to exceptionally high price levels, not seen in any other European country.[25] The city administration – having lost the position of Governatorate – recognized the need to bring services to where they were most needed, but was rendered impotent by a growing budgetary deficit and did

Figure 15.4. A model of the E42 exhibition site, later known as the EUR.

Figure 15.5. An original sketch of the main piazza of E42 (which was built in a similar manner).

not succeed in managing the urban expansion. The building boom of the first twenty years was strictly conditioned by land speculation: plans for private subdivisions became practically the only development control tool used. Expelled from the most lucrative locations by public and private office developments, inhabitants of central areas moved to more recently developed subdivisions. The housing problem, the inadequate transportation facilities and the scarcity of public services explained why urban planning remained one of the most debated topics in city council and in the press. Once again, speculators and builders were accused of favouring an inefficient and easily corruptible municipal administration.[26] But, above all, two opposing political strategies became apparent: the Right against the plan and the Left in favour of it.

The usefulness of the 1931 plan ran its course, and by 1950 the need for a new urban planning tool was generally accepted. This started a long process that finally concluded more than a decade later with the adoption of a new regulatory plan. A first sketch was presented in 1955 by a Technical Drafting Committee (*Comitato di Elaborazione Tecnica* – CET). The plan's fundamental idea was the displacement of the central functions to the eastern quadrant of the city, along a main road axis, later called the Eastern Directional System (*Sistema Direzionale Orientale* – SDO) where new public and private office buildings were supposed to locate. It also directly connected to the highway linking northern Italy to the south. By directing growth to the east, the centre would be relieved from through-traffic and, at the same time, there would be an alternative to the concentric growth of land values. This provoked opposition on the part of those who had an interest in the parallel development of the west side.

In the meantime the 1960 Olympic Games affected the city's development. Thanks to substantial national investment, a series of crucial infrastructure works were completed: the international airport (which is in the adjacent municipality of Fiumicino); the Olympic Avenue which feeds the city in the west, connecting the old and the new sport facilities of the Foro Italico (the former Foro Mussolini) to the EUR (Esposizione Universale di Roma, formerly E42), which got a new Palace of the Sports, a velodrome, a swimming pool, athletics facilities, and the passages along the Lungotevere and the Muro Torto. This gave a boost to the development of the western sector, totally contradicting what was suggested in the municipal planning offices, and triggering vehement controversies in the press. After a decade of debate, a revised plan was eventually adopted by city council in 1962. At that time the city had 2.2 million inhabitants and the plan foresaw a population, twenty-five years hence, of 4.5 million to be located in self-sufficient neighbourhoods.[27]

Planning Paradoxes: Illegal Developments, Legal Failures and New Hopes

The year the plan was adopted saw a record number of new homes built. Unfortunately, that was exactly when the economic cycle saw a downturn; many new dwellings remained unsold due to the high prices needed to cover the cost of land. The plan was accused of causing the crisis because of the numerous environmental and functional limitations it imposed. In reality, the market was unable to respond to the demand for housing for lower-income people. This caused informal settlement to develop, initially without streets or services, on land designated for agriculture or plainly off limits for environmental reasons. Cheap land – outside the plan – was the first to be developed by 'self-build'; an informal building industry emerged, as well as an underground real estate market.

In 1976, after thirty years of uninterrupted centre-right government, the left gained power in the municipality, and kept it until 1985. It was estimated that, at that time, the population of such settlements reached 800,000, which was one-third of the total population. The legalization of these irregular neighbourhoods – official registration, provision of infrastructure and social services – was a major task of successive left-wing administrations from 1976 to 1985.[28] In the meantime measures were taken to safeguard historic parks – Villa Doria Pamphili, Villa Ada, Castelfusano and Castelporziano, Veio and the entire Via Appia Antica area, while a General Amendment established the ways and phases of future development. Work continued around one of the main features of the plan, the new eastern business centre, but the project was regularly scaled back. The relocation of ministries that was supposed to make the centre take off was constantly postponed. The conversion of the historic centre and nineteenth-century buildings into offices, and the space in the well-managed EUR was enough to satisfy the growing demand for office space.

The EUR, an example of planning outside the plan, was one of the few success stories of Roman post-war urban planning. Its status as an independent agency and landlord allowed its head official, Virgilio Testa, to set up a development policy for offices and up-market residences, which proved timely and economically viable. Until recently, when it became part of the City of Rome, the EUR provided independent police services, maintenance of public spaces and, above all, it kept strict control over the quality of projected buildings. This environment attracted several ministries (Finance, Postal Services, Merchant Marine), and the headquarters of some big corporations (Esso, Telecom, Alitalia), which confirmed it as the only appropriate location for executive activities. Two other factors contributed to the EUR's success: the opening of the first subway line, linking it directly to the Stazione Termini, the city's main railway station, and the construction of remarkable sports facilities for the 1960 Olympics.

The city had profoundly changed in 1993, when the Left returned to power. The old plan proved useless: the city, which was forecast to reach 4.5 million inhabitants, never passed 2.8 million, and kept losing population to adjacent municipalities. Only one of four subway lines had been completed, and private car use exploded turning urban traffic into a daily nightmare. Once again, the preparation of a new plan was slow. In 1997, a scheme was adopted, with the intent of excluding the non-urban areas from development. It was the first step towards a future plan.

The Holy Year of 2000 was dreaded by many for the huge influx of tourists it would entail – a phenomenon less and less accepted by the Romans. Even though the new plan (see figure 15.6) emphasized rail transit, part of the national funding went to the rehabilitation of road infrastructure, and part of it was used to restore and maintain historic buildings of varying value. In December 2001 the new plan, designed for the medium-term (fifteen years) with a stabilized population of approximately 2.5 million, was presented to the mayor,[29] and was eventually adopted by the city council in March 2003.

In the absence of a strategic plan setting social and economic objectives, the 2003 regulatory plan explicitly limits itself to defining a functional organization of the city adapted to its various urban fabrics, and looks for public-private partnerships. Great emphasis is placed on rail

Figure 15.6. The synthetic image of the latest version of the Master Plan (2003) shows greater emphasis on green areas and infrastructure, as well as the increase of the areas considered historic.

transport (regional and metropolitan) which would ensure that 50 per cent of the inhabitants would be within a 500 metre radius of a station – modest objectives, if one compares them to other European capitals, but very innovative by Roman standards. The already substantial stock of parks, public gardens and natural and agricultural areas is somewhat increased, bringing the area of open space to 21 m² per inhabitant and making Rome one of the greenest European cities. On the whole, the plan is intended to serve as a flexible policy instrument that must be managed openly and efficiently in the years to come.

Final Observations

Let us try to draw some conclusions. Rome is obviously an exception insofar as it was the capital city of a long standing Roman Empire and subsequently of the Vatican, itself for long time a powerful state and a world cultural centre. Unlike the situation in other countries, Rome's role as a capital city was never under discussion. A city that always had a municipal government – even, in various forms, under the Pope's rule – had to compete, at the beginning, with the national government on decisions such as the location and the typology of basic capital facilities. Since the break with the Italian government in 1870 – that resulted in the excommunication of its members – the Church had very little say in the development of the new capital city. The new government found no shortage of prestigious buildings in which to house its high officials. The King took one of the Pope's main residences – Palazzo del Quirinale – (a masterpiece of Renaissance architecture); the Parliament finally adopted a baroque building – Palazzo di Montecitorio – with an interesting addition at the beginning of the twentieth century. Some ministries – Finance, War, Agriculture – were built anew along Via XX Settembre, east of the historic core, where, according to some, the new capital city should have gone (but did not), according to the idea of differentiating from an inglorious clerical past. However, no Haussmann ever came to substitute the old core with new developments. Here and there some demolitions took place, but that never developed into a coherent new grand design. This was attempted only during the twenty fascist years: recovering the archaeological core right into the central city, developing some grandiose new complexes, and imposing some kind of architectural style.

The democratic republic born after World War II behaved the opposite way. In fact the municipality – left alone to control the city's growth – found itself, in the first decades, at a loss when confronted with the overwhelming forces of land speculators. And, when growth stopped, it had to rehabilitate large portions of the municipal territory where extensive illegal developments had taken place. Only one attempt was made to redesign the city's structure in the plan prepared between 1955 and 1962 – the Eastern Directional System; an interesting idea in 1955, it was already obsolete by the 1960s, due to great changes in the context and trends of development.

On the whole, the city's planners (or, better, the planning machine) always had difficulties in adjusting to social and economic change. Nowadays Rome is a rather well-off metropolis whose income comes mostly from the tourist and service sector (with a relevant share of high-tech firms). Its influence covers a large metropolitan area where urban population and jobs tend to move. Mobility is possibly its main problem, which calls for more efficient public transport, as current plans indicate. Because plans took a long time to be adopted their forecasts were frequently overtaken by events. With one exception in 1883, special financial provisions to support the capital city's structures were independent of the official plan. Changes and new developments were introduced through 'great events' (such as the 1911 National Exhibition, 1942 World Exhibition, 1960 Olympic Games, 1990 World Cup, 2000 Holy Year) rather than through regular planning, but this is often true elsewhere. The extraordinary amount of public art, heritage buildings and archaeological sites makes it difficult for Rome to function as any other modern city. The care that is devoted to the heritage, however, seems appreciated by its citizens, considering the amount of passionate debate arising whenever some major development is proposed.

NOTES

1. The Kingdom of Italy was officially born in 1861, to endorse a political design that conjugates the expansionist wishes of Piemont (whose capital is Turin, but is called the Kingdom of Sardinia) with

the objectives of modernization of a bourgeoisie that does not feel represented by the constellation of small regional states that were created at the conference of Vienna. On 27 March 1861 Rome was declared the capital of Italy.

2. Caracciolo (1974), pp. 27–34. Also see Bartoccini (1985), pp. 433–473.

3. Luigi Pianciani (1810–1890) was a liberal progressive Mayor who held office from November 1872 to July 1873, and then for a short period in 1881.

4. Alessandro Viviani (1825–1905), a railway and civil engineer, political exile, and a member of the *Commissione per l'ingrandimento della città* (Commission for the Expansion of the City). In 1871, he was in charge of writing the Regulatory Plan for Rome.

5. Insolera (1962), pp. 29–39

6. *Ibid.*, p. 36.

7. Via Nazionale is probably the first example of land speculation in modern Rome. Belgian cardinal De Mérode, acquired the lands of the Terme di Diocleziano from Englishman Billingham. In 1867 he proposed that the Commune purchase the land of Piazza dell'Esedra, and is given free an area to build a road connecting the Piazza to the city. The Italian administration accepted the proposal and in 1871 Viviani proposed connecting the new street with the Piazza Venezia. See Tafuri (1959), pp. 95–108.

8. Quintino Sella (1827–1884) was a politician and Minister of Finance with various governments from 1862 to 1873. He was also a member of the communal Council of Rome, a mathematician, geologist, speleologist and mountain climber, and skillful parliamentarian. He was always a supporter of political economic austerity by courageous balanced budgets.

9. Numerous as they are in the first decades, political declarations contrary to industrialization of the capital, were feared for the popular manifestations that could condition the freedom of Parliament. See Caracciolo (1974), pp. 240–267.

10. Cuccia (1991).

11. Insolera (1962), pp. 44–53; Sanfilippo (1992), pp. 52–60.

12. Insolera (1962), p. 63.

13. Ernesto Nathan (1845–1921) born in London of Italian mother Sara Levi and German father Meyer Moses Nathan. Both father and son were followers of Giuseppe Mazzini, an outstanding political figure in Italian history. Ernesto Nathan became an Italian citizen in 1888. He was Mayor of Rome between 1907 and 1913, and one of the founders of the Society Dante Alighieri, a major national institution for the diffusion of Italian culture abroad.

14. Edmondo Sanjust di Teulada (1858–1936), an expert in hydraulics and head of the civil engineering department of the city of Milan between 1903 and 1908, visited many European countries (including Russia, where in St. Petersburg he met Mayor Nathan), and he also visited the United States for conferences or governmental missions.

15. The building regulations attached to the plan, launched in 1912, previewed isolated mansions where the area covered by the building was only one-twentieth of the total plot.

16. On the origins of public building in Italy see Piccinato (1987), pp. 115–133; for Rome see Cocchioni and De Grassi (1984); for Garbatella and Aniene see Fraticelli (1982).

17. Piantoni (1980); See in particular, Valeriani (1980), pp. 305–326. There were also foreign pavilions. Amongst the most remarkable were that of Austrian J. Hoffmann, and that of Englishman E. Lutyens, that with some modifications two years later became the centre of the British School in Rome, and that of the United States in the 'colonial American' style, of New York office of Carrere and Hastings.

18. Published in *Il Popolo d'Italia*, quoted from the National Fascist Party on 22 November 1928.

19. The role of architects can be seen in: Cederna (1979); Manacorda and Tamassia (1985), pp. 16–31.

20. They are the quarters of Santa Maria del Soccorso, Primavalle, and Val Melaina. See Rossi (2000).

21. On Marcello Piacentini (1881–1960) see Lupano (1991). Another outstanding figure is Gustavo Giovannoni (1873–1947) an architect and prominent architectural historian. Giovannoni represented the traditional tendencies in contrast to the new ones embodied by Piacentini.

22. Governatorato di Roma (1931).

23. The Esposizione Universale was to be held in 1941, but the date was moved to 1942 so that it would coincide with the twentieth anniversary of the 'Fascist Revolution'.

24. For more recent studies on this topic see Quilici (1996). Also see Ciucci (1989).

25. Sanfilippo (1992), pp. 21–42.

26. Cederna (1956); Della Seta and Della Seta (1988).

27. The history is documented in numbers 27 and 28–29 in *Urbanistica* (1959), then reunited in one document: *Roma. Città e piani* (1959) and then in successive editions of Italo Insolera (1962). The recent text of Vidotto (2001) is opposed to one interpretation that privileges the protagonist role of land speculation.

28. Clementi and Perego (1983); Piazzo (1982).

29. The first version of the plan is illustrated in *Urbanistica*, 116, June 2001. Previous analyses are in *Urbanisme*, 302, September–October 1998 and *Capitolium*, **III**, 11–12, December 1999 and **IV**, 13, March 2000.

Chapter 16

Chandigarh: India's Modernist Experiment

Nihal Perera

Chandigarh, one of the first state capitals built in independent India, is also the first Modernist capital to follow the CIAM model. The need for it was created by the division of the Province of Punjab between India and Pakistan, at their separation in 1947, and the allocation of its magnificent capital, Lahore, to Pakistan (see figure 16.1). Although designed as the state capital of Indian Punjab, symbolically Chandigarh acquired the attention of the national leaders from the beginning and several world-renowned planners and designers were involved in the project.

The factors that played a major role in its planning include the independence of India, the partitioning of the British colony and imagined state of India, the allotment of several major cities to Pakistan, nostalgia for lost places, the flow of refugees, national goals, and postcolonial imaginations. National aspirations, especially those represented in the first Prime Minister, Jawaharlal Nehru's ideas of India and the notions of modernity promoted by Punjabi officials, particularly A.L. Fletcher, T.N. Thapar, and P.L. Varma, had profound effects on the plans and planning. At the immediate level, the authorities remained divided on the location, character, and size of the city. Adding more complexity to the process, two plans were prepared for the city. The project was initially awarded in January 1950 to the American firm Mayer and Whittlesey; and Albert Mayer and Matthew Nowicki were the primary designers of the plan. The second plan was made by Le Corbusier, supported by Pierre Jeanneret, Maxwell Fry, and Jane Drew. The death of Nowicki in a plane crash and the increase in the value of the US Dollar are cited as reasons for replacing the Mayer team in November 1950.[1] While Nehru championed the project, the plan was negotiated by many social agents, especially those mentioned above and the inhabitants of the site.

With the approval for a new capital of Punjab in 1949, the first phase of Chandigarh Master Plan area of 70 km^2 (in 1951) was acquired and a Periphery Control Act of 1952 enacted to control development within an 8 km (5 mile) periphery, expanded to 16 km in 1962. Chandigarh, a

CHANDIGARH: INDIA'S MODERNIST EXPERIMENT

Figure 16.1. The location of Chandigarh.

Union Territory which falls directly under the central government, was governed by the Chief Commissioner. The estimated expenditure of the first phase was Rs. 167.5m (£10.5m) and the project was well-funded.[2] Among the new towns built in India at that time, Chandigarh reports the highest per capita government expenditure and the highest maintenance cost.[3] The sources of funds were:[4]

Loans from the Rehabilitation Ministry, Government of India (1950–1953)	Rs. 30m
Grant from the Government of India (1953–1956)	Rs. 30m
Contribution by the Government of Punjab	Rs. 30m
Loan from the Government of India for housing	Rs. 4.4m
Estimated receipts from the sale of plots	Rs. 86m

The city was formally inaugurated on 7 October 1953. In 1966, due to the further division of (Indian) Punjab into Haryana and Punjab and both states claiming it, the city and some area on its periphery were converted into a Union Territory administered by the central government with the city functioning as the capital of both states. Since 1984, the Governor of Punjab has been functioning as the Administrator of the Union Territory, assisted by the Advisor to the Administrator. Most policy issues are currently settled by senior administrative and technical officers under the overall charge of the Administrator, although for the past ten years an elected municipal corporation has been demanding a greater role in such matters.

Chandigarh is a well studied city. The leading scholars of the city include Norma Evenson, Kiran Joshi, Ravi Kalia, and Madhu Sarin,[5] most of whose work has been largely architect-centric and focused on the fame of Le Corbusier,[6] who was able to realize an overall form for the city and designed some magnificent buildings in the capital. The following pages will provide a brief overview of the politics of planning, a comparison of plans, and how the plan was adapted to ground realities.

Indian Aspirations

As Kalia points out, the staggering desire of its leaders to establish India as an independent and modern nation in many ways shaped the new city.[7] The defining ambiguity of the planning process is represented in the conflicting notions of 'independence' and 'modernity'. While independence demanded a future marked by national prosperity, the dominant goal of development was to 'catch up with the West'. Until the 1970s, almost all prestigious capital building projects across the world were designed by architects from the 'West'.[8] Only a few, like Brasília, were designed by local architects and planners; even these designs followed the Western-Modernist idiom.

According to Sarin, during the Freedom Movement, national leaders and intellectuals had been searching for an art and architecture which would serve as an expression of independent nationhood. Many staunch nationalists favoured a search in the past, especially in the Mughal style and historic architectural treatises, for example, *Mansara Shilpa Shastras*. Unlike the case of Bhubaneswar – another capital built during the same time – the choice in Chandigarh was in favour of moving 'forward'; anything to do with the 'tradition' was too easily associated with backwardness.[9] The competition in Chandigarh was largely between a European-type modernity and one which represented a developed India – an 'Indian modern'.[10]

The Western-trained civil servants and professionals largely desired a rational and efficient city. In addition to building new industrial towns in the region, they were involved in large-scale development projects (for example, dam building projects) and the establishment of industrial and refugee towns on sound engineering principles with little preoccupation with symbolic and aesthetic aspects. In Chandigarh, the Punjabi officials desired to build a city on the scale of Lahore, accentuated by function and efficiency, and were more inclined towards 'European modernity'. In its recommendations in August 1948, the Cabinet Sub-Committee (New Capital) not only made direct references to garden city principles, but also specified parameters largely based on Western notions of appropriate development.[11] These included self-contained, use-specific neighbourhoods, which were not the norm in India but found in colonial New Delhi. This created a paradox: a country coming out of European colonialism looking up to Europe to establish itself as an independent and developed nation.

Nehru followed a third path. His ideas about architecture and planning dovetailed his views on economic development.[12] He saw Chandigarh 'as a showpiece of economic development and national aspiration based on his conviction that India must industrialize to survive and prosper'.[13] He was inspired by the USA and the USSR but did not desire to emulate either.[14] His ideas were strongly rooted in the history of India, but his focus was on its changing spirit: 'From age to age she has produced great men and women, carrying on the old tradition and yet ever adapting it to changing times'.[15] Instead of Westernizing India, Nehru's choice was to Indianize the foreign inputs: 'It was India's way in the past to welcome and absorb other cultures. That is much more necessary to-day'.[16] He was in search of a city which would display a modernity distinct from and free of the colonial version.[17] Nehru wanted to build community life on a 'higher scale' without breaking the old foundations of India.[18]

A further source of conflict was the question of spatial scale. The literature on Chandigarh points to the conclusion that the site was selected primarily at the national scale, to satisfy national priorities. Many politicians in the region favoured building a small administrative town with a population of about 40,000, adjoining an existing city near their hometowns. The Punjabi officials wanted to build a larger town with an initial population of 150,000 which could replace the material and psychological loss of Lahore, a magnificent city which had been the hub of Punjab's commercial and cultural activities prior to separation. This responded to, and fed off, the popular Punjabi nostalgia for Lahore. Indian leaders and Punjabi officials were in agreement in this regard. When diplomacy failed, Varma resorted to the national government to resolve this conflict and Nehru's intervention confirmed the Chandigarh site suitable for a new capital. The site, bounded by two riverbeds, at the foothills of the Shivalik Range of the Himalayas, with a picturesque backdrop, was chosen in March 1948. The Surveyors' Report of 1949 'unequivocally' endorsed the physical suitability of the site.[19]

The desire to create an Indian (national) identity in Chandigarh is further evident in the displacement of over 6,000 families from the area. There was considerable opposition to the acquisition of the site from existing villagers, and they were allowed to remain temporarily 'as tenants of the government' until the land was needed for building purposes.[20] By acquiring the villagers' land and directing the refugees to settlements built in other places, the authorities created a site 'unfettered by existing encumbrances'.

Punjabi officials did not believe that Indian designers could accomplish what they desired, and wanted to search for designers from Europe. Nehru consistently disagreed. He feared that 'The average American or English town planner will probably not know the social background of India'.[21] He suggested two Western planners already working in India and therefore perhaps conversant with it: Otto Koenigsberger and Albert Mayer. The Punjabi officials were unimpressed by them; they were too Indianized for the task the officials had imagined. But Nehru's power at the time was too strong for Punjab officials to challenge and Mayer was given the job.

The appointment of Mayer did not end the conflicts and not all Indian leaders shared Nehru's views. The Indian government's overwhelming desire was to create a great monumental city,[22] which was also functional and efficient. Although Nehru won the initial contest, the Punjabi officials won the larger battle when Nehru authorized them to visit Europe in search of new designers after the death of Nowicki and Mayer's expressing his inability to manage the task from the USA. However, the documentation suggests that Mayer was systematically displaced because he no longer represented the desires of the Indian government. The turning point was the death of Nowicki up to which Nehru held strong to his position.[23]

The Designer Ambitions

The physical plans for Chandigarh were developed within two different imaginations of what makes a good city and what is good for India (see figure 16.2). The differences between the plans are also ideological; the planners adopted significantly different approaches. Le Corbusier was an architectural Modernist and Mayer was influenced by garden city principles prevalent in the USA during that time and by his own experience in India. The advocates of garden cities sought a spatial escape out of the industrial city; a way to create cleaner living environments with hybridized urban-rural characteristics away from the problem-city. The Modernists imagined

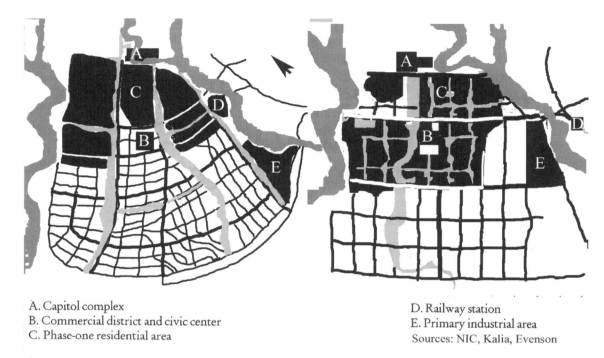

A. Capitol complex
B. Commercial district and civic center
C. Phase-one residential area
D. Railway station
E. Primary industrial area
Sources: NIC, Kalia, Evenson

Figure 16.2. The Mayer (left) and Le Corbusier (right) Chandigarh plans.

a temporal escape into a 'post-industrial' future; the Modernist city represents a future that is radically different from the industrial present and the European past. Both approaches responded to the problems facing the industrial city in Europe and the USA, but their validity in a 'pre-industrial' India was never questioned.

Although he was hired to execute the Mayer-Nowicki plan, Le Corbusier substantially revised it. He straightened the curving roads, created a grid, added more levels of separation to the circulation system, increased the size of the residential block, combined the civic centre and central business district and moved it further north, relocated the railroad station beyond the river, increased the amount of open and recreational space, removed the dependence of the urban form on the natural features such as streams (although he did accommodate a seasonal riverbed running through the master plan area as a 'leisure valley' of green space), and removed the use of landmarks. Despite some significant similarities, the plans have fundamental differences.[24]

The garden city model then popular in the United States was based on the 'Radburn ideal' of 'decentralized, self-contained settlements organized to promote environmental considerations by conserving open space, harnessing the automobile, and promoting community life'.[25] This influence is evident in the residential 'superblocks' in the Mayer-Nowicki plan (see figure 16.3). Mayer used Los Angeles' Baldwin Hills to explain the superblock idea and used Radburn (NJ) and Greenbelt (MD) to explain the proposed system of internal pedestrian paths.[26] The Modernist city to which Le Corbusier subscribed was developed in the CIAM (*Congrès internationaux d'architecture moderne*) manifesto. The group's goal was social transformation; a main premise of the larger ideology, which James Holston calls Architectural Modernism, is that the transformation of the built environment can instigate social change.[27] For Le Corbusier, the city 'should be free from the "inhibiting restraints" of the past'.[28] The Architectural Modernists undertook to transform the inhabitants' daily practices through an unfamiliar environment that would direct them to a better future (see figure 16.3).

The second area of difference concerns the inhabitants of the planned city. While Le Corbusier conceived the city from the Capitol complex, the Mayer-Nowicki team began their planning process from the neighbourhood. The neighbourhood, for them, was the basic generative unit, its strength, unity, and identity: 'We did not plan down to [the neighbourhoods] but up from them'.[29] Le Corbusier's intention was solving urban problems in general.[30] Similar to New Delhi and Canberra, therefore, the city represented an abstract future and identity at the expense of immediate cultural compatibility.

Thirdly, the experience the planners had in India, their views about it, and the degree of 'Indian culture' they opted to accommodate in

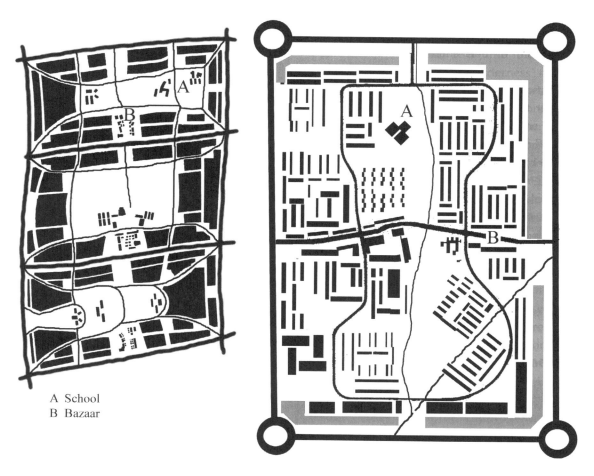

Figure 16.3. The Mayer (right) and Le Corbusier (left) neighbourhoods: superblock and sector.

their plans were radically different. Starting with a programme for new model villages, Mayer had helped develop master plans for Kanpur, Bombay, and Delhi. By the time he undertook the Chandigarh assignment, Mayer had significantly Indianized. A strong concern for the Indian culture is evident in the work of Nowicki who did most of the design development work. In contrast, Le Corbusier's trip to the site in 1951 was his first to India. His views of both Chandigarh and India were not based on any substantive study or experience of the Indian society.[31] Evenson asserts that '[Corbusier's] fondness for Baroque expansiveness combined with his long-term obsession with the industrialized city had rendered him unsympathetic to the functional workings and aesthetic subtlety of the traditional Indian environment'.[32] According to Kenneth Frampton, 'Because of the fixation on the Athens Charter, Le Corbusier and his colleagues were unable to arrive at a more intimate residential fabric'.[33]

In his own view, Le Corbusier knew what India's problem was and had a solution for it:

India had, and always has, a peasant culture that exists since a thousand years! India possessed Hindu . . . and Muslim temples [Maharaja palaces, and gardens] . . . But India hasn't yet created an architecture for modern civilization (offices, factory buildings) . . . We must begin at the beginning.[34]

Instead of familiarizing himself with Indian social and environmental conditions, he opted to familiarize the Punjabi officials, who visited him in France, with architecture appropriate for a modern civilization. Le Corbusier sent them to Marseilles to see his *Unité d'Habitation*.[35] Chandigarh was thus imagined from a European vantage point. This view of Le Corbusier is reflected in the straight geometries, the uniformity of components of the city, and distances and proportions which were familiar to him from France (for example, the relationship of the monumental axis and the rhythm of bus stops in Paris).[36] A rare reference to the distinctiveness of place is found in the notion of 'Tropical Architecture', which Maxwell Fry and Jane Drew began practicing in Africa – prior to their arrival in India – and conceptualized later.[37] This reference to the climate was, however, 'a subtle objectification of the subjects [the "Climatic Other"] referring to more impersonal, material, and scientific factors than the culture'.[38]

While Mayer viewed industry as a catalyst for development, Le Corbusier saw it as inappropriate and argued that it would be erroneous to introduce industry in Chandigarh.[39] The Cabinet Sub-Committee wanted an industrial area to be built at the third stage of development.[40] Following a prevalent Western concept of the time, both plans separated the industrial area from residential areas (see figure 16.2). The elimination of traditional ties between work and living through single purpose zoning would marginalize a sizeable population. Similar defamiliarization can be seen in regard to the location of the railway station, which is the transportation hub of Indian cities served by it. The Railway Department of India requested that the station be located about 1.5 km from the city centre. In the Mayer-Nowicki plan, it is located on the city-side of the Sukhna Cho ('Cho' in local dialect means a seasonal riverbed which floods during the monsoons but is dry the rest of the year), at a distance of about 2.5 km from the civic centre, and connected by a direct 'green-way' and footpath. The Le Corbusier team located the station on the opposite side of the river at a distance of about 6.4 km from the commercial centre.

Finally, both plans emphasize parks and open areas, and the park system is repeatedly looked upon in a favourable manner by the critics. These were common planning strategies employed in industrial cities of the time. In contrast to connectivity, the focus of the park system in the Mayer-Nowicki plan, the Le Corbusier plan put emphasis on representation. While Nowicki provided identity by varying sizes and shapes of buildings, the second plan contains a few monuments derived from Le Corbusier's earlier work. The highlight is the 'Open Hand' monument which has become the symbol of Chandigarh:[41] 'Open to Give. Open to Receive'. The second is the Martyrs' Memorial which makes Chandigarh unique; according to Tai and Kudaisya, it is the only memorial to the victims of partitioning in the entire subcontinent. The two monuments are evocative of the circumstances in which Chandigarh came about and symbolized the city's genesis and preoccupations.[42]

The lack of monumentality is the reason given for not implementing the Mayer plan. The principal actors who concentrated more on the visual appeal of the site most likely envisioned a monumental city rising from it. According to Evenson, the Mayer plan 'does not read as a monumental capital – as that positive act of possession by which a capital may symbolize the control of a people over their destiny'.[43] 'Although the Indian officials of Chandigarh had originally been completely satisfied with the Mayer plan . . . [possibly] the added qualities of monumental urbanity . . . moved them to accept . . . the changes proposed by the second group'.[44] Le Corbusier not only isolated the Capitol area (*la tête*) for himself and designed some magnificent buildings (see figure 16.4), but was also able to materialize his dream to design a city in India.

Figure 16.4. The Assembly.

The City Lived

As Suneet Paul points out, Chandigarh did change the architectural morphology at national and international levels.[45] Nehru is more graphic: 'it hits you on the head and makes you think. You may squirm at the impact but it makes you think and imbibe new ideas'.[46] 'The residents are boastful of the city and enjoy a pattern of living which many Indianized cities just don't offer', adds Paul.[47] At the same time, the city is being Indianized in multiple ways. What the designers and administrators failed to see is that as rapidly as people are assigned to a space, the subjects tend to familiarize their own spaces through daily practices.[48] Chandigarh has been changing and is advertised for the tourist as a very different place from that which its designers would have anticipated. In place of the image of an administrative city and a famous one-man wonder, the Capitol, the city promoters highlight the Rock Garden, the Sukhna Lake, the Cactus Garden, and the Mansa Devi Temple.[49]

The comparatively low level of roadside commercial activity in Chandigarh speaks to the impact that an unfamiliar and unviable place might have on such practices. Yet the familiarization of space and discovering viable places within this new city was seen in the early development of self-built markets principally at the locations where the original villages existed, for example, Bajwada Village and Nagla Village (see figure 16.5). Shastri Market consisted of narrow lanes reminiscent of traditional Indian bazaars, and, in the words of Prakash and Prakash, 'The place is popular, crowded, and alive with the noises of bargaining and haggling'.[50]

In addition to the daily activities of the inhabitants, the city administration itself has breached the plan from the outset. Because the poor were excluded from the city, by not providing affordable housing for them, dealing with the 'unauthorized' settlements they created to house themselves has remained a serious problem for Chandigarh until today. Creating their own solution, the construction workers were the first to build 'non-planned' settlements; the principal ones were adjacent to Bajwada Village, near the Capitol complex construction site and in Sector 17. Yielding to the pressure of the residents, in 1959, the authorities demarcated the sites of 'non-planned' settlements as 'temporary' locations for 'labour colonies'[51] (figure 16.6). The separation of mono-functional land-uses, particularly the separation between housing and industry, has also been difficult to enforce. As a result of the unsuccessful struggle to move such industrial activities to phase II of the industrial area, the administration amended the regulations permitting household industries in residential areas in 1975.[52]

In summation, planning Chandigarh was a complex process participated by a whole group of enthusiastic players, highlighted by national

Figure 16.5. Chandigarh's markets, from high income (left) to low income (right).

Figure 16.6. Residential segregation in Chandigarh, from high income (left) to low income (right).

leaders, Punjabi officials, two teams of designers, and the inhabitants. Planning continues to be a contested process and the resulting city is a hybrid. Yet the ability of the players to influence the outcome was uneven: the city was planned largely as a national representation, and very little attention was paid to its inhabitants and the social and cultural context of north-west India. It does not bear the stamp of a single person, but the magnificent contributions of Nehru and Le Corbusier are quite evident. Highlighting the compromises he made and the shortcoming of the plan, Le Corbusier himself emphasized the need to increase the population (density) in the initial area of development (Phase I), before undertaking any physical expansion of the city suggested in the Phase II plan.[53] Urbanization, familiarization, and Indianization are precisely what Chandigarh has been going through ever since it was built. As much with the plan, the existing city is also shaped through various violations of the plan and negotiations between various agencies mediated by the administration. The plan and the designer provide a stable reference point to hang onto within a constantly changing discourse of the city.

NOTES

1. The Mayer team's fee for completing the plan in America and detailing it in India was $30,000 (Rs. 126,000). This included $10,000 (Rs. 42,000) for consulting experts outside of the team. Le Corbusier was paid an annual salary of £2,000 (Rs. 32,000) and a per diem of £35 (Rs. 560) when in India, subject to a maximum limit of £4,000 (Rs. 64,000). Kalia (2002), pp. 32, 43.

2. More than half of this amount (Rs. 86m) was for development of the town and the provision of civic amenities and the rest (Rs. 81.5m) was for government buildings and the water supply system. Sarin (1982), p. 61.

3. Compared to Rs. 2,352 in Pimpri, which was the lowest, Chandigarh's per capita goverment expenditure was Rs. 4,384. Prakash (1969), pp. 48, 59.

4. Sarin (1982), p. 61; Prakash (1969), p. 61. See also Joshi (1999).

5. Kalia (1987); Sarin (1982); Evenson (1966).

6. For a critical analysis, see Perera (2004). In his recent book Vikramaditya Prakash (2002) also provides a criticism from a design standpoint.

7. Kalia (1987).

8. Vale (1992).

9. Sarin, (1982), p. 25.

10. Perera (2004).

11. Kalia (1987).

12. Speech given at the Seminar and Exhibition of Architecture, on 17 March 1959, in New Delhi. India (1957).

13. Giovannini (1997), p. 41.

14. Nehru (1946), p. 548.

15. Nehru (1946), p. 563.

16. Nehru (1946), p. 566.

17. Khilnani (1997), p. 130.

18. Letter to Mayer, Nehru in Kalia (1987).

19. Kalia (1987), pp. 3–4, 17.

20. Kalia (1987), p. 12; Evenson, (1966), p. 7.

21. Kalia (1987), p. 26.

22. Kalia (1987), p. 33.

23. Kalia (2002), p. 38.
24. Furore (2000).
25. Birch (1997), p. 123.
26. Evenson (1966), p. 17.
27. Holston (1989), pp. 31, 41.
28. Curl (1998), p. 383.
29. Mayer (1950).
30. Sarin (1982), p. 37.
31. Sagar (1999), p. 120.
32. Frampton (2001), p. 38.
33. *Ibid.*
34. In Kalia (2002), p. 87.
35. Le Corbusier (1955), p. 115; Evenson (1966), p. 25.
36. Frampton (2001), p. 31.
37. Perera (1998), pp. 70–79.
38. Perera (1998), p. 73.
39. Kalia, (2002); Evenson (1966).
40. Kalia (1987), p. 18.
41. Prakash (2002).
42. Prakash (2002).
43. Evenson (1966), p. 18.
44. Evenson (1966), p. 38.
45. Paul (1999), p. 44.
46. Nehru (1959), p. 49.
47. Paul (1999), p. 114.
48. Perera (2002).
49. Huet (2001), p. 167.
50. Prakash and Prakash (1999), p. 33.
51. Sarin (1982), pp. 110–111.
52. Sarin (1982), pp. 95, 97.
53. Sarin (1982), p. 75.

Chapter 17

Brussels – Capital of Belgium and 'Capital of Europe'

Carola Hein[1]

Brussels's history and urban form have been shaped by the history of the Low Countries and the Duchy of Brabant.[2] Continuously occupied since the Roman period, the city grew around a fortified French encampment established by Charles of France, Duke of Lorraine in 979 as Bruocsella – 'settlement in the marshes'. Eventually the city expanded to the higher ground on the east.[3] By the thirteenth century, the city started to thrive due to its position on the trading routes between Cologne and Bruges, and established itself as a centre for the manufacture of textiles, tapestries, and other luxury goods. A regional capital under a variety of foreign occupants, following the Congress of Vienna when Belgium was united with the northern Netherlands, Brussels became the second capital of the Dutch kings. Royal successions and warfare among the major European empires led to the sequential occupation of Brussels by France, Spain, the Habsburg Empire, and Germany. After the Belgian revolution of 1830, Brussels became the capital of the new nation.

The new state integrated earlier symbols of capital ambition, such as parts of the ensemble of the Quartier du Parc (late eighteenth century), including the Place Royal, the Parc Royal, and the ring boulevard created in the early nineteenth century on the site of the former fortification walls.[4] The second Belgian King, Léopold II (1865–1909), in particular, tried to give the city metropolitan and national character, stimulating major urban transformation financed with private money.[5] During his tenure, municipal initiatives and royal interventions transformed Brussels to create the framework of a national capital. Under the Mayor Jules Anspach, the city realized the central boulevards (1868–1871) over the meandering river Zenne, cutting through the old city to connect the northern and southern train stations. These interventions complemented the King's projects that focused on Brussels's suburbs. In tune with the comprehensive road development proposed by Victor Besme, surveyor of the roads of the suburbs of Brussels in 1863 and 1866,[6] Léopold II introduced a complete plan for beautifying the city, introducing major parks and green spaces, broad avenues and a uniform design for private buildings. Radical large-scale transformation and destruction characterized Léopold's time, and the word *architect* became a curse in traditional areas such as the Marolles, where several densely-built blocks were expropriated to construct the enormous Palace of Justice, inaugurated in 1883. The royal plans were sometimes at odds with city government projects, and some national projects, such as the creation of a central station linking the north and south stations, proposed by Besme in 1858, dragged on long after Léopold's reign with the new train link opening only in 1952.

Since Léopold II, no authority in Brussels has sponsored significant changes to beautify the capital. During both world wars, German forces occupied the city and briefly created a Greater Brussels administration in the Second World War, which was dismantled thereafter. Although Brussels did not suffer destruction during World War II, many neighbourhoods were torn apart in the post-war period. High-rise buildings and modern construction bordering decaying buildings and empty sites became characteristic. Investors bought entire blocks, one by one, let them decay, and were finally granted demolition and rebuilding permits when the old buildings could no longer be saved. Masterpieces, including major architectural works such as Victor Horta's Maison du Peuple, were demolished. Particularly in the 1960s, new office buildings rose quickly and *'bruxellisation'* became a term for urban destruction. Recent interest in at least superficial history has created a new trend that is shaping Brussels: 'façadisme', meaning the preservation of the façades while the interiors are completely rebuilt.[7]

After World War II, disputes among the two major cultural and language groups led to the establishment of Flemish and French community organizations that address cultural issues beyond regional spheres and the creation of three distinct regions inside Belgium. The full regionalization of 1989 equipped Flanders, Walloon, and Brussels-Capital, with important powers. Among the five organizations created in the context of regionalization, all but the Walloon Region – which opted for Namur – chose Brussels as their headquarters, making it the capital of the Flemish regional government (joined with the Flemish community), of the Brussels-Capital Region, and the French community.[8] The regional and community organizations constructed their government and administrative buildings throughout the city. So far they have had less impact on Brussels's urban form than the nineteenth- and early twentieth-century national capital designs or the post World War II transformation of Brussels as one – although the most important – of three official European headquarters of the European Union (EU) (the others being Strasbourg and Luxembourg).[9]

The saga of Brussels, 'capital of Europe', began with Belgium's refusal to host the first European organization, the European Coal and Steel Community (ECSC) created in 1952. The other five member states agreed on the choice of Brussels, however parts of the Belgian government rejected the choice for intragovernmental reasons. The Belgian negotiator offered Liège, a provincial city, instead, but the other member countries refused. After three days of intense discussion, Luxembourg's president and foreign minister Joseph Bech offered his tiny capital as the temporary seat of the new European organization. While the member nations chose Luxembourg as the ECSC's provisional seat, Strasbourg, for pragmatic reasons, became home to the European Parliament. The Council of Europe, an earlier, larger but less powerful European body was headquartered in Strasbourg, which had the only non-national plenary hall that could house the new assembly. This decision effectively decentralized the different ECSC institutions and laid the foundation for the current polycentric capital.[10]

Chosen in 1958 as the third of the supposedly temporary European capitals, Brussels became home to the European Economic Community (EEC) and the European Atomic Energy Community (Euratom) – the two new European organizations created by the 1957 Treaty of Rome. With the fusion of the European communities in 1967, the city lost the European Investment Bank (EIB) to Luxembourg, but became the main headquarters of two of the three most important institutions of what is today the EU: the main location of the Commission, a supranational body independent of the governments, and the only office for the Council, the decision-making institution representing the governments and until today the most powerful European organ. Brussels also fought successfully for years to obtain a presence of the third major institution, the European Parliament (which, since 1979, is selected through direct elections). Each of the three institutions requested and obtained its own headquarters building over the last five decades. Their conception and construction followed in turn and each took a decade or more to finish. The history of their planning and construction mirrors the administrative, political, economic and urban transformations of the three periods discussed in this chapter.

The Berlaymont, today the headquarters building of the EU Commission, began the transformation of the former upper-class residential Quartier Léopold into an administrative district for the European institutions at a time when the national government had largely unchallenged planning power in Brussels. The evolving projects and the failed construction for the Council building – known as 'Justus Lipsius' (after one of the streets on the site)[11] – reflects the ups and downs of the economic boom and bust of the 1960s and 1970s. It also represents the emergence of public opposition to the encroachment of the European institutions in the district. The construction of Justus Lipsius, the Paul-Henri Spaak parliament building and its adjacent administrative offices named after Altiero Spinelli[12] demonstrates the economic revival of the 1980s, the filling in of the European district, and the emergence of regional political representation in Brussels. The challenge of accommodating the ten new members of the European Union from 2004 – each requiring 200,000 m^2 of office space with supporting housing and services – and the construction of a new venue for the European Council, the regular meetings of the heads of states,[13] requires better planning than the city has traditionally enjoyed, and challenges the main stakeholders to develop new ideas for European integration in Brussels.

Brussels needs to develop a capital concept that offers a solution to the integration of European, national, and regional capital city functions and permits the assimilation of a large foreign population while respecting local citizens and their way of life. Brussels, the largest among the three European capital host cities, with almost a million inhabitants, has a foreign population of nearly 30 per cent, including 140,000 from EU countries. By 2005 the EU employed more than 34,000 permanent staff and about 2,100 temporary workers; of these an estimated 27,000 are located in Brussels,[14] where EU institutions occupy 1,600,000 m^2 of office space, about one-fifth of all office space in the city.[15] Furthermore, the city is home to the North Atlantic Treaty Organization (NATO) and to Belgian, regional and community capital functions. Over recent decades, ordinary citizens have begun to exert some influence on the way in which the city is developing, making Brussels into a centre of citizen initiatives. But traditional planning patterns, particularly the intimate collaboration between public and private sectors continues. The city officials are tacitly expected to work with property developers to produce profitable buildings. Sites are made available and building permit exemptions are given to the developers without much public participation in terms of neighbourhood (grass-roots) organizations and the like.

Analysis of the history of the European presence in Brussels provides an example of global-local interactions.[16] As supranational organizations and multi-national corporations, with their vast size and economic might, play an increasingly important part in the design and planning of cities, their impact on the quality of urban life and on local representation grows. This situation demon-strates how European government is built to the detriment of local citizens if they do not have adequate political representation. The officially 'temporary' presence of the European organizations in Brussels aggravated the situa-tion because the headquarters cities had to compete for European functions. The European organizations could not develop a headquarters policy and the host nations were largely bur-dened with providing the necessary buildings and infrastructure, while not being allowed to provide structures that had capital allure. As long as a unanimous vote was necessary to select a headquarters city, no single capital could be selected,[17] and the three presumed temporary headquarters, Brussels, Strasbourg, and Luxembourg, revealed themselves as the definitive polycentric European capital, confirmed in that position by the European Council in Edinburgh of 1992.

The process of integrating the EU and its predecessors in Brussels was different from that in Strasbourg and Luxembourg. Brussels had to provide multiple buildings for the rapidly growing European organizations and their most important institutions, the Commission and the Council, whereas Strasbourg and Luxembourg hosted specialized organizations with fewer personnel. Brussels opted for the large-scale transformation of an inner-city district, whereas Strasbourg, hindered by French neglect, built little and focused on symbolic construction and Luxembourg developed a European district on the formerly agricultural Kirchberg plateau.

Brussels's particular political and administrative structure partially explains the situation. Political differences between Fleming and Walloon communities have disturbed the smooth functioning of the European organizations from as early as 1952. Despite these domestic quarrels, all national governments have supported the European organizations, and did more than officially required to facilitate private investment that would provide necessary buildings.

Another drawback to the city's European function has been its particular regional organization. The Brussels agglomeration has two official languages and consists of nineteen independent municipalities, including the City of Brussels. Until 1989, when the Brussels-Capital Region government was elected directly for the first time, regional planning was in the hands of a national minister. The national government, holding extraordinary planning powers in the Brussels's agglomeration, had no interest in architectural preservation or even in Brussels's votes, so it promoted the rapid transformation of the city to the detriment of its traditional structure and largely against the wishes of the inhabitants. Competition among the communes allowed many decisions to be made by the state. With support from the City of Brussels and in close collaboration with corporate business, the government transformed the city from a regional centre into a metropolis and the capital of the EU.

The creation in 1989 of a directly elected regional government, the Brussels-Capital Region, has brought about some changes, including the approval of a Regional Plan in 1998. These steps were the belated consequence of popular agitation that began in the late l960s.

Building the Berlaymont and Accommodating Europe to Promote Brussels as a Metropolis

In early 1958, the EEC and Euratom occupied recently constructed private office buildings in Brussels's Quartier Léopold. The Belgian government used the European presence issue to boost Brussels's urban development. Following the 1954 government change, Expo 58 became the occasion for a profound and radical transformation. In its name, the gov-ernment carried out major urban projects such as the ring and central highway network, and the development of Zaventem Airport. In record time, Brussels transformed itself into a modern city that could host the Fair and also provide ideal accommodation for the European organizations.[18] Although the road administration (*Fond des Routes*) insisted that there would be no major demolition or urban transformation due to the new streets, its work led to tertiary sector concentration in the centre and population flight to the suburbs.[19] In the early years, political and economic leaders as well as citizens welcomed the innovations. At the time, the regional plan by the architecture and planning office, Group Alpha, proposed new traffic infrastructure and urban development to accommodate growth to 2 million inhabitants and offered several sites for a future European quarter (see figure 17.1).[20] It was in this euphoric context that the EEC and Euratom came to Brussels. As the European presence was labelled temporary, the Belgian government offered recently erected office buildings close to the city centre (in the Quartier Léopold), reserving large empty sites for the construction of a European district to be built once a final decision on the site of the European capital had been made.

The choice of the Quartier Léopold – an area close to the city centre with first-rate accessibility – as the temporary site for the European organizations was not accidental. It was intrinsic to transforming the city into a metropolis. A private organization, that counted King Léopold I among its investors, planned (in 1838) and financed the district as the first extension of the City of Brussels (see figure 17.2).[21] The Quartier's wealthy residents left for the suburbs in the 1920s, vacating large residences that were

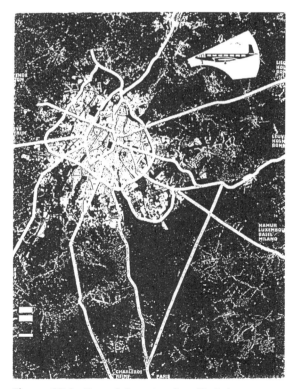

Figure 17.1. Brussels's regional traffic infrastructure as proposed in the 1958 application for the capital of Europe.

Figure 17.2. Aerial view of Rue de la Loi in 1939 with the Résidence Palace, a prestigious apartment complex built in the Quartier Léopold to prevent the affluent population of the area from leaving for the suburbs. The triumphal arch of the Parc du Cinquantenaire is in the background.

easily assembled for redevelopment. In 1948, Group Alpha identified the area as an ideal site for national and international organizations.[22] The provisional location there of the European organizations was thus in tune with the transformation of this prominent residential district into an office area.

Drawing on the tradition of public and private collaboration, Belgium responded to the precarious European presence in a particularly capable manner. The government worked politically to win a European presence for its capital, because it would improve its economy and image. It generally limited its investment to structures that simultaneously served the European presence and Brussels's urban development. But the government did not support building features that were intended purely to increase the symbolic character of the European organizations. It has followed this practice whenever new European institutions have moved to the city: it provides access to the site and basic urban infrastructure and uses political and administrative power to facilitate private construction of buildings. In order not to jeopardize the Quartier Léopold's transformation into an office area, the national government therefore did not control development and, in fact, gave private entrepreneurs too much freedom.[23] The latter frequently failed to honour local planning resolutions. The City of Brussels supported government decisions and building construction with rapid approval. Citizen opposition was non-existent. The Berlaymont complex stands as the major example of this period.

The EEC and Euratom experienced rapid and unpredictable growth after 1958, pushing them to search constantly for new buildings. A construction company suggested building on the site of the former Berlaymont monastery at the edge of Quartier Léopold. This outstanding site allowed consolidation of the agencies in a grand new building, named the 'Berlaymont'. Situated on the Rond-point de la Loi, the building's main entrance faces a boulevard which connects to the centre of the city. The only planning restriction was to limit the height of the new building to 55 m. Over time, a large number of Brussels's citizens have come to criticize the design of the structure.

The Berlaymont was financed by the Belgian Office of Overseas Social Security (*Office de Sécurité Sociale Outre-Mer* (OSSOM)/*Dienst voor de Overzeese Sociale Zekerheid* (DOSZ)) after the European organizations had expressed interest in using it, and the state provided extensive site

Figure 17.3. Aerial view of the x-shaped Berlaymont with the Rue de la Loi in the foreground. To the right, the circular façades bordering the rond-point Schuman; to the left, part of the Charlemagne, first headquarters of the Council; and in the background, the typical low-rise single-family Brussels' row-houses of the Quartier des Squares.

infrastructure. Because the building had to be convertible into a Belgian government office, it was never conceived as a symbol of Europe. After the 1962 Council decision to delay once more a definitive decision on the location of a European capital, construction began with the east wing in 1963. The architects designed the current x-shaped building, for use by all European organizations,[24] including a plenary hall, and other imposing spaces for the Parliament. The Berlaymont's offices were conceived with an open-plan layout like that of the Belgian ministries, just in case the European organizations left Brussels and the host city had to use the building (see figure 17.3).

The Berlaymont became the key to the creation of a new street and subway system in Brussels, continuing the transformation initiated for Expo 58. Instead of a north-south subway line as originally planned, an east-west axis was built, connecting the centre of the city to the European quarter and the well-to-do residential areas in the south-east of the capital.

Meanwhile, the needs of the European communities constantly increased and the organizations rented premises without architectural pretensions from the private sector in the Quartier Léopold, promoting office construction by private developers. The demands of European institutions have strongly affected the office building sector in Brussels, which evolved from construction on demand to speculative building. Other international organizations moved to Brussels due to its central geographic position, modern roads and the lack of urban restrictions. The city experienced the greatest office-building boom on the continent, attracting British developers in particular.

The requirements of the European organizations changed further with the amalgamation of the EEC, ESCS and Euratom in 1965–1967 and their subsequent regrouping. The number of EC personnel in Brussels was already larger than the capacity of the Berlaymont building, still under construction. The Europeans heavily criticized the Berlaymont and threatened to use only parts of it or erect a new building. The temporary status of the capital, however, made it difficult for the organization to make demands. Belgium rejected any solution other than a complete occupation of the Berlaymont, while the European Commission wanted to be the sole occupant of the building for reasons of prestige. By then, Belgium had concentrated too much money and material in the area to let the Europeans go elsewhere. The country's decision to pay part of the rent persuaded the Commission to occupy the Berlaymont with its formal areas designed for the Parliament but unusable by the Commission.[25]

With the presence of the Berlaymont building and the new road construction, the Quartier Léopold became the permanent home of the European communities in Brussels and Europe's central executive district.[26] The Belgian state never seriously considered a different location, even if additional or larger buildings were needed. Following its heavy investment in the site, the government was even ready to facilitate demolition to satisfy the demand of the EC. The European presence pulled major international organizations into the area. It also determined that all other European organizations would build their headquarters in the Quartier Léopold, as the Council and the European Parliament did.

Economic Unification, Megalomaniacal Projects, Citizen Protests and Projects for the Council Building

The Commission had barely occupied the Berlaymont when the Council asked for its own building. The site and the design, discussed from the late 1960s to the mid-1980s, pitted technical and functional urban planning under central control against aesthetic and social ideas based on public debate. While the national government tried to pursue its tested policy of co-operating with the private sector, societal changes, citizen awareness and local government opposition led to the downfall and delay of early grandiose projects.

The Council could not design and finance its own building in the absence of a definitive headquarters decision. Like the Berlaymont, the new building was to be constructed by the Belgian state and rented to the institution. The Council rejected several sites as too small or lacking a prestigious approach and the search widened to suburban sites.[27] But various interests blocked any suggestion of decentralization. The French and Luxembourg delegations opposed decentralization, because the limited extension space in the Quartier Léopold prevented expansion of the European institutions and thus preserved their interests.[28] Faced with this opposition and acknowledging its earlier investments in the Quartier Léopold, the Belgian government revived the initial project for a site opposite the Berlaymont (see figure 17.4). For the first time the government could not realize its project unchallenged. The planning context had changed radically since the late 1960s. Government instability limited political power, and the City of Brussels withdrew its support after experiencing public criticism in the elections.[29]

The early steps of the regionalization process in Brussels in 1968 created the first directly elected regional body, the Brussels Agglomeration (*Agglomération de Bruxelles / Agglomeratie Brussel*) which had limited powers. The Agglomeration's interventions largely concurred with the requests

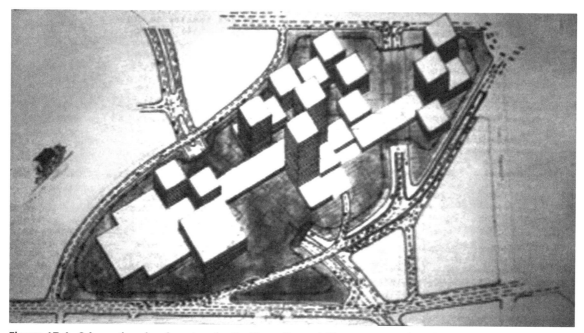

Figure 17.4. Scheme for a headquarters for the Council on a 6.4 hectare site with a platform over the Chaussée d'Etterbeek.

of residents' groups opposing the principles that had guided urban and regional planning since the 1950s. The interdisciplinary group ARAU (*Atelier de Recherche et d'Action Urbaines*),[30] together with two other initiatives: *Inter-Environnement Bruxelles* (IEB) and the Flemish group *Brusselse Raad voor het Leefmilieu* (BRAL), and the active architectural school La Cambre, became the focus of citizen activities in the Brussels agglomeration.[31] They requested open planning processes and democratization of decision-making, instead of the functionalist concepts defended by most public authorities. The community groups supported preservation of inner-city housing, a mix of functions and priority for public transportation.

ARAU used public events, guided tours showing examples of building speculation, pamphlets, press conferences, and counter-projects to advance their cause (see figure 17.5). Faced with opposition and diminished support from the city, the national government had to reduce its projects.

After other European member states refused to finance the Council building in 1978, the Belgian government decided to have a private financier launch the construction and erect it on the planned, but slightly reduced site.[32] This solution avoided an architectural competition that had been recommended for the prestigious building. Even though a competition does not guarantee

Figure 17.5. The façade of the Council building alongside the Rue de la Loi in a counter-project by the Brussels's citizen initiative Inter-Environment Bruxelles (IEB). The subtitle to this proposal reads: '*Construire l'Europe en detruisant la ville*' (Building Europe by destroying the city).

architectural quality, it would have reminded the public that the EU has cultural not just economic and political aims. But a competition was not in the interests of the Belgian government, because it would have slowed the process by arousing public debate and requiring the study of entries from architects and urban planners throughout Europe.[33] Instead, the government opted for a request for proposals from private developers. During this time of economic crisis, a major construction project could stimulate the national building industry.

This request for proposals emphasized economic, technical and functional requirements rather than urban or architectural quality. The Council proposed functional, constructive and security aspects, while the Belgians, under the pressure of the committees, desired urban integration. The results satisfied no one, and no final choice could be made. The futile process provoked lively reactions from local citizen groups, who prepared counter proposals featuring traditional streets, small-scale structures, urban diversity and integration, and also sketched out projects for a new European city on the site of the former Josaphat station.[34] These counter-proposals were not implemented, but from that point, the pressure groups could no longer be completely ignored.

Many of the problems in the design of the Council building related to the absence of a responsible authority. The refusal of the Council to lead the design of the building necessitated control by the Belgian government, which was generally more inclined to favour functionality and economic advantages over imposing and aesthetic design. Other member states would have objected to any clear position taken by the Belgian government. There was no official role for citizen groups and their voices were heard only as loud protest. Private developers with the tacit support of national and local government, however, guaranteed the transformation of the Quartier Léopold into Brussels's European district, even though the major projects for the Council failed in this period.

New Actors in Brussels and the Parliament Complex

Years of delay and opposition came to an end in the mid-1980s. The national government's attitude changed following the economic revival and the threat of strong regional opposition after the directly elected regional government, the Brussels-Capital Region, had been established. The understanding that Belgium had to provide decent offices for the Council, if it wanted to house the European Parliament, accelerated the process. The Belgian government took extraordinary measures and in rapid succession three major building complexes were developed: Justus Lipsius, the headquarters building of the Council; the Paul-Henri Spaak building with the parliamentary assembly hall and the adjacent Altiero Spinelli building housing parliamentary office facilities.

In 1983, after fifteen years of wrangling over the site and the design of its headquarters building, the Council decided to finance the construction. Another two years passed before the Minister for the Brussels Region signed the necessary permits and an accord with the EC.[35] The urban layout and architectural design of the Council building was finally achieved by means of a typical Brussels compromise, with urban design guidelines by an independent Brussels architectural and urban planning firm, Group Planning, and architectural design by the twenty-one architects who had participated in the earlier contested request for proposals.[36] In order to distinguish the new building from the nearby

Figure 17.6. The Justus Lipsius building for the Council. The corner of the building was cut to blend better with the circular rond point Schuman. The 'E' (for Europe) sitting on its legs is somewhat visible in the façade alongside the Rue de la Loi.

banks and business buildings, and to identify it as a key European building, the architects found the simplest solution: they inscribed the letter 'E' – for Europe – on the façade, sitting sideways on its legs and formed by the concrete of columns and beams (see figure 17.6). This is hardly adequate as an architectural symbolism for Europe.

The construction of Justus Lipsius took place simultaneously with the planning and development of a complex for the European Parliament. The Belgian desire to unite all three institutions – the Commission, Council, and Parliament – in Brussels coincided with the repeated request by the Parliament for regrouping its activities on a single site. A site for a parliament building existed on the edge of the Quartier Léopold alongside the train tracks and the Luxembourg station. The Belgian government could not intervene openly in the construction of a European Parliament complex, as that meant confronting France and Luxembourg.[37] To prevent disruption of the existing political balance, government and business representatives suggested a privately funded project in the form of an international conference centre with 750 seats – a parliament in disguise.[38] In order to allow for the rapid construction of the conference centre, the Brussels Region Minister bypassed requirements to amend the official plan. Luxembourg and France complained loudly, reminding Belgium that it had no right to erect a parliament building and that the question of the European capital required a unanimous decision.[39]

The 1987 announcement of the project aroused

widespread local protest. Citizen groups, who had been limited to a powerless consulting committee, criticized the procedure, the speculative operation, and the likely impact of the construction and of future offices. Nevertheless, in 1988 the contending parties signed an agreement that promised consultation with the local population, building renovation, and support for culture in the neighbourhood. The citizens had no way to enforce the agreement, and nothing major has been done so far. In 1988, the EU Parliament agreed to rent the semi-circular building. In 1992, even before the Council meeting in Edinburgh later in the year made Brussels, Strasbourg, and Luxembourg, the three temporary headquarters, the definitive capitals of Europe, the Parliament also rented an area in its vicinity alongside the railway line on which 300,000 m^2 of office space could be constructed, and further extensions are in the planning stage.

Aesthetically, the building is problematical and the scale of the whole ensemble shows a lack of regard for the social cohesion and the absorption capacity of the infrastructure in this district (see figure 17.7). It is a typical example of Brussels urban planning and of the weakness of the EU

Figure 17.7. Aerial view of the European Parliament complex in Brussels. The oval hemicycle building with the half-barrel shaped cupola is the Paul-Henri Spaak building. The administrative offices in the Altiero Spinelli Building are alongside the covered railway tracks. The Gare du Luxembourg is on the axis of the Parliament, with the neoclassic Place du Luxembourg beyond. In the background are the Council building Justus Lipsius and the Berlaymont, headquarters of the Commission, under reconstruction.

siting policy. It is also the first European building in Brussels after the Berlaymont to catch the interest of architectural critics – the underlying current in the architectural analysis is one of criticism and complaint of failed integration.[40] Once again, we can regret the absence of a design competition and the lack of democratic procedure that might have led to the appropriate expression of a building that is central to the European Union and politics.

In spite of its late entry into the battle of the sites, Brussels has succeeded in bringing together the three major European institutions in the Quartier Léopold before the last phase of regionalization and the establishment of the Brussels-Capital Region. The new regional government clearly entered too late to control the European projects. Nonetheless, it has started to affect the future of the area and since the 1990s the new regional government has opposed the national government several times on issues in the European quarter. Collaboration among investors, the public sector, and local organizations is increasing, but investors tend to address aesthetic requirements only as window-dressing on economically-motivated large projects.

Knowing that some businesses have left the European district because of its poor environmental quality, investors and politicians have recently started to recognize the importance of architectural and urban form for the quality of life and work, particularly in the Quartier Léopold. For the Brussels-Capital Region that means embellishment of the Quartier through urban furniture, improvement of the appearance of streets and public spaces and increased public transport. Improvement of the urban environment is even more important because Strasbourg, the direct competitor of Brussels in the fight for the Parliament seat, uses architecture and urban planning as a conscious means to gain publicity, while Luxembourg has invested heavily in the development and the transformation of the Kirchberg plateau, home to the EU in the Grand Duchy.

To rival its competitors and their urban and architectural efforts, Brussels needs a regional plan to achieve balanced development. The plan finally arrived after the direct election of the Brussels-Capital Region government in 1989. The new *Plan Régional d'Aménagement du Sol* (PRAS) is designed to develop Brussels into a metropolis of tertiary functions, while controlling office buildings and making them contribute to improving the urban landscape rather than destroying it. A long-term presence of the European organizations in the Quartier Léopold seems guaranteed, even though 3300 staff members of the European Commission left the Berlaymont in 1991 following the detection of asbestos. They occupied new office buildings in Auderghem in south-east Brussels, an area with a well-developed network of public transportation and streets[41] opening a new growth pole. The government for the Brussels-Capital Region wanted the Berlaymont to be rebuilt for the Commission, arguing that the Berlaymont has become a European and urban symbol. The private owners of the Berlaymont had to accept the Brussels-Capital Region's requirements and began expensive and lengthy renovations.[42] Meanwhile the Commission's departments were split over fifty-seven buildings in 1992. Even after the Commission returned to the Berlaymont in October 2004, this de-concentration has remained in order to accommodate the needs of the now twenty-five members of the EU. As of 2004 the Commission was housed in fifty-two buildings with 792,000 m² and is currently considering a decentralized location in Brussels.[43]

Europe in the Quartier Léopold

After forty years of national and local actions, the Quartier Léopold, a symbol of the *laissez-faire* tradition and the Belgian bourgeoisie, has been largely transformed into a European quarter. Brussels has become the *de facto* European capital despite the lack of planning. Recent efforts show that business people, the city government, and ordinary citizens now realize that their city is sometimes considered unappealing to both business and tourism, and that neither the city nor the Quartier Léopold evoke positive images. The results so far are not convincing. This is dangerous to Europe as a concept, because an iconic building such as the US Capitol or the British Houses of Parliament is a useful symbol in encouraging public identification with government institutions.

The perception of Brussels as the place of a faceless bureaucracy, and as a fictive place is reflected in social theorist Jean Baudrillard's statement 'Brussels is such an abstract place, it is not to Brussels that one is going to feel in debt, no one will feel a relation of reciprocity, of obligation, of responsibility toward Brussels'.[44] This feeling is not limited just to the architecture and design; it also reflects the lack of a feeling of responsibility that European citizens have towards their capital and the EU organizations. However, some pride and responsibility is needed to counter the expedient forces that have governed Brussels's European fate so far, to develop a clear vision and reinvent the Quartier Léopold and other European poles as liveable sites and positive symbols of Europe.

Brussels seems the appropriate space to test the future of Europeanization and its impact on urban and regional form, as well as the concrete interaction of Europe with its citizens. The present analysis clearly shows that the urban and architectural design of the Quartier Léopold and future European poles in Brussels or other European cities, need to be designed with both global and local perspectives, taking into account the interests of all participants.[45] If the residents of Brussels can establish networks and connections with other European citizens, they may be able to balance the already existing European financial and political networks, and initiate a new culture that seriously addresses the social, economic and cultural problems that stem from the city's Europeanization.

NOTES

1. This chapter is based on primary research reported in Hein (2004*a*). See also Hein (forthcoming, 2006).
2. English language references on Brussels are limited, see for example: Billen, Duvosquel and Case (2000); Hein (2004*a*); Jacobs (1994); Papadopoulos (1996). The general and urban history of Brussels is relatively well documented in French and Dutch. See for example: Abeels (1982); Aron (1978); Demey (1992); Lambotte-Verdicq (1978).
3. With its working-class areas to the west and the upper-class districts to the east, Brussels is an exception in Western Europe; since prevailing winds in Europe are from the west, most European cities have industrial and working-class areas situated to the east, but in Brussels the highlands on the east side provide better living conditions.
4. On Brussels architectural history see particularly Aron, Burniat and Puttemans (1990) and des Marez (1979).
5. Hall (1997*b*); Ranieri (1973); Therborn (2002).
6. Besme was *Inspecteur voyer dans les faubourgs de Bruxelles* from 1859 to 1903.
7. For a discussion of these two terms see also Käpplinger (1993).
8. On Brussels's multiple capital functions see also Lagrou (2000).
9. On the history of the European presence in Brussels see particularly Demey (1990); Hein (1987; 1993; 1995; 2004*a*; 2005*a*, *b*).
10. On the larger context of the creation of a capital for Europe, visionary projects for single and monumental capitals and symbolic buildings, as well as the history of the implementation in the three headquarters cities see Hein (2004*a* and *b*).

11. The Council building was named after the Belgian scholar Justus Lipsis (1547–1606) and in tribute to the street that disappeared to make room for the Council building.
12. The oval hemicycle building Paul-Henri Spaak owes its name to the Belgian politician (1899–1972) who was President of the General Assembly of the ECSC from 1952 to 1953 and Secretary General of the NATO Council after 1956. Its adjacent office facilities were named after Altiero Spinelli, the founder of the European federalist movement (in 1943) (1907–1986), and major promoter of a European Defense Community (EDC) and a European Political Community (EPC).
13. Held in Brussels since 2003 according to the Treaty of Nice.
14. Swyngedouw and Baeten (2001). See also www.europa.eu.int
15. Christiane (forthcoming, 2006).
16. On the topic of glo-cal interaction in Brussels see Swyngedouw (1997); Swyngedouw and Baeten (2001).
17. The Treaty of Rome requires that the seat of the European government be selected by a unanimous decision. The original six EEC nations could never agree, and since the membership is now twenty-five prospects for a unanimous decision are probably unlikely.
18. See also *Bauwelt* (1958).
19. Ministère des Travaux Publics et de la Reconstruction (1956).
20. Gourvernment Belge (1958).
21. A private society, the *Société Civile pour l'aggrandissement et l'embellissement de la Capitale de la Belgique* created the Quartier Léopold in 1837. On the Quartier Léopold see Burniat (1992).
22. Ministère des Travaux Publics (1966).
23. The impact of the 1962 Belgian Town and Country Planning Act on the transformation of Belgian cities reveals how planning legislation was designed for developers. See Laconte (2002).
24. The architects commissioned, Lucien de Vestel, Jean Gilson, André and Jean Polak, were well established in Belgium and Brussels, and have since designed numerous buildings in the European district. The Polak brothers designed the Atomium, the symbol of the Expo 58. The UNESCO headquarters in Paris is often cited as model a for the Berlaymont. The Berlaymont architects, however, say that the reference is more likely the Nestlé administrative building in Vevey, Switzerland. Jean Polak interview by Carola Hein, 25 June 1993.
25. The big halls could not be rented out as the translators refused to work there.
26. This term was coined by the geographer Papadopoulos (1996).
27. Hein (2004a), chapter 7.
28. *Ibid.*
29. Belgian Prime Ministers changed rapidly between 1958 and 1985, with eleven changes in government over this period.
30. ARAU was founded in 1968 and led by Maurice Culot and René Schoonbrodt.
31. On ARAU and La Cambre see *wonen-TA/BK* (1975); ARAU (1984); Culot (1974; 1975); Culot, Schoonbrodt and Krier (1982); Krier, Culot and AAM (1980); Schoonbrodt (1979); Strauven (1979).
32. On the history of the Council building see also Laporta (1986) and Schoonbrodt (1980).
33. Competitions do not have a long-standing history in Brussels. In 2000 the Belgian architectural magazine *A+* tried to change that situation by publishing two special issues on competitions. See *A+* (2000).
34. Employees of the Council criticized the project: Vaes (1980); Vantroyen (1984). On the projects by the opposition groups and the projects for the Josaphat station see Hein (2004a; 2005b) and Inter-Environnement Bruxelles Groupement des Comités du Maelbeek (1980).
35. Vantroyen (1984).
36. Nicaise (1985). Resistance to demolition of the Résidence Palace delayed the construction. Vaes (1980).
37. Fear that the accession of East European countries to the EU might bring new worthy candidates for the capital functions appears to have been a major factor in this decision.
38. See detailed discussion Demey (1992).
39. Fralon (1987).
40. Dubois (1994); *Arca* (1993); Kähler (1995); Wislocki (1996).
41. On the various sites used by the Commission see *Bâtiment* (1992).
42. Designed by the Belgian architect Steven Beckers with Pierre Lallemand as aesthetic advisor.
43. For the current discussion on EU integration in Brussels and possible deconcentration see also Hein (forthcoming, 2006).
44. Sassatelli (2002).
45. Baeten (2001) similarly argues in this context the necessity to develop a global and local view of the city's problems, to 're-empower the "victims" of the current "world-city regime"'.

Chapter 18

New York City: Super Capital – Not by Government Alone

Eugenie L. Birch

New York qualifies as a capital city under two categories: a Former Capital (a one-time United States's political capital that retains an important urban role) and a Super Capital (being the headquarters of the United Nations (UN), the international governmental organization).[1] New York is a Super Capital for reasons that are distinct from those of many of the other cities in this book. New York was a national capital for only one year (1789–1790) and, today, the city is neither its country's, nor even its state's, political capital. It became a kind of world capital after winning the competition to host the United Nations in 1947.

While capitals share many features with other cities, they have distinct design and development characteristics that set them apart. In the New York case, many urban design elements had first-time application and became exemplars of their type. With regard to the context, New York demonstrates the workings of a tri-partite governmental (city, state and federal) structure where each level has sharply defined powers. And in implementation, New York shows how the public and private sectors partner to fashion creative funding and administrative structures. This added up to a 'chemistry' of design, politics and finance that catalyzed New York's emergence as a Super Capital.

New York reached its Super Capital status because it possesses a singular combination of activities. Its prominence devolves not only from the UN presence but also from its function as the premier home to international business and culture. Four large-scale developments, including the United Nations, Rockefeller Center, World Trade Center and Lincoln Center for the Performing Arts, embody this phenomenon. This narrative focuses on these projects. It identifies the leaders who built them and examines the political, design and implementation strategies employed in their execution.

New York: Provincial Capital Lays the Foundation for the Super Capital

Many scholars have documented the Big

Apple's emergence from a former national capital (1789–1790) to a national city prior to World War II, attributing the city's rise to three phenomena: enormous population growth, economic dominance, and leadership in culture, communications and style.[2] They have shown how federal policy encouraged the city's success. For example, relaxed immigration rules spurred population growth – between 1890 and 1940 foreign-born residents ranged from 29 to 41 per cent of the total (peaking in 1910) at a time when the city experienced a more than 400 per cent overall population increase. The creation of the Federal Reserve system (1913) stabilized the nation's banking, making the New York branch dominant due to its oversight of foreign exchange transfers and open market operations. These policies yielded the large labour force/consumer market that stimulated the city's economic growth and assured a smooth flow of capital that encouraged the proliferation of banking, manufacturing, services (legal, accounting and insurance) and the headquarters of the nation's key corporations in Manhattan. Standard Oil, for example, moved from Cleveland to New York City in 1884 to coordinate its enormous international and domestic businesses.

The rest of the growth story flows from the needs and interests of the population. Bankers and industrialists required excellent communications, demands met by the mounting number of newspaper and periodical publishers and telecommunications providers – telegraph and telephone – and, later, radio and television. Upper- and middle-income citizens supported culture, leading to the establishment of internationally ranked museums, universities, learned societies and performing arts institutions. The unmet needs of low-income residents and newcomers inspired public and private involvement in social reform, where innovative models for housing, public health, education, including a tuition-free city university system, emerged.

By the 1940s, this 7.5 million-inhabitant, 322 square-mile city had become the 'capital of capitalism' and dominated its surroundings and the nation in terms of economic strength and population.[3] It had two strong central business districts in Manhattan and many residential precincts throughout its five boroughs.

Six factors determined the shape of New York, laying the stage for the later Super Capital projects. They were: geography; topography; the gridded street pattern (introduced 1811); housing code (1901); comprehensive zoning ordinance (1916) and public/private infrastructure investments in the port; and a 722-mile mass transit network and surrounding system of commuter rail, highways, bridges and tunnels (early nineteenth century to mid twentieth century).

Furthermore, the public authority, a type of public benefit corporation that came into use in the 1920s, played a critical role in developing the infrastructure projects that under-pinned New York City's Super Capital identity. Enabled under state (or in the case of bi-state arrangements, federal) legislation, these entities financed construction with tax-exempt revenue-anticipation bonds. In the first half of the century, the Port Authority of New York (1921), charged with shipping activities, and the Triborough Bridge Authority (1933), created to oversee the construction of a major river crossing, successfully demonstrated the effectiveness of this type of quasi-governmental unit.

Comprehensive planning had no role in these arrangements. It emerged late in the game with two efforts that had mixed results in guiding the city to its Super Capital position.[4] The first was the emergence of the Regional Plan Association of New York, a foundation-sponsored organization composed of civic-minded business leaders. It employed the nation's most talented planners to produce the *Regional Plan for New York and its Environs* (1922–1931), a twelve-volume assessment and development strategy for a 5,528 square mile region. Having no legal status and appearing on the eve of the Depression, followed by a world war, the Plan had no hope of immediate implementation. Nonetheless, it persuasively advanced ideas that would become postwar policy, including advocating Manhattan as the region's economic engine; strengthening vehicular transportation routes; and promoting residential decentralization.

The second effort was the creation of the New York City Planning Commission (1938), with a seven-member board appointed by the mayor, operating under a mandate to develop a master plan and its implementation tools. While its draft plan (1940) met with defeat, the Commission's management of zoning and capital budgeting encouraged private construction suitable for a Super Capital.

Finally, with a large population of workers and residents concentrated in a limited land area, especially Manhattan, the city became the home of the nation's highest density development. The skyscraper, invented in Chicago in the late nineteenth century, became its icon. By the 1930s, the Manhattan skyline was synonymous with New York and, later, it became the emblem of this Super Capital.

New York becomes a Super Capital

Moving to a Super Capital was a complicated process associated with the growing importance of the United States in global affairs after World War II. (In this context, the city's prominence in finance, culture and communications played out in an international arena.) New York benefited from America's political stability, extraordinary economic strength and renewed openness to foreign immigration.[5] By the 1970s, New York was a Super Capital; it was a leader of international government, finance and culture.

But did New York City become a Super Capital according to a comprehensive plan? No, it did not. The Regional Plan Association issued two plans in 1968 and 1996, but as in the past, these efforts had no legal standing.[6] The city did not have a comprehensive city plan until 1969 and even then, this plan did not receive official sanction from the City Council but was approved only by the City Planning Commission.

New York's rise was due to the efforts of small groups of public and private leaders who advanced large, city-shaping projects that would, over time, add to the collective strength of the city. As has been related earlier, this tradition had deep roots. It was born of an aggressive, entrepreneurial, brashness that has characterized New Yorkers in the last two centuries, making them unlike other US citizens and often the subject of the nation's derision. It resulted in important investments in infrastructure, commercial facilities, and cultural venues, all efforts that encouraged or facilitated the concentration of population and activities. However, in about fifty years, primarily in the decades following World War II, this behaviour intensified, resulting in four exemplary projects – not plans – that helped meld the city into a Super Capital. These projects, alone, did not establish the city as a Super Capital, but they did provide the skeleton and the icons of a new type of capital, one that combined political, economic and cultural hegemony.

Winning the bid for the United Nations headquarters in 1947 made the city the symbolic centre of the postwar world. However, its dominance in the global economy, physically represented by Rockefeller Center (1931–1973) and a revitalized Lower Manhattan downtown/ World Trade Center complex (1947–1987), and its leadership in worldwide cultural activities, symbolized by Lincoln Center for the Performing Arts (1955–1992) solidified the city's position.

Public Servants and Wealthy Families Provide Leadership

Between the 1920s and 1970s, three public servants, Robert Moses, Austin Tobin and Nelson Rockefeller, played critical roles in the development of these projects. Their long tenures, political acumen, risk-taking behaviour, vision and substantial accomplishments set them apart. Boldly accumulating and using political power to achieve their ends, they made extensive use of public authorities and other devices to undertake and streamline large-scale efforts.[7]

Robert Moses began his career in public service as an analyst in the Municipal Research Bureau, a Rockefeller-endowed reformist institution, but soon entered government employ, remaining for forty-four years. He simultaneously held multiple state and city jobs, including head of the Triborough Bridge Authority (1933–1968), New York City Parks Commissioner (1933–1959), member of the New York City Planning Commission, City Construction Co-ordinator and Chair, Mayor's Committee on Slum Clearance (1949–1960). Through these positions he facilitated two of the four projects, the United Nations and Lincoln Center. He was also responsible for major enhancements to the metropolitan infrastructure that sustained the large population required of a Super Capital and for thousands of units of low- and moderate-income housing.[8] In all, he directly oversaw $27 billion in public works.[9]

Austin Tobin, as head of the Port Authority of New York and New Jersey (1942–1972) directed the construction of the World Trade Center (WTC), starting in 1962. Under his leadership, the Port Authority produced the facilities necessary to support the Super Capital: it assumed operating responsibility for the region's three airports (1947); built the regional bus terminal (1947); and constructed the container port (1950) to handle New York's shipping.

Nelson Rockefeller, governor of New York State (1958–1973), began his New York City-building activities in 1931 as a renting agent and, later, president of the Rockefeller Center Corporation. His enduring interest in architecture and development stemmed from this work. He was deeply involved in the three other projects, the United Nations, Lower Manhattan downtown/World Trade Center and Lincoln Center. In addition, as governor, he oversaw the creation of state agencies supportive of Super Capital growth, the Metropolitan Transportation Authority (MTA, 1968), State Housing Finance Agency, Urban Development Corporation (UDC, 1968) and the United Nations Development Corporation (UNDC, 1968). These authorities facilitated the investment of billions of dollars in transit improvements, middle-income housing construction, and United Nations-related office/ hotel/residential construction.

These public servants would not have been successful without the contributions of wealthy families who were willing to tie their ambitions (and pour money, time and effort) to the city. Emblematic are the Rockefellers. John D. Rockefeller, the nation's first billionaire, moved Standard Oil's headquarters to New York City in 1884 and founded Rockefeller Institute (1901), the city's internationally distinguished scientific research institute. His son, John D. Jr built Rockefeller Center and underwrote the land costs for the United Nations. In the third generation,

John D. III guided the development of Lincoln Center for the Performing Arts and David led the revitalization of Lower Manhattan with the consequent construction of 60 million square feet of office space.[10] Nelson, of course, was involved in all of the efforts either as a representative of the family or as a public servant.[11]

Two sets of interpersonal relations contributed to the appearance and effectuation of the Super-City projects. The first was the long involvement between the Rockefellers and architect Wallace K. Harrison. Harrison began at Rockefeller Center as a junior associate of Harvey Wiley Corbett, an important early skyscraper designer and would go on to form his own, highly successful firm. He was chairman, International Committee of Architects, United Nations; member, Architectural Advisory Board, World Trade Center; and director, Board of Architects, Lincoln Center. Harrison was a close personal friend and mentor of Nelson Rockefeller, related to him by marriage.[12] The second was the intertwined affairs of the implementers, especially the Rockefellers and Robert Moses. In the 1920s, Moses closely collaborated with John D. Rockefeller, Jr in setting up the Cloisters Museum in Fort Tryon Park and Palisades Parkway; in the 1940s, he worked with Nelson on the United Nations (although in 1968 Nelson would fire him by creating the MTA, absorbing the Triborough Bridge and Tunnel Authority); in the 1950s and 1960s, he collaborated with John D. III on Lincoln Center and with David, on downtown redevelopment, especially the planned, but never executed, Lower Manhattan Expressway. These relationships (between patron and architect and implementer and sponsor), built on mutual respect and shared visions, fostered more than a billion dollars in investment in projects that were strategically located and had impacts disproportionate to their size.

Rockefeller Center Sets the Pace

Built between 1931 and 1939 on 12 acres of leased land, the first phase of Rockefeller Center accidentally became the world's first 'skyscraper city'. Its success led to the acceptance of a similar arrangement for the United Nations. And it was the prototype for such later Super Capital commercial expressions as London's Canary Wharf and Paris's La Défense. These mixed-use complexes would supply the offices, retail spaces and amenities supportive of economic activities with national and global reach.

The design of Rockefeller Center was accidental, a result of financial necessity. Originally driven by plans to relocate the Metropolitan Opera House there, it evolved into a commercial project when the Depression forced the Met's withdrawal. John D. Rockefeller Jr, who had signed a twenty-year, $3,300,000 per annum lease with the site's owner, Columbia University, found himself holding land yielding only $300,000 in annual revenues. This obligation caused an about-face by the Metropolitan Opera Corporation (the name changed to the Rockefeller Center Corporation later), a private company acting as the project's management entity. Its leaders ordered the already-hired architectural teams[13] to produce a project whose revenues covered the lease.

The seedy site was not ideal for high-end commercial use since it was bordered by an elevated subway and housed bars, houses of ill-repute and other undesirable uses (see figure 18.1a). However, the teams responded with a dramatic fourteen-building group of office, retail, entertainment and open space woven into the city's grid and linked together by an intricate underground pedestrian and service network (see figure 18.1b). Of particular note was its architectural unity, exceptional open space and extensive use of public art.[14]

Figure 18.1(a). The site of Rockefeller Center featured densely packed, low-scale tenements, seen here immediately above St Patrick's Cathedral. (**b and c**) Rockefeller Center included fourteen buildings including its centrepiece, the RCA building, open space and public art.

Between 1931 and 1940, the Corporation implemented the plans, building 5.5 million rentable square feet. The lead commercial leaseholder, Radio Corporation of America (RCA), represented the rapidly expanding entertainment and news sectors. Surrounding tenants included soon-to-be media giants: Time-Life, Associated Press and RKO. Rockefeller enterprises, Standard Oil, Esso, Sinclair Oil and Eastern Airlines, rounded out the anchors. Rockefeller's sales force courted international retail and commercial tenants for the British Empire, La Maison Française and International buildings, located on fashionable Fifth Avenue frontage.[15] (The conscious marketing of the space in Europe marked important international outreach at a time when most of America was firmly isolationist.) Finally, they included the Radio City Music Hall, a movie theatre, the trademark outdoor skating rink and several restaurants. In the words of architectural historian Carol Krinsky, Rockefeller Center represented 'the anticipated city of the future'.[16]

This design had three sources: the École des Beaux Arts tradition with its formal, symmetrical building/open space site planning; city regulations, especially the zoning and building code ordinances governing towers that allowed the movement of development rights to permit construction above theatres; and the real estate market that determined the amount of rentable square footage.

Costing $125 million (in 1929 dollars), the complex was fully rented by 1940, eliminating its operating deficits in 1941.[17] It retired its mortgage in 1950.[18]

The United Nations Establishes New York as a World Capital

In 1945, the United Nations elected to build its headquarters in the United States, after a good deal of manoeuvring among the postwar victors. Russia cast the deciding vote, counterbalancing a French/British desire to retain the UN in Europe.[19] Immediately, San Francisco, Boston, Philadelphia and New York competed for the honour.[20] New York City Parks Commissioner Robert Moses and a seven-member, blue-ribbon committee (including Nelson Rockefeller and Winthrop Aldrich, Nelson's uncle and president of Chase National Bank) developed the Big Apple's bid. Offering 350 acres in Flushing Meadow Park, the site of the 1939 World's Fair (see figure 18.2), they envisioned a world capital with impressive ceremonial spaces (a 750,000 square foot terrace bordered by fifty-one massive pylons – one for each UN country – arranged along a 2,000 foot axis; a large, bone-white amphitheatre), four principal buildings (with enough space for 15,000 workers), parking (for 2,200 vehicles), a new commuter rail station, and, minimally, 800 units of off-site housing (see figure 18.2b). They touted the area's proximity to Manhattan and its five-minute distance from LaGuardia Airport. The proposal carried an $85 million price-tag.[21]

The site selection committee rejected this offer (as well as another in Westchester County) as being too suburban. At the last minute, Rockefeller, working with Moses and Wallace Harrison, saved the day. They proposed six blocks between East 42nd and 48th Streets from First Avenue to the East River Drive in the slaughterhouse district on Manhattan's midtown waterfront (see figure 18.3). The site, about 17 acres, owned by real estate speculator William Zeckendorf, already had a schematic design for 'X City' conceived by Harrison and anchored by new halls for the Metropolitan Opera House and New York Philharmonic. In a tense, harried five days, Harrison re-labelled the buildings,

Figure 18.2. Robert Moses's and his team presented the New York City bid for the United Nations, a massive model accompanied by renderings by Hugh Ferris.

Rockefeller convinced his father to underwrite the $8.5 million purchase price and Moses smoothed the way through city and state barriers, including securing $2 million for additional land purchases, tentative permissions for street closings, zoning changes, re-routing First Avenue and extending bulkhead into the East River.[22]

On 15 December 1946 New York won the bid in a landslide vote, forty-six to seven. The editors of the *New York Times* gleefully celebrated, declaring that 'New Yorkers feel themselves citizens of the world through the very process of their growth' and predicted that the headquarters 'will not remotely resemble the group of classic buildings that housed the League of Nations' but will 'be a modern skyscraper . . . perhaps something like Radio City'.[23] They were correct.

In the following months, the city, state and federal governments tidied up the loose ends. New York City re-zoned the area and made substantial infrastructure improvements. The state passed legislation giving the United Nations jurisdiction over the property and making it tax-exempt. It also authorized the city to relocate on-site residents and businesses. The US Congress ratified the UN immunity and released John D. Rockefeller from the federal gift tax on the funds used to purchase the site.[24]

The United Nations quickly appointed Harrison as Director of Planning. It also selected a Board of Design, sixteen architects drawn from the member-nations and carefully screened for their adherence to the International Style and *Congrès internationaux d'architecture moderne* (CIAM) principles.[25] The UN believed that as representative of a new world order, the headquarters should be forward-looking and innovative. This desire translated into a desire for Modernist architecture. The Board of Design, which included Le Corbusier, met forty-five times from February to June 1947 before reaching an

Figure 18.3. The site selected for the United Nations was smaller than originally envisioned (top). The United Nations rose dramatically from an area that had recently been the home of slaughter houses (bottom).

agreement on the site plan and buildings. In the end, they recommended a superblock scheme conceived by Le Corbusier and detailed by his young disciple, Brazilian Oscar Niemeyer. After UN acceptance, Harrison supervised the plan's execution, completing construction in 1953.[26]

Although owing a debt to Rockefeller Center, the plan was thoroughly cast in the International Style. Eliminating five streets, the designers transformed the site into a large superblock. They anchored the southern end with the complex's major office building, the Secretariat, a thin, unadorned thirty-nine storey rectangle with shimmering blue-green curtain glass walls framed in white marble.[27] At its north side, they placed the low-slung General Assembly building having an upwardly curving roofline topped by a small dome (see figure 18.3). East of the Secretariat, they built an office block. They left the remainder of the site as a park. A quick turnoff from First Avenue, edged with a row of flagpoles flying the colours of the member-nations, provided officials with a dignified entrance to the Secretariat and the delegates' entrance to the General Assembly. The public had a separate entrance adjacent to the park. The exterior and interior spaces featured public art focusing on themes of peace.

This complex cost about $93 million (about $683 million in 2000 dollars) inclusive of site acquisition, improvements and construction. The United States government gave the UN a $65 million interest-free loan, the city put in $20 million for improvements and additional land purchases, and John D. Rockefeller contributed $8.5 million for site acquisition.[28]

In later years, the UN and its members added structures to the site and its surroundings. In 1963 the UN built the Dag Hammarskjold Library located south of the Secretariat. Between 1976 and 1987, it also constructed five buildings along First Avenue for office and hotel space, employing tax-exempt bonds issued by the United Nations Development Corporation (UNDC) to finance these new facilities. This New York state development corporation facilitated fund raising, but, when the 1986 federal tax reform laws curtailed the use of bonds for these purposes, UNDC development activities evaporated.[29] Finally, some countries converted opulent residences adjacent to the United Nations for their missions, forming an ill-defined diplomatic district on Manhattan's East Side.

Lower Manhattan/World Trade Center provide Space for Global Finance

In 1955, David Rockefeller, an officer of the nation's second largest bank, Chase Manhattan Bank, became active in Lower Manhattan, the city's once dominant downtown then in decline. Its most recent office building dated from the 1920s and Midtown, with its substantial regional locational advantages, superior office space and amenities as at Rockefeller Center, was rapidly becoming the first choice for new commercial construction. To arrest this trend, Rockefeller convinced Chase not only to build a new $140 million downtown headquarters but also to commission the design from Skidmore Owings and Merrill (SOM), a firm noted for Modernism. The SOM architectural firm combined the 2½ acre site as a superblock that featured a two-million square foot tower rising 800 feet from a huge plaza. When completed in 1960, it was the first International-Style building in Lower Manhattan. Chase's bold decision to invest in downtown would stimulate 60 million square feet of new office construction in the next three decades.

Rockefeller led this boom by rallying the business elite to form the Downtown Lower Manhattan Association (DLMA), an advocacy

Figure 18.4. David Rockefeller (centre) worked with Mayor Robert Wagner (left) in Lower Manhattan to revitalize the area for offices.

organization (see figure 18.4). Between 1958 and 1963, the DLMA commissioned two SOM-authored land-use and transportation plans that set the agenda for the following decades. For example, to enhance circulation, SOM specified a Lower Manhattan Expressway, a pet project of Robert Moses that citizen activist Jane Jacobs would ultimately help defeat. SOM also resurrected an idea for a World Trade Center (WTC), first articulated by Governor Thomas Dewey in 1947, to stimulate economic development.

Nelson Rockefeller, by now governor of New York, readily adopted the World Trade Center concept, commissioning a Port Authority of New York and New Jersey (PA) study. The PA appointed an architectural advisory board, including Wallace Harrison, Gordon Bunshaft and Edward Durrell Stone, who reported favourably. Extremely controversial, the WTC proposal became mired in state and local politics. Only after agreeing in 1962 to locate it on 16 acres on the West Side and to take over an ailing New Jersey commuter rail line, did the PA receive the required legislative go-ahead. That same year, the PA selected the WTC designer, Michigan architect Minoru Yamasaki, giving him programmatic but not aesthetic direction. When the PA inaugurated the WTC in 1973, it was well on its way to completion with 10 million square feet of offices, ½ million square feet of retail and ½ million square feet of hotel.

In designing the complex, Yamasaki, an International-Style renegade, blocked off five through streets to create a superblock, but added stylistic embellishments, called 'New Formalism', to the individual buildings. He cast the silver aluminum-clad Twin Towers, each of whose one hundred and ten storeys was an acre in area, as the dominant feature. As with Rockefeller Center, a multilevel underground retail and transportation concourse provided pedestrian circulation and links to regional mass transit.[30] Visible for miles, the towers transformed the skyline and ultimately became so associated with global finance and politics that terrorists

Figure 18.5. Nelson Rockefeller (left) was also intimately involved in the revitalization of Lower Manhattan. Both he and his brother believed that with re-investment in modern office space it would continue to be the centre of world capital.

targeted their destruction, first in 1993 and later in 2001. These attacks heightened the symbolic importance of the World Trade Center and New York City as a Super Capital. Fittingly, the redesign of the World Trade Center reinforces this theme in featuring the world's tallest skyscraper and expanding former uses in a defiant assertion of the city's Super Capital status.

In building the WTC, the PA employed a complex implementation strategy, featuring administrative and financial wizardry. For example, to avoid the national government's oversight of the whole agency that under the US Consitution would be triggered if the PA assumed responsibility for a bi-state commuter rail, the PA created a subsidiary corporation. To gain city permissions for street-closings necessary for the superblock, it agreed to make payments in lieu of taxes and land-fill the excavation debris – ultimately providing the territorial nucleus for the 92 acre Battery Park City (see figure 18.6).[31] To finance the construction, it floated low interest revenue-anticipation bonds. In addition, it worked out special leasing arrangements to provide such tourist-attracting amenities as the Windows on the World restaurant and Observation Deck. In the end, the WTC cost $700 million (1970s dollars), a sum far exceeding its original estimates.

Operating under a legislative mandate that restricted tenants to those having demonstrable links to international trade, the PA had a rough

Figure 18.6. The twin towers of the World Trade Center rose 110 storeys and were visible for miles around. The landfill from the excavation (outlined in photo) would serve as the site for Battery Park City, housing more offices, housing and public amenities.

time with leasing, especially since the space became available during a recession. Through the 1970s and 1980s the PA and state agencies rented 40 per cent of the WTC space. By the 1990s, however, the government presence declined as domestic banks, stockbrokers and insurance companies, now part of a global economy, met tenant qualifications.

Lincoln Center establishes Cultural Prominence

Lincoln Center had its origins in Robert Moses's urban renewal efforts undertaken according to the 1949 Housing and Slum Clearance Act and succeeding legislation. Moses employed the federal programme to modernize the city, especially the slum-ridden areas of Manhattan, with new housing (as required by law) and educational and cultural facilities. Among his schemes was a seventeen block (80 acre) area called Lincoln Square (see figure 18.7). He sketched out the housing, lined up Fordham University and then turned to the Metropolitan Opera, long in search of a new home. Using Wallace Harrison as an intermediary, he offered a one block site to the Opera's board. Realizing the enormous amount of fund raising that a new facility would necessitate, the Board enlisted John D. Rockefeller III's help. Rockefeller, whose interests had been in Asian affairs, was attracted for two reasons. First, through his international work, he had seen that the United States lacked a strong identity for its cultural activities. Second, he wanted to give something back to his home city.

After a few months of exploratory meetings, Rockefeller, Harrison and the leadership of the Metropolitan and the New York Philharmonic (participating because of the threatened loss of its lease in Carnegie Hall), persuaded Moses to undertake a revolutionary and more ambitious endeavour. They called for a performing arts centre, the first of its kind in the United States, encompassing the Metropolitan Opera, Philharmonic, Julliard School of Music, ballet, repertory theatre, a library and a museum.[32] Moses readily agreed, added three more blocks and, by 1954, unveiled a preliminary plan. While this scheme would evolve over time, it envisioned a grandiose cultural centre worthy of a Super Capital.

Lincoln Center's promoters were quick to claim its significance, citing its role in establishing the city's national and international cultural hegemony: 'Lincoln Center will add another capital [to New York] as important to the performing arts as the United Nations is to world affairs, Wall Street to finance and Fifth Avenue to fashion'.[33] An official of Milan's La Scala observed: 'Up to now music has two world capitals. When Lincoln Center is built there will be only one'.[34]

As with the United Nations and Rockefeller Center, Lincoln Center's plan resulted from collaboration among prominent architects, but in this case, each selected by the independent, constituent organizations to design their respective buildings. They formed a Board of Architects, chaired by Harrison (see figure 18.8). In this instance, Harrison co-ordinated the design, but he did not wield the same decision-making authority as he did for the United Nations.

This group settled on a modified superblock oriented on an east-west axis. They consolidated three blocks to make room for a central plaza surrounded by the principal buildings: with the

Figure 18.7. The original site of Lincoln Center was designated a slum under the Housing and Slum Clearance Act of 1949 (left). The plan for Lincoln Center had classical antecedents but was also influenced by CIAM principles (right).

Opera in the centre and the State Theater and Philharmonic Hall to south and north respectively. Moses had required the inclusion of a Damrosch Park for free concerts which the designers placed on the south-west corner of the site, balancing it with the Vivian Beaumont/New York Public Library complex on the north-west. For the fourth block, linked by a pedestrian bridge, they placed the Julliard School and, later, a building for the Film Society and residential uses (see figure 18.7). The complex had International Style and historic features, making it an early prototype for postmodern design. The use of the superblock and sleek, white travertine-clad, minimally decorated structures combined with a classical building arrangement and plaza treatment made it a twentieth-century Campidoglio, Rome's renowned civic/cultural plaza complex.[35]

A private entity, the Lincoln Center for the Performing Arts Corporation (LCPAC), undertook implementation. It acted as a general leaseholder and manager for the centre. Its board, consisting of representatives of the constituent organizations, elected Rockefeller, president. A powerful group, it enlisted President Dwight D. Eisenhower for the 1959 groundbreaking. In the early years, it worked closely with Robert Moses, first in his urban renewal capacity and later in his World's Fair chairmanship, to ensure favourable conditions for securing land and financing. The last building was completed in 1992.

The complex cost more than $185 million or

Figure 18.8. The designers of Lincoln Center for the Performing Arts included architects of international fame. From left to right: Edward Matthews (SOM), Philip Johnson, Joseph Mielziner, Wallace Harrison, John D. Rockefeller (patron) Eero Saarinen, Gordon Bunshaft, Max Abramowitz and Peitro Belluschi.

over a billion dollars in contemporary terms. The private sector raised $144.4 million, 70 per cent of this attributable to Rockefeller's fund-raising efforts.[36] The public sector contributions included the land write-down derived from the urban renewal formula. (To minimize costs, the LCPAC purchased only the land under the building footprints, leaving the remainder in city ownership.) The city allotted $12 million towards the State Theater and public library and built the parking garage. Under Governor Nelson Rockefeller, the state gave $15 million for the State Theater, financing it as part of the 1964 Worlds Fair, an interesting concept because the fairgrounds were several miles distant.

Diffusion vs Originality

Taken together, the four projects, Rockefeller Center, United Nations, Downtown Manhattan/World Trade Center and Lincoln Center for the Performing Arts, constitute the key physical elements of the Super Capital of New York. The distinguishing characteristic of these projects is the unity of their leadership. For four decades, a small group promoted them. They held a bold, but not particularly co-ordinated vision. They were united in their desire to build big, symbolically important complexes. They were singularly dedicated to bolstering New York's central place in the region, nation and world, transforming failing or undeveloped precincts of the city to achieve this goal. They were opportunistic in their choices of sites and activities. They had no plan, other than to modernize a nineteenth-century city with, what was considered at the time, the best of urbanism. They invented or expanded existing tools to implement their schemes, crafting solutions as they went along.

Each project represents a place-specific solution to a design/economic/political challenge. However, on the level of design, the leadership adopted the International Style, especially the postwar use of the superblock. They believed that this type of site plan, in erasing the street grid, allowed the freedom to design unique large-scale complexes whose scale and architecture distinguished them as emblems of a new world order, one in which New York City was supreme in the 1950s.

As the leaders conceived the projects, they consciously looked to other people or cases for instruction. For example, Harrison insisted that Le Corbusier be included on the UN team, despite Corb's reputation as being difficult to work with. Also, Harrison led two trips to Europe to view theatres and performance halls in hopes of emulating them for Rockefeller and Lincoln Centers.

Perhaps the most startling feature of the projects was how collaborative they were. The use of boards or committees of design, the cooperation between the different levels of government, the invention of public/private partnership arrangements to master financing are representative.

Finally, these projects did not take a long time to execute, once the decision was taken to build them. Although on average, they were not entirely completed for about forty years, their respective cores were done in six to eight years. The first phases of Rockefeller Center took eight years; the UN, six years; World Trade Center, six years; and Lincoln Center, seven years.

New York's emergence as a Super Capital occurred rapidly and was not the result of an articulated, grand master plan. While individual site or precinct plans guided development, no written document existed to plot the whole course. Nonetheless, a planning framework, one that included providing for transportation and housing and an acceptance of large, transformative projects, was ever-present in the thinking of the leadership, a unique partnership of public and private sector participants. They assumed that New York City could and would become increasingly important in the postwar period and incrementally contributed the features that would make the city a Super Capital.

Epilogue

Super Capitals are dynamic places – they must be to retain their status. Today, all four prototype projects are changing. Rockefeller Center is undergoing a full-scale modernization and the United Nations will also be renovated, funded by a revitalized United Nations Development Corporation. The rebuilt World Trade Center, destroyed by terrorists in 2001, will re-emerge as a defiant symbol of New York's Super Capital status. Finally, Lincoln Center has recently embarked on an ambitious plan to rejuvenate its ageing structures.

NOTES

1. See Hall, chapter 2 in this volume.
2. Jackson (1984); Hall (this volume), chapter 2.
3. Jackson (1984), p. 319.
4. Strictly speaking, the Commissioners' Plan (1811) and the New York City Improvement Plan (1907) preceded these efforts. The state-authorized 1811 plan laid out the Manhattan grid. The 1907 report identified the need to connect the city's land area with bridges and to have a coordinated street system among the five boroughs.
5. The United States would lift its restrictive immigration policies dating from the 1920s in the 1960s resulting in New York again becoming an ethnic melting pot. By 1990, 28 per cent of its population would be foreign-born.
6. New York City Planning Commission and Richards (1970); Regional Plan Association of New York (1968); Yaro and Hiss (1996).
7. New York was not a pioneer in the use of public authorities for these kinds of activities, London employed the Port of London Authority in 1909; New York, however, made extensive use of the device and demonstrated to the world its effectiveness in efficiently accomplishing large scale projects.
8. See Caro (1975); Schwarz (1993).
9. Newhouse (1989), p. 187.
10. David Rockefeller also played a leadership role in other activities ranging from the Rockefeller University expansion (1950s) and the development of the 1,000-unit Morningside Gardens (1957), the city's first integrated, middle-income housing project.
11. These are the major projects. There are many more including Sloan-Kettering Memorial Hospital, International House, Museum of Modern Art, Riverside Church.
12. Harrison's wife, Ellen Milton Harrison was the sister of David Milton, the husband of Abby Rockefeller, Nelson's sister. From the Rockefeller Center days, Nelson and Harrison had a close relationship, enjoying daily morning coffee and other personal interchanges. Sixteen years older than Nelson, Harrison was a mentor, friend and employee. When President Franklin Delano Roosevelt named Nelson, Co-ordinator for Inter American Affairs, Nelson persuaded Harrison to reduce his practice, move to Washington DC and serve as his director of cultural affairs, a position he held from 1940 to 1946.
13. The architectural teams consisted of: Reinhard & Hofmeister; Corbett, Harrrison and MacMurray; and Hood, Godley & Fouilhoux.
14. In the postwar period working with Wallace Harrison, the corporation added five new buildings, including new headquarters for Time-Life and associated open space on Sixth Avenue. These new buildings also included: Sperry Corporation (Emery Roth, 1961); Exxon (Harrison & Abramowitz, 1971); McGraw-Hill (Harrison & Abramowitz, 1972); Celanese (Harrison and Abramowitz, 1973).
15. According to Krinsky (1978), the Corporation went so far as to secure favourable customs treatment for their tenants with passage of federal legislation enabling payment of duties on products not as they entered the country as was customary practice, but on their sale.

16. Krinsky (1978), p. xxiii.

17. According to Krinsky (1978), the author of the definitive work on Rockefeller Center, Met Life extended a $65 million, 5 per cent mortgage but the Corporation only used $45 million. John D. Rockefeller Jr added other funds.

18. The design of the later expansion was much less successful. The public spaces were so dysfunctional that the Rockefeller Brothers Foundation would underwrite public space research by urbanist William H. Whyte. His influential findings (Whyte, 1980) became the basis for the city's zoning ordinance reform in the 1970s and transformed thinking on this topic. In financing the expansion, the Rockefeller Center Corporation did not underwrite it but entered into separate partnership agreements with each building's sponsor.

19. Dudley (1994).

20. 'Interim UNO Site Here is Discussed', *New York Times*, 9 January, 1946, p. 10.

21. Berger, Meyer, 'City Sets Up Scale Model World Capital for the United Nations', *New York Times*, 9 October, 1946, p. 3.

22. Warren Moscow, 'Rockefeller Offers UN $8,500,000 Site on the East River for Skyscraper Center', *New York Times*, 12 December, 1946, p. 1; Special Subcommittee of the United Nations Headquarters Committee, 'Text of Headquarters Committee Report on the U.N.', *New York Times*, 13 December, 1946, p. 4.

23. 'Capitol of Nations', *New York Times*, 16 December, 1946, p. 22.

24. 'City Sets Hearings on U.N. Zoning Plan', *New York Times*, 9 January, 1947, p. 3; 'Dewey Signs Bill for U.N. Site Here', *New York Times*, 1 March, 1947, p. 8; 'U.N. to Help Find Homes', *New York Times*, 1 March, 1947, p. 8; 'U.N. Building Plan Approved by the City', *New York Times*, 28 March, 1947, p. 1.

25. Harrison worked with the previous owner, William Zeckendorf on plans for the site, 'X City', and thus knew the conditions and possibilities.

26. Dudley (1994).

27. According to Newhouse (1989), Pietro Belluschi's Equitable Building (Portland, Oregon) provided a model for the Secretariat.

28. New York City Bar Association (2001), p. 1.

29. *Ibid.*, p. 8.

30. He arranged six buildings around a 5 acre marble-paved plaza on which he centred a 90-foot-diameter fountain, holding Fritz Koenig's bronze 'Globe' sculpture – the one that miraculously survived the 2001 attack.

31. Gordon (1996).

32. The performing arts centre idea was emerging in other American cities. Washington, DC started one, later named the Kennedy Center, at about the same time.

33. 'Text of Statement by Lincoln Center for the Arts,' *New York Times*, September 12, 1957, p. 28.

34. *Ibid.*

35. The architects were: the Opera (Harrison, 1996); Theatre (Johnson 1964); Concert Hall (Abramowitz 1962); Park (Eggers and Higgins 1969); Beaumont (Saarinen/Bunshaft 1965); Julliard (Belluschi, 1968); and Film Society (Brody and Associates and Abramowitz Kings-land Schiff, 1992).

36. The Rockefellers, their agents and relatives directly provided at least $35.4 million.

Chapter 19

What is the Future of Capital Cities?

Peter Hall

What is the future for capital cities? It all depends on the capital city. As Chapter 2 showed, they come in many shapes and sizes: nation-state political capitals, supra-national political capitals, sub-national or provincial capitals, commercial capitals. What will happen to each category will depend on global trends, mediated through features that are specific to countries or continents.

There are two key global trends, independent but closely related: the *globalization* of the world economy, and what can only be called its *informationalization* (an ugly but necessary word): the shift in the economies of advanced economies, away from manufacturing and goods-handling and towards service production, particularly into advanced services that handle information.[1] Neither is new: there was a species of globalization at the time of the Renaissance in Florence, whose bankers played a principal role in it, and again in the nineteenth century when London played the central role.[2] Nor is the shift to the informational economy, which was already recognized over half a century ago;[3] by the 1990s, in typical advanced countries, between three-fifths and three-quarters of all employment was already in services, while between one-third and one-half was in information handling: there can be little doubt that by 2025 80–90 per cent of employment will be in services, and up to 60–70 per cent will be in information production and exchange.[4] Its most significant expression is the emergence of the so-called *Advanced Business Services*: a cluster of activities that provide specialized services, embodying professional knowledge and processing specialized information, to other service sectors.[5]

Globalization and informationalization together result in the increasing importance of cities at the very top of the hierarchy, the so-called *world cities* or *global cities*. These, too, are by no means new: ancient Athens or renaissance Florence could be regarded as examples,[6] and they were recognized in academic literature throughout the twentieth century.[7] A study of four world cities – London, Paris, New York and Tokyo – distinguished four key groups of advanced service activity: *Finance and Business Services*, 'Power and Influence' (or 'Command and Control'), *Creative and Cultural*

Industries and *Tourism*. All are service industries processing information in a variety of different ways; all demand a high degree of immediacy and face-to-face exchange of information, so that strong agglomeration forces operate; all are synergistic, with many key activities – hotels, conference centres and exhibition centres, museums and galleries, advertising – overlapping from one category to another and so operating critical interstitial spaces. Thus, strong agglomeration tendencies apply not only within each sector, but also between them.[8]

The most important advance in our understanding of the new global hierarchy of cities has come from the *Globalization and World Cities* (GaWC) Study Group and Network at Loughborough University, led by Peter Taylor. They argue that previous approaches – even key contributions as

Table 19.1. The Loughborough Group 'GaWC' inventory of world cities. (Cities are ordered in terms of world city-ness values ranging from 1 to 12. Capital cities are italicized.)

A. Alpha World Cities

12: *London, Paris,* New York, *Tokyo*

10: Chicago, Frankfurt, Hong Kong, Los Angeles, Milan, *Singapore*

B. Beta World Cities

9: San Francisco, Sydney, Toronto, Zürich

8: *Brussels, Madrid, Mexico City,* São Paulo

7: *Moscow, Seoul*

C. Gamma World Cities

6: *Amsterdam*, Boston, *Caracas*, Dallas, Düsseldorf, Geneva, Houston, *Jakarta*, Johannesburg, Melbourne, Osaka, *Prague*, Santiago, *Taipei, Washington*

5: *Bangkok, Beijing, Rome,* Stockholm, *Warsaw*

4: Atlanta, Barcelona, *Berlin, Buenos Aires, Budapest, Copenhagen,* Hamburg, Istanbul, *Kuala Lumpur, Manila,* Miami, Minneapolis, Montreal, Munich, Shanghai

D. Evidence of World City Formation

D(i) Relatively strong evidence

3: Auckland, *Dublin, Helsinki, Luxembourg,* Lyon, Mumbai, *New Delhi,* Philadelphia, Rio de Janeiro, Tel Aviv, Vienna

D(ii) Some evidence

2: *Abu Dhabi*, Almaty, *Athens*, Birmingham, *Bogotá, Bratislava,* Brisbane, *Bucharest, Cairo,* Cleveland, Cologne, Detroit, Dubai, Ho Chi Minh City, *Kiev, Lima, Lisbon,* Manchester, Montevideo, *Oslo,* Rotterdam, *Riyadh,* Seattle, Stuttgart, The Hague, Vancouver

D(iii) Minimal evidence

1: Adelaide, Antwerp, Århus, *Athens,* Baltimore, Bangalore, Bologna, *Brasília,* Calgary, Cape Town, *Colombo,* Columbus, Dresden, Edinburgh, Genoa, Glasgow, Göteborg, Guangzhou, *Hanoi,* Kansas City, Leeds, Lille, Marseille, Richmond, St Petersburg, *Tashkent, Tehran,* Tijuana, Torino, Utrecht, *Wellington*

Sources: Beaverstock, Taylor and Smith (2000); Taylor *et al.* (2002); Taylor (2004).

those of Friedmann and Sassen – concentrate on measuring urban *attributes*, ignoring the mutual relationships, the *interdependencies*, between cities.[9] The Loughborough team do not attempt to measure these relationships directly, since data on flows of information are scarce; instead, they use a proxy, the internal structures of large Advanced Producer Services firms, expressed by the relationship between head office and other office locations. The result (table 19.1) shows that only two in five of the so-called Global Cities – those italicized in the table – are national capital cities. This is particularly interesting since it appears to confirm what many critics of globalization have suggested: that there is at least a partial disconnect between the operation of the twenty-first-century global economy and that of the political system of nation states that arose between the Middle Ages and the nineteenth century.

To some extent, this is an artificial construct, arising from details of the political organization of federal states: at the very head of the table, New York is not a capital because of the decision of the Founding Fathers to create a political capital on federal territory (although, down to that point, Philadelphia not New York was the capital); in "the second rank, only one of the six, Singapore, is a national political capital (and that a city state) and only one, Milan, is a provincial capital, while Hong Kong, until 1997 a city state, now has the same status in relation to Beijing; in the third," all four cities belong to federal states with special political capitals. Going further down the table, it becomes increasingly evident that the same principle constantly repeats itself: many of the world's top cities are in effect provincial capitals in large, economically-advanced federal states, in Europe (Germany, Spain), North America (Canada, the United States), Australia, Brazil or South Africa. But also, important provincial cities in centralized countries appear: Osaka, Istanbul, Shanghai, Lyon, Manchester, Birmingham, Rotterdam. These represent genuine independent centres of commercial power.

For the future, two contradictory tendencies seem likely to do battle. The first is the principle of increasing *centralization* of power in relatively few cities at the top of the global hierarchy, that will increasingly manage the key Advanced Producer Services through which the global economy is controlled. But this power will be commercial in character, only coincidentally corresponding to the distribution of political power through national capitals. This particularly applies to such cities in the large federal countries, but also to some European countries that are similarly organized. However, the main focus of the powerful all-purpose capital city will continue to be the national capitals of Europe, particularly since the rejection in 2005 of the proposed European constitution has dealt a severe blow to any hope of a more federal European Union. Indeed, in Europe the national capitals seem destined to enjoy an enhanced role, since the European Union's policy of encouraging a more polycentric pattern of urban development Europe-wide, contained in the 1999 *European Spatial Development Perspective*, is having the paradoxical result of encouraging a more *monocentric* form of urban development at the national level, in and around the capital cities which form the magnets for immigration of local labour and inflow of international capital.[10] This process, very evident in the 1980s and 1990s in such capitals as Dublin, Lisbon and Madrid, is now equally evident in and around Eastern European capitals like Riga, Tallinn, Warsaw and Budapest, even in advance of their countries' formal accession to the EU in May 2004. One key reason, again underlined by the votes on the constitution, is that linguistic and cultural divides seem certain to continue to play a much larger role in Europe between countries (and, in Belgium and perhaps Spain at least, within them) than in more newly-settled countries with more homogenous backgrounds like the United States, Australia or Brazil – or, to take the opposite extreme of the world's largest and oldest homogenous nation-state – China.

Within such a global framework and its local variations, there will of course be dynamic shifts. The most important, without doubt, will be the rise of Beijing to its appropriate place in the top range of global cities. But, given its historic position as political capital as against the great commercial cities of the Chinese seaboard, it will continue to share its global functions with Shanghai – still, on the GaWC system, astonishingly low in the hierarchy – and Hong Kong (not to mention the latter city's local rivals, Shenzhen and Guangzhou). Similar rises up the hierarchy may occur for other East and South Asian capitals, above all New Delhi (but there, too, sharing global roles with Mumbai and Kolkata), but also Bangkok, Kuala Lumpur, Jakarta and Hanoi. Cities on the Pacific Rim of the dynamic East Asian economies will similarly see their roles enhanced – though this will apply more to their commercial non-capitals or provincial capitals (Sydney, Melbourne, Auckland) than to Canberra or Wellington.

Equally, however, there may well be slippages. The biggest uncertainty of all is the future roles of the capitals of countries that are – actually or potentially – failed states. Most of these are located in sub-Saharan Africa or the Middle East; a few are located elsewhere, including the borders of the former Soviet Union. Paradoxically, especially in Africa, such cities are growing because of in-migration induced by civil war or other disturbances in their rural hinterlands – a source of potential weakness rather than strength. Others face the risk of internal disintegration through disturbances, terrorism and civil war: Beirut and Sarajevo in the 1990s, Baghdad and Monrovia in the 2000s. The first two cases, at least, offer a hope that such disintegration, drastic as it may seem at the time, may be reversed.

As against that, some new capitals may come into existence: one – perhaps provisional, until resolution of the vexed status of East Jerusalem – for a new Palestinian state; others if Iraq, an artificial British invention during the post-World War I disintegration of the Ottoman Empire, in turn ceases to exist. African rulers with grandiose ideas or acute political problems, or both, may decide to establish new capitals on the model of Abuja; the Republic of Korea may at last take the drastic step, so many times considered, of relocating out of Seoul. But such cases are likely to be few and far between. The great historic era of the ends of empires, which began with the British handover of India in 1947 and ended with the disintegration of the Soviet Union in 1991, is over. And most states with ambitions to build new capitals lack the money to realize go with them, unless – an unlikely event – their rulers decide to raid their own numbered Swiss bank accounts for the purpose.

So a certain stability in the list of the world's capital cities is likely to be the order of the day – at least, in comparison with the hectic changes of the last half-century. That of course assumes corresponding stability in the world order, a stability even greater than obtained in the century between the Napoleonic Wars and World War I: a stability, it must be recalled, that included the unrolling of the modern map of Europe and the era of Empire-building in much of Asia and Africa. History surprised us before and will doubtless surprise us again. But the preconditions for another major upheaval seem faint. A new edition of this book, in the year 2055, might not surprise the ghosts of its authors over-much.

NOTES

1. Castells (1989, 1996); Hall (1988, 1995); and Hall and Pain (2006).

2. Hall (1998); Kynaston (1994, 1995, passim).

3. Clark (1940).

4. Castells (1989), p. xxx.

5. Wood (2002).

6. Taylor (2004), pp. 8–15.

7. Geddes (1915); Hall (1966, 1984); Friedmann (1986); Friedmann and Wolff (1982); Sassen (1991).

8. Department of the Environment and Government Office for London (1996).

9. Taylor (2004), p. 8.

10. Hall and Pain (2006).

Bibliography

A + (2000) Concours. *A +*, nos 166 and 167.

Abeels, Gustave (1982) *Pierres et rues: Bruxelles, croissance urbaine, 1780–1980: Exposition organisée par la Société Générale de Banque en collaboration avec la 'Sint-Lukasarchief'*. Brussels: Société Générale de Banque, St.-Lukasarchief.

Abercrombie, P. (1910) Washington and the proposals for its improvement. *Town Planning Review*, **1**(July), pp. 137–147.

Abercrombie, Patrick (1945) *Greater London Plan 1944*. London: HMSO.

Abercrombie, Patrick and Forshaw, J.H. (1943) *County of London Plan 1943*. London: Macmillan.

Aberdeen, Ishbel Gordon, Marchioness of (1960) *The Canadian Journal of Lady Aberdeen*. Toronto: Champlain Society.

Adams, T. (1916) Ottawa-Federal Plan. *Town Planning and the Conservation of Life*, **1**(4), pp. 88–89.

Albuquerque, José Pessôa Cavalcanti de (1958) *Nova metrópole do Brasil: relatório geral de sua localização*. Rio de Janeiro: Imprensa do Exército.

Allen, W. (2001) *History of the United States Capitol: A Chronicle of Design, Construction, and Politics*. Washington DC: US Government Printing Office.

Almandoz, Arturo (ed.) (2002) *Planning Latin America's Capital Cities,1850–1950*. London: Routledge.

Ambroise-Rendu, M. (1987) *Paris – Chirac: Prestige d'une ville, ambition d'un homme*. Paris: Plon.

Andreu, P. and Lion, R. (1991) L'Arche de la Défense: a case study. *RSA Journal*, **139**, pp. 570–580.

Anonymous (1908) Wettbewerb um einen Grundplan für die Bebauung von Gross-Berlin (1908) [announcement of the competition for Greater Berlin]. *Der Baumeister*, p. 18B; also in *Wochenschrift des Architekten-Vereins zu Berlin*, p. 275.

Anomymous (1910) *Beurteilung der zum Wettbewerb 'Gross-Berlin' eingereichten 27 Entwürfe durch das Preisgericht*. Berlin: Wasmuth.

Anonymous (1911) *Wettbewerb Gross-Berlin 1910. Die preisgekrönten Entwürfe mit Erläuterungsberichten*. Berlin: Wasmuth.

Anonymous (1930) Der Berliner Platz der Republik. *Wasmuths Monatshefte für Baukunst & Städtebau*, **14**, pp. 51–56.

Anonymous (1946) Zur Ausstellung 'Berlin plant'. Aus der Ansprache des Stadtrats Professor Scharoun zur Eröffnung. *Neue Bauwelt*, **10**, pp. 3–6.

Anonymous (1958) Ideenwettbewerb zur sozialistischen Umgestaltung des Zentrums der Hauptstadt der Deutschen Demokratischen Republik, Berlin. *Deutsche Architektur*, **7**(10), special attachment.

Anonymous (1960) Ideenwettbewerb zur sozialistischen Umgestaltung des Zentrums der Hauptstadt der Deutschen Demokratischen Republik, Berlin. *Deutsche Architektur*, **9**(1), pp. 3–36.

Anonymous (1988) Rebuilding the capital. *Japan Echo*, **15**(2), pp. 5–7.

APUR (Altelier Parisien d'Urbanisme) (1980) *Schéma Directeur d'Aménagement et d'Urbanisme. Paris Projet*, nos 19–20. Paris: Les Éditions d'Imprimeur.

ARAU (1984) Construire l'Europe en détruisant la ville! in *Bruxelles vu par ses habitants*. Brussels: ARAU, pp. 116–123.

Arca (1993) Il Parlamento Europeo a Bruxelles = The European Parliament HQ. *Arca*, no. 74, pp. 42–47.

Arndt, Adolf (1961) *Demokratie als Bauherr*. Berlin: Verlag Gebrüder Mann.

Arndt, Karl, Koch, Georg Friedrich, and Larsson, Lars Olof (1978) *Albert Speer. Architektur*. Berlin: Propyläen Verlag.

Aron, Jacques (1978) *Le tournant de l'urbanisme bruxellois 1958–1978*. Brussels: Fondation Joseph Jacquemotte.

Aron, Jacques, Burniat, Patrick and Puttemans, Pierre (1990) *Guide d'architecture moderne, Bruxelles et environs, 1890–1990, Itinéraires*. Bruxelles: Didier Hatier.

Association of Urban Management and Development Authorities (2003) *Land Policy for Development in the National Capital Territory of Delhi*. New Delhi: AMDA.

Astaf'eva-Dlugach, M.I. *et al.*, (1979) *Moskva*. Moscow: Stroiizdat.

Åström, Sven-Erik (1957) *Samhällsplanering och regionsbildning i kejsartidens Helsingfors. Studier I stadens inre differentiering 1810–1910*. Social Planning and the Formation of Social Areas in Imperial Helsingfors. Studies on the Inner Differentiation of the City 1810–1910. Helsinki: Mercators Tryckeri.

Australia, Minister for Home Affairs (1908) Instructions from Minister for Home Affairs in 'Yass-Canberra Site for Federal Capital General (1908–09) Federal Capital Site Surrender of Territory for Seat of Gov-ernment of the Commonwealth. National Archives of Australia (NAA: A110, FC1911/738 Part 1).

Babad, Michael and Mulroney, Catherine (1989) *Campeau: The Building of an Empire*. Toronto: Doubleday.

Backheuser, Everardo (1947–1948) Localização da nova capital. *Boletim Geográfico*, nos. 53, 56, 57, 58.

Bacon, Edmund N. (1967) *Athènes à Brasília*. Lausanne: Edita.

Baeten, Guy (2001) The Europeanization of Brussels and the urbanization of 'Europe': hybridizing the city-empowerment and disempowerment in the EU district. *European Urban and Regional Studies*, 8(2), pp. 117–130.

Banham, M. and Hillier, B. (eds.) (1976) *A Tonic to the Nation: The Festival of Britain*. London: Thames and Hudson.

Baran, B. (1985) Office automation and women's work: the technological transformation of the insurance industry, in Castells, M. (ed.) *High Technology, Space, and Society*. Beverly Hills and London: Sage.

Barlow Report (1940) *Report of the Royal Commission on the Distribution of the Industrial Population*, Cmd 6153. London: HMSO.

Barreto, Frederico Flósculo Pinheiro (ed.) (1996–99) *Historiografia da gestão urbana do Distrito Federal: 1956 a 1985*, 6 volumes. Brasília: PIBIC, FAU/UnB.

Bartholomew, H. (1950) *A Comprehensive Plan for the National Capital and Its Environs*. Washington DC: US National Capital Parks and Planning Commission.

Bartoccini, F. (1985) *Roma nell'Ottocento*. Bologna: Cappelli Editore.

Bastié, Jean (1975) Paris: Baroque elegance and agglomeration, in Eldredge, H.W. (ed.) *World Capitals: Towards Guided Urbanization*. Garden City, NY: Anchor/Doubleday.

Bastié, Jean (1984) *Géographie du Grand Paris*. Paris: Masson.

Bater, J.H. (1980) *The Soviet City: Idea and Reality*. Beverly Hills, CA: Sage Publishers.

Bâtiment (1992) L'administration européenne à Bruxelles doit encore s'aggrandir. *Bâtiment*, no. 228.

Bauwelt (1958) Petite Ceinture – Schnellverkehrsstraße in Brüssel. *Bauwelt*, no. 24 pp. 568–569.

Bauwelt (1993) In der ECU-Hauptstadt. *Bauwelt*, no. 84.

Beard, Charles (1923) *A Memorandum Relative to the Reconstruction of Tokyo*. Tokyo: Tokyo Institute for Municipal Research.

Beaverstock, J.V., Smith, R.G. and Taylor, P.J. (2000) World-city network: a new metageography? *Annals of the Association of American Geographers*, **90**, pp. 123–134.

Ben-Joseph, Eran and Gordon, David L.A. (2000) Hexagonal planning in theory and practice. *Journal of Urban Design*, **5**(3), pp. 237–265.

Beers, D. (1987) Tomorrowland: we have seen the future and it is Pleasanton. *Image (San Francisco Chronicle/Examiner Sunday Magazine)*, 18 January.

Berg, Max (1927) Der neue Geist im Städtebau auf der Grossen Berliner Kunstausstellung. *Stadtbaukunst alter und neuer Zeit*, **8**(3), pp. 41–50.

Berger, Martine (1992) Paris et l'Ile de France: rôle national et fonctions internationals, in Berger, Martine and Rhein, Catherine (eds.) *L'Ile de France et la Recherche Urbaine*, Vol. 1. Paris: STRATES/Université de Paris I.

Berger, Meyer (1946) City sets up scale model world capital for the United Nations, *New York Times*, October 9, pp. x, 3.

Berlinische Galerie (ed.) (1990) *Hauptstadt Berlin*.

Internationaler städtebaulicher Ideenwettbewerb 1957/58. Berlin: Berlinische Galerie.

Bernhardt, Christoph (1998) *Bauplatz Gross-Berlin. Wohnungsmärkte, Terraingewerbe und Kommunalpolitik im Städtewachstum der Hochindustrialisierung 1871–1918*. Berlin: de Gruyter.

Berton, K. (1977) *Moscow: An Architectural History*. New York: St. Martin's Press.

Bierut, Bolesław (1951) *The Six-Year Plan for the Rebuilding of Warsaw*. Warsaw: Ksiazka i Wiedza.

Billen, Claire, Duvosquel, Jean-Marie and Case, Charley (2000) *Brussels, Cities in Europe*. Antwerp: Mercatorfonds.

Birch, Eugenie L. (1984) Observation man. *Planning Magazine*, March, pp. 4–8.

Birch, Eugenie L. (1997, 1983) Radburn and the American planning movement: the persistence of an idea, in Krueckeberg, D. (ed.) *Introduction to Planning History in the United States*. New Brunswick, NJ: The Center for Urban Policy Research, Rutgers University, pp. 122–151.

Blau, Eve and Platzer, Monika (eds.) (1999) *Shaping the Great City. Modern Architecture in Central Europe 1890–1937*. Munich: Prestel.

Bleecker, Samuel (1981) *The Politics of Architecture: A Perspective on Nelson A. Rockefeller*. London: Routledge.

Blomstedt, Yrjö (1963) *Johan Albrecht Ehrenström, kustavilainen kaupunkirakentaja*. Helsinki: Helsingfors stads publikationer 14.

Bloom, Nicholas D. (2001) *Suburban Alchemy: 1960s New Towns and the Transformation of the American Dream*. Columbus, OH: Ohio State University Press.

Bonatz, Karl (1947) Der neue Plan von Berlin. *Neue Bauwelt*, **48**, pp. 755–762.

Boyer, Jean-Claude and Deneux, J.-F. (1984) Pour une approche géopolitique de la region parisienne. *Hérodote*, 33(4).

Brasil, Presidência da República (1960) *Coleção Brasília*, 18 volumes. Rio de Janeiro: Serviço de Documentação.

Brasil, Presidência da República (1977) *Plano Estrutural de Organização Territorial do Distrito Federal*, 2 volumes. Brasília: Secretaria de Planejamento.

Brumfield, W.C. (1991) *The Origins of Modernism in Russian Architecture*. Berkeley, CA: University of California Press.

Brumfield, W.C. (1993) *A History of Russian Architecture*. Cambridge: Cambridge University Press.

Brunet, R. *et al*. (1989) *Les villes 'européenes'. Rapport pour la DATAR*. Paris: La Documentation Française.

Brunila, Birger and af Schulten, Marius (1955) Asemakaava ja rakennustaide, in *Helsingin kaupungin historia*, IV:I. Helsinki: SKS/Suomalaisen Kirjallisuuden Seuran kirjapaino.

Brunn, Gerhard and Reulecke, Jürgen (eds.) (1992) *Metropolis Berlin. Berlin als deutsche Hauptstadt im Vergleich europäischer Hauptstädte 1871–1939*. Bonn: Bouvier.

Buck, N., Gordon, I., and Young, K. (1986) *The London Employment Problem*. Oxford: Oxford University Press.

Buck, N., Gordon, I., Hall, P., Harloe, M. and Kleinman, M. (2002) *Working Capital: Life and Labour in Contemporary London*. London: Routledge.

Bundesminister für Wohnungsbau Bonn und Senator für Bau- und Wohnungswesen Berlin (ed.) (1957) *Berlin. Planungsgrundlagen für den städtebaulichen Ideenwettbewerb Hauptstadt Berlin*. Berlin: Ernst.

Bundesminister für Wohnungsbau Bonn und Senator für Bau- und Wohnungswesen Berlin (ed.) (1960) *Hauptstadt Berlin. Ergebnis des internationalen städtebaulichen Ideenwettbewerbs*. Stuttgart: Krämer.

Bundesministerium für Raumordnung, Bauwesen, Städtebau (ed.) (1989) *40 Jahre Bundeshauptstadt Bonn 1949–1989*. Karlsruhe: Müller.

Bundesministerium für Verkehr, Bau- und Wohnungswesen (ed.) (2000) *Demokratie als Bauherr. Die Bauten des Bundes in Berlin 1991–2000*. Hamburg: Junius.

Burg, Annegret and Redecke, Sebastian (eds.) (1995) *Kanzleramt und Präsidialamt der Bundesrepublik Deutschland. Internationale Architekturwettbewerbe für die Hauptstadt Berlin (Chancellery and Office of the President of the Federal Republic of Germany. International Architectural Competitions for the Capital Berlin)*. Basel: Birkhäuser.

Burlen, K. (ed.) (1987) *La Banlieue Oasis: Henri Sellier et les Cités-Jardins, 1900–1940*. Saint-Denis: Presses Universitaires de Vincennes.

Burnham, Daniel H. and Bennett, Edward H. (1909) *Plan of Chicago*. Chicago: Commercial Club.

Burniat, Patrick (1981–1982) Le Quartier Léopold à Bruxelles, création, transformation, perspectives d'avenir. Mémoire de licence en urbanisme et aménagement du territoire. Université Libre de Bruxelles.

Burniat, Patrick (1992) Die Erosion eines Stadtteils: das Leopold-Viertel in Brüssel. *Werk, Bauen + Wohnen*, **79**(46), pp. 10–21.

Caracciolo, Alberto (1974) *Roma capitale*, 2nd ed. Roma: Editori Riuniti.

Caro, Robert A. (1975) *The Power Broker, Robert Moses and the Fall of New York*. New York: Vintage Books.

Carpenter, Juliet, Chauviré, Yvan and White, Paul

(1994) Marginalization, polarization and planning in Paris. *Built Environment*, **20**(3), pp. 218–230.
Castells, M. (1989) *The Informational City: Information Technology, Economic Restructuring and the Urban Regional Process*. Oxford: Basil Blackwell.
Carter, Paul (1987) *The Road to Botany Bay: An Essay in Spatial History*. London: Faber and Faber.
Carter, Paul (1995) Landscapes of Disappearance. Paper presented at the Asia-Pacific Workshop on Associative Cultural Landscapes. World Heritage Convention and Cultural Landscapes, Sydney Opera House, Australia ICOMOS Report.
Castells, M. (1989) *The Informational City: Information Technology, Economic Restructuring and the Urban-Regional Process*. Oxford: Basil Blackwell.
Castells, M. (1996) *The Information Age: Economy, Society and Culture, Vol. I: The Rise of the Network Society*. Oxford: Blackwell.
Castro, Cristovam Leite de (1946) A nova capital do Brasil. *Revista de Emigração e Colonização*, **7**(4).
Cauchon, Noulan (1922) A Federal District Plan for Ottawa. *Journal of the Town Planning Institute of Canada*, **1**(9), pp. 3–6.
Cederna, A. (1956) *I vandali in casa*. Roma-Bari: Laterza.
Cederna, A. (1979) *Mussolini urbanista*. Roma-Bari: Laterza.
Chakravarty, S. (1986) Architecture and politics in the construction of New Delhi. *Architecture and Design*, **11**(2), pp. 76–93.
Chaslin, F. (1985) *Les Paris de François Mitterrand*. Paris: Gallimard.
Cherry, Gordon and Penny, Leith (1986) *Holford: A Study in Architecture, Planning and Civic Design*. London: Mansell.
Chevalier, Louis (1977) *L'Assassinat de Paris*. Paris: Calmann-Lévy.
Ciucci, G. (1989) *Gli architetti e il fascismo*. Torino: Einaudi.
Clark, C. (1940) *The Conditions of Economic Progress*. London: Macmillan.
Clementi, A. and Perego, F. (eds.) (1983) *La metropoli 'spontanea'. Il caso di Roma 1925–1881*. Roma-Bari: Laterza.
Cobbett, William (1830) *Rural Rides*. London.
Cocchioni, C. and De Grassi, M. (1984) *La casa popolare a Roma*. Roma: Edizioni Kappa.
Cohen, E. (1999) *Paris dans l'imaginaire national de l'entre-deux-guerres*. Paris: Publications de la Sorbonne.
Collier, R.W. (1974) *Contemporary Cathedrals: Large-scale Developments in Canadian Cities*. Montreal: Harvest House.
Colton, T.J. (1995) *Moscow: Governing the Socialist Metropolis*. Cambridge, MA.: Harvard University Press.
Congrès Internationaux d'Architecture Moderne (1943) *La Charte d'Athènes*. Paris: Plon.
Connah, Roger (ed.) (1994) Tango Mäntyniemi. The Architecture of the Official Residence of the President of Finland. Helsinki: Painatuskeskus.
Conner, James (1993) Canberra: The Lion, the Witch and the Wardrobe, in Freestone, Robert (ed.) *Spirited Cities: Urban Planning, Traffic and Environmental Management in the Nineties*. Sydney: The Federation Press.
Costa, Cruz (1967) *Contribuição à história das idéias no Brasil*. Rio de Janeiro: Civilização Brasileira.
Costa, Lúcio (1995) *Registro de uma vivência*. São Paulo: Empresa das Artes and EDUnB.
Coulter, Charles (1901) An Ideal Federal City, Lake George, N.S.W. National Library of Australia, PIC R134 LOC 2596.
Creese, Walter L. (1985) *The Crowning of the American Landscape: Eight Great Spaces and Their Buildings*. Princeton: Princeton University Press.
Cruls, Luis (1894) *Relatório da Comissão Exploradora do Planalto Central do Brazil*. Rio de Janeiro: H. Lombaerts.
Cuccia, G. (1991) *Urbanistica, edilizia, infrastrutture di Roma capitale 1870–1990*. Roma-Bari: Laterza.
Cullen, Michael S. (1995) *Der Reichstag. Parlament, Denkmal, Symbol*. Berlin: Bebra Verlag.
Culot, Maurice (1974) The rearguard battle for Brussels. *Ekistics*, **37**(219), pp. 101–104.
Culot, Maurice (1975) ARAU Brussels. *Architectural Association Quarterly*, **7**(4), pp. 22–25.
Culot, Maurice, René Schoonbrodt, and Krier, L. (1982) *La Reconstruction de Bruxelles: recueil de projets publiés dans la revue des Archives d'architecture moderne de 1977 à 1982, augmenté de trente pages inédites*. Bruxelles: Éditions des Archives d'Architecture Moderne.
Curl, James Stevens (1998) Review of the experience of modernism. Modern architects and the future city, 1928–1953. *Journal of Urban Design*, **3**(3).
Curtis, William J.R. (1990) Grands projets. *Architectural Record*, March, pp. 76–82.
Cybriwsky, Roman (1991) *Tokyo: The Changing Profile of an Urban Giant*. London: Belhaven.
Dagnaud, M. (1983) A history of planning in the Paris region: from growth to crisis. *International Journal of Urban and Regional Research*, **7**.
Davis, Timothy (2002) Preserving nature or creating a formal allée: two schemes, in Miller, Iris, *Washington in Maps,1606–2000*. New York: Rizzoli.

de Swaan, A., Olsen, D.J., Tenenti, A., de Vries, J., Gastelaars, R. v. E. and Lambooy, J.G. (1988) *Capital Cities as Achievement: Essays.* Amsterdam: Centrum voor Grootstedelijk Onderzook, University of Amsterdam.

Delhi Development Authority (1962) *Master Plan for Delhi.* New Delhi: Ministry of Works and Housing.

Delhi Development Authority (1990) *Master Plan for Delhi. Perspective 2001.* New Delhi: Ministry of Urban Affairs and Employment.

Delhi Development Authority (2003) *Guidelines for the Master Plan of Delhi, 2021,* ddadelhi.gov.in.

Delhi Improvement Trust (1956) *Interim General Plan for Greater Delhi.* New Delhi: Ministry of Health.

Della Seta, Piero and Della Seta, Roberto (1988) *I suoli di Roma. Uso ed abuso del territorio nei cento anni della capitale.* Roma: Editori Riuniti.

Delorme, Jean-Claude (1978) Jacques Gréber: urbaniste francais. *Metropolis,* **3**(32), pp. 49–54.

Demey, Thierry (1990) *Bruxelles. Chronique d'une capitale en chantier,* 2 volumes. Brussels: Paul Legrain.

Demosthenes, M. (1947) *Estudos sobre a nova capital do Brasil.* Rio de Janeiro: Agir.

Department of the Environment and Government Office for London (1996) *Four World Cities: A Comparative Study of London, Paris, New York and Tokyo.* London: Llewelyn Davies Planning.

des Cars, Jean and Pinon, Pierre (1992) *Paris – Haussmann.* Paris: Picard/Pavillon de l'Arsenal.

des Marez, Guillaume(1979) *Guide illustré de Bruxelles. Monuments civils et religieux.* Brussels: Touring Club Royal de Belgique.

Deutscher Bundestag (ed.) (1991) *Berlin – Bonn. Die Debatte. Alle Bundestagsreden vom 20. Juni 1991.* Cologne: Kiepenheuer & Witsch.

Dolff-Bonekämper, Gabi (1999) *Das Hansaviertel. Internationale Nachkriegsmoderne in Berlin.* Berlin: Verlag Bauwesen.

Donald J. Belcher and Associates (1957) *O relatório técnico sobre a nova Capital da República.* Rio de Janeiro: Departamento Administrativo do Serviço Público.

Draper, Joan (1982) *Edward Bennett Architect and City Planner, 1874–1954.* Chicago: Art Institute of Chicago.

DREIF (1994) *Schéma Directeur de l'Ile-de-France.* Paris: DREIF.

DREIF (1995) *Les Bureaux en Ile-de-France.* Paris: Observatoire Régional de l'Immobilier d'Entreprise en Ile-de-France.

DREIF/APUR/IAURIF (1990) *Le Livre Blanc de l'Ile-de-France.* Paris: DREIF/APUR/IAURIF.

Dubois, Marc (1994) Chi difende la qualità dell'architettura? = Who defends the quality of architecture? *Domus,* no. 758, pp. 78–79.

Dudley, G.A. (1994) *Workshop for Peace: Designing the United Nations Headquarters.* Cambridge, MA: MIT Press.

Durth, Werner (1989) Hauptstadtplanungen. Politische Architektur in Berlin, Frankfurt am Main und Bonn nach 1945, in Baumunk, Bodo-Michael and Brunn, Gerhard (eds.) *Hauptstadt. Zentren, Residenzen, Metropolen in der deutschen Geschichte.* Cologne: DuMont.

Durth, Werner, Düwel, Jörn and Gutschow, Niels (1998) *Architektur und Städtebau der DDR, 2 vols.* Frankfurt and New York: Campus Verlag.

Düwel, Jörn (1998) Am Anfang der DDR. Der Zentrale Platz in Berlin, in Schneider, Romana and Wang, Wilfried (eds.) *Moderne Architektur in Deutschland 1900 bis 2000. Macht und Monument.* Stuttgart: Hatje.

Eggleston, W. (1961) *The Queen's Choice: A Story of Canada's Capital.* Ottawa: National Capital Commission.

Enakiev, F.E. (1912) *Tasks for the Reform of St. Petersburg.* St. Petersburg, pp. 19–22.

Engel, Helmut and Ribbe, Wolfgang (eds.) (1993) *Hauptstadt Berlin – Wohin mit der Mitte? Historische, städtebauliche und architektonische Wurzeln des Stadtzentrums.* Berlin: Berliner Wissenschaft Verlag.

Epstein, David G. (1973) *Brasília, Plan and Reality.* Berkeley: University of California Press.

Escher, Felix (1985) *Berlin und sein Umland, Zur Genese der Berliner Stadtlandschaft bis zum Beginn des 20. Jahrhunderts.* Berlin: Colloquium Verlag.

Eskola, Meri and Eskola, Tapani (eds.) (2002) *Helsingin helmi. Helsingfors pärlä. The Pearl of Helsinki.* Helsinki Cathedral 1852–2002. Helsinki: Kustannus Oy Projektilehti.

Espejo, Arturo L. (1984) *Racionalité et formes d'occupation de l'espace: le projet de Brasilia.* Paris: Anthropos.

Evenson, Norma (1966) *Chandigarh.* Berkeley, CA: University of California Press.

Evenson, Norma (1973) *Two Brazilian Capitals.* New Haven, CT: Yale University Press.

Evenson, Norma (1979) *Paris: A Century of Change, 1878–1978.* New Haven, CT: Yale University Press.

Evenson, Norma (1984) Paris, 1890–1940, in Sutcliffe, Anthony (ed.) *Metropolis 1890–1940.* London: Mansell.

Federal Plan Commission for Ottawa and Hull (1916) *Report of the Federal Plan Commission on a General Plan for the Cities of Ottawa and Hull 1915.* Ottawa: Federal Plan Commission.

Ficher, Sylvia and Batista, Geraldo Sá Nogueira (2000) *GuiArquitetura Brasília*. São Paulo: Abril.

Fils, Alexander (1988) *Brasilia*. Dusseldorf: Beton-Verlag.

Fischer, Karl F. (1984) *Canberra: Myths and Models, Forces at Work in the Formation of the Australian Capital*. Hamburg: Institute of Asian Affairs.

Flagge, Ingeborg and Stock, Wolfgang Jean (eds.) (1992) *Architektur und Demokratie. Bauen für die Politik von der amerikanischen Revolution bis zur Gegenwart*. Stuttgart: Hatje.

Ford, George B. (1913) The city scientific. *Engineering Record*, No. 67, 17 May, pp. 551–552.

Forsyth, Ann (2002) Planning lessons from three U.S. new towns of the 1960s and 1970s: Irvine, Columbia, and The Woodlands. *Journal of the American Planning Association*, **68**(4), pp. 387–416.

Foster, S. G. and Varghese, M.M. (1996) *The Making of the Australian National University: 1946–96*. Sydney: Allen & Unwin.

Fourcaut, A. (2000) *La Banlieue en Morceaux. La crise des lotissements défectueux dans l'Entre-deux-Guerres*. Paris: Créaphis.

Frampton, Kenneth (2001) 'Keynote Address', in Takhar, Jaspreet (ed.) *Celebrating Chandigarh: Proceedings of Celebrating Chandigarh: 50 Years of the Idea, 9–11 January 1999*. Chandigarh: Chandigarh Perspectives, pp. 35–41.

França, Dionísio Alves de (2001) *Blocos residenciais de seis pavimentos em Brasília até 1969*. Brasília: FAU/UnB.

Fralon, José-Alain (1987) Bataille pour un hémicycle. *Le Monde*, 2 July.

Fraticelli, V. (1982) *Roma 1914–1929*. Rome: Officina Edizioni.

Freestone, R. (1989) *Model Communities: the Garden City Movement in Australia*. Melbourne: Thomas Nelson Australia.

French, R.A. (1995) *Plans, Pragmatism and People: the Legacy of Soviet Planning for Today's Cities*. Pittsburgh: The University of Pittsburgh Press.

Friedmann, J. (1986) The world city hypothesis. *Development and Change*, **17**, pp. 69–83.

Friedmann, J. and Wolff, G. (1982) World city formation: an agenda for research and action. *International Journal of Urban and Regional Research*, **6**, pp. 309–344.

Fujimori, Terunobu (1982) *The Planning History of Tokyo in the Meiji Era*. Tokyo: Iwanami Shoten.

Fukuda, Shigeyoshi (1918) Shin Tokyo (New Tokyo). *Journal of the Institute of Japanese Architects*, **32**(380), pp. 86–124.

Fukuoka, Shunji (1991) *Reconstruction Planning in Tokyo: The System of Administration for Urban Redevelopment*. Tokyo: Nihon Hyoronsha.

Furore, Angela M. (2000) Could Mayer Have Made It Work? An Evaluation of the Mayer Plan for Chandigarh, India. Unpublished Thesis, Muncie, IN: Ball State University.

Gaffield, Chad (1997) *History of the Outaouais*. Montreal: Les Presses de l'Université Laval.

Galantay, Ervin Y. (1987) *The Metropolis in Transition*. New York: Paragon House.

Gallagher, Patricia, Krieger, Alex, McGill, Michael and Altman, Andrew (2003) Rethinking fortress federalism. *Places*, **15**(3), pp. 65–71.

Gastellars, R. v. E. (1988) Revitalising the city and the formation of metropolitan culture: rivalry between capital cities in the attraction of new urban elites, in de Swaan, A., Olsen, D.J., Tenenti, A., de Vries, J., Gastelaars, R. v. E. and Lambooy, J.G., *Capital Cities as Achievement: Essays*. Amsterdam: Centrum voor Grootstedelijk Onderzook, University of Amsterdam, pp. 38–43.

Gaudin, J.-P. (1985) *L'avenir en plan; techniques et politique dans la prévision urbaine (1900–1930)*. Seyssel: Champ Vallon.

Geddes, P. (1915) *Cities in Evolution*. London: Williams and Norgate. (Reprinted (1998) in LeGates, R. and Stout, F. (ed.) *Early Urban Planning 1870–1940*, Vol. 4. London: Routledge.)

Gelman, J. (1924) Town Planning in Russia. *Town Planning Review*, **14**(July).

George, J. (1998) *Paris-Province: de la révolution à mondialisation*. Paris: Fayard.

George, P. (1967) Un difficile problème d'aménagement urbain: l'évolution des noyaux historiques – centres de villes, in Sporck, J.A. and Schoumaker, B. (eds.) *Mélanges de Géographie Physique, Humaine, Economique, Appliquée Offerts à M. Omer Tulippe*, Vol II. Gembloux: J. Duculot.

Gibbney, Jim (1988) *Canberra: 1913–1953*. Canberra: Australian Government Publishing Service.

Gilbert, A. (1989) Moving the capital of Argentina: a further example of utopian planning? *Cities*, **6**, pp. 234–242.

Gillespie, A.E. and Green, A.E. (1987) The changing geography of producer services employment in Britain. *Regional Studies*, **21**, pp. 397–412.

Gillespie, Angus Kress (1999/reprinted 2001) *Twin Towers, The Life of New York City's World Trade Center*. New Brunswick: Rutgers University Press.

Gillette, Howard Jr. (1995) *Between Justice and Beauty: Race, Planning, and the Failure of Public Policy in*

Washington, D.C. Baltimore: Johns Hopkins University Press.

Giovannini, Joseph (1997) Chandigarh revisited, architecture. *The AIA Journal*, **86**(7), pp. 41–45.

Glushkova, V.G. (1998) Economic transformations in Moscow and the socio-cultural environment of the capital, in Luzkov et al. (ed) *Moscow and the largest Cities of the world at the Edge of the 21st Century*. Moscow: Committee of Telecommunications and Mass Media of Moscow Government.

Godard, F. et al. (1973) *La renovation urbaine à Paris: structure urbaine et logique de classe*. Paris: Mouton.

Goldman, Jasper (2005) Warsaw: reconstruction as propaganda, in Vale Lawrence J. and Campanella, Thomas J. (eds.) *The Resilient City: How Modern Cities Recover from Disaster*. New York: Oxford University Press, pp. 135–158.

Goodsell, Charles (2001) *The American Statehouse*. Lawrence, Kansas: University Press of Kansas.

Gordon, David L.A. (1998) A City Beautiful plan for Canada's capital: Edward Bennett and the 1915 plan for Ottawa and Hull. *Planning Perspectives*, **13**, pp. 275–300.

Gordon, David L.A. (2001a) From noblesse oblige to nationalism: elite involvement in planning Canada's capital. *Journal of Urban History*, **28**(1), pp. 3–34

Gordon, David L.A. (2001b) Weaving a modern plan for Canada's capital: Jacques Gréber and the 1950 plan for the National Capital Region. *Urban History Review*, **29**(2), pp. 43–61.

Gordon, David L.A. (2002a) Ottawa-Hull and Canberra: implementation of capital city plans. *Canadian Journal of Urban Research*, **13**(2), pp. 1–16.

Gordon, David L.A. (2002b) William Lyon Mackenzie King, town planning advocate. *Planning Perspectives*, **17**(2), pp. 97–122.

Gordon, David L.A. (2002c) Frederick G. Todd and the origins of the park system in Canada's capital. *Journal of Planning History*, **1**(1), pp. 29–57.

Gordon, David L.A. and Gournay, Isabelle (2001) Jacques Gréber, urbaniste et architecte. *Urban History Review*, **29**, pp. 3–5.

Gordon, David L.A. and Osborne, B. (2004) Constructing national identity: Confederation Square and the National War Memorial in Canada's capital, 1900–2000. *Journal of Historical Geography*, **30**(4), pp. 618–642.

Gottmann, J. (1983a) The Study of Former Capitals. *Ekistics*, **50**, pp. 315, 541–546.

Gottmann, J. (1983b) Capital Cities. *Ekistics*, **50**, pp. 88–93.

Gournay, Isabelle and Loeffler, Jane C. (2002) A tale of two embassies: Ottawa and Washington. *Journal of the Society of Architectural Historians*, **61**, pp. 480–507.

Gouvernment Belge (1958) *Bruxelles E*. Brussels: Gouvernment Belge.

Governatorato di Roma (1931) *Piano Regolatore di Roma 1931-IX*. Milano-Roma: Treves-Treccani-Tumminelli.

Government of India (1951) *District Census Handbook*. Delhi: Directorate of Census Operations.

Government of India (1971) *District Census Handbook*. Delhi: Directorate of Census Operations.

Government of India (1991) *District Census Handbook*. Delhi: Directorate of Census Operations.

Government of the Commonwealth of Australia (1911) *Information, conditions and particulars for guidance in the preparation of competitive designs for the Federal Capital City of the Commonwealth of Australia*. Melbourne: Printed and published for the Government of the Commonwealth of Australia by J. Kemp, Government Printer for the State of Victoria.

Government of the National Capital Territory of Delhi (1997) *Report of Vijay Kumar Malhotra Committee regarding Amendments in the Unified Building Bye-Laws*. New Delhi: Government of NCT Delhi.

Governo do Distrito Federal (1970) *Plano Diretor de Águas, Esgotos e Controle de Poluição*. Brasília: SVO and Caesb.

Governo do Distrito Federal (1976) *Análise da estrutura urbana do Distrito Federal*. Brasília: GDF and Seplan.

Governo do Distrito Federal (1984) *Atlas do Distrito Federal*, 3 volumes. Brasília: GDF.

Governo do Distrito Federal (1985) *Plano de Ocupação do Território*, 2 volumes. Brasília: SVO and Terracap.

Governo do Distrito Federal (1986) *Plano de Ocupação e Uso de Solo*. Brasília: SVO and Terracap.

Governo do Distrito Federal (1991) *Relatório do Plano Piloto de Brasília*. Brasília: ArPDF, Codeplan and DePHA.

Governo do Distrito Federal (1992) *Plano Diretor de Ordenamento Territorial*. Brasília: IPDF.

Governo do Distrito Federal (1997) *Segundo Plano Diretor de Ordenamento Territorial*. Brasília: IPDF.

Governo do Distrito Federal (2001) *Anuário Estatístico do Distrito Federal*. Brasília: Codeplan.

Gravier, Jean-François (1947) *Paris et le Désert Français*. Paris: Le Portulan.

Greater London Authority (2002) *Draft London Plan*. London.

Greater London Council (1969) *Greater London Development Plan*. London: County Hall.

Gréber, Jacques (1950) *Plan for the National Capital: General Report Submitted to the National Planning Committee*. Ottawa: National Capital Planning Service.

Griffin, Walter Burley (1912) The plans for Australia's new capital city. *American City*, **7**(July), p. 9.

Griffin, Walter Burley (1914) *The Federal Capital: Report Explanatory of the Preliminary General Plan, Dated October 1913*. Melbourne: Albert J Mullett, Government Printer.

Guglielmo, R. and Moulin, B. (1986) Les grands ensembles et la politique. *Hérodote*, **43**, pp. 39–74.

Guimarães, Fábio de Macedo Soares (1946) O planalto central e o problema da mudança da capital do Brasil. *Revista Brasileira de Geografia*, **11**(4).

Gupta, Narayani (1988) *The City in Indian History, Report of the National Commission on Urbanisation*, Vol. 4. New Delhi: Ministry of Urban Development.

Gutheim, Frederick (1977) *Worthy of the Nation: the History if Planning for the National Capital*. Washington: Smithsonian Institution Press.

Gwyn, Sandra (1984) *The Private Capital: Ambition and Love in the Age of Macdonald and Laurier*. Toronto: McClelland and Stewart.

Hain, Simone (1998) 'Von der Geschichte beauftragt, Zeichen zu setzen'. Zum Monumentalitätsverständnis in der DDR am Beispiel der Gestaltung der Hauptstadt Berlin, in Schneider, Romana and Wang, Wilfried (eds.) *Moderne Architektur in Deutschland 1900 bis 2000. Macht und Monument*. Stuttgart: Hatje.

Hakala-Zilliacus, Liisa-Maria (2002) *Suomen eduskuntatalo. Kokonaistaideteos, itsenäisyysmonumentti ja kansallisen sovun representaatio*. Helsinki: SKS/Suomalaisen Kirjallisuuden Seura toimituksia 875.

Hall, P. (1966) *The World Cities*. London: Weidenfeld and Nicolson.

Hall, P. (1984) *The World Cities*, 3rd ed. London: Weidenfeld and Nicolson.

Hall, P. (1987) The anatomy of job creation: nations, regions and cities in the 1960s and 1970s. *Regional Studies*, **21**, pp. 95–106.

Hall, P. (1988) Regions in the transition to the information economy, in Sternlieb, G. (ed.) *America's new Market Geography*. Piscataway, NJ: Rutgers University, Center for Urban Policy Research.

Hall, P. (1995) Towards a general urban theory, in Brotchie, J., Batty, M., Blakely, E., Hall, P. and Newton, P. (ed.) *Cities in Competition: Productive and Sustainable Cities for the 21st Century*. Melbourne: Longman Australia, pp. 3–31.

Hall, P. (1998) *Cities in Civilization*. London: Weidenfeld and Nicolson.

Hall, P. (1999) Planning for the mega-city: a new Eastern Asian urban form? in Brotchie, J., Newton, P., Hall, P. and Dickey, J. (ed.) *East West Perspectives on 21st Century Urban Development: Sustainable Eastern and Western Cities in the New Millennium*. Aldershot: Ashgate, pp. 3–36.

Hall, P. (2000) The changing role of capital cities. *Plan Canada*, **40**(3), pp. 8–11.

Hall, P. (2002) *Cities of Tomorrow*, 3rd ed. Oxford: Blackwell.

Hall, P. and Pain, K. (2006) *The Polycentric Metropolis: Mega-City Regions in the European Space of Flows*. London: Earthscan.

Hall, Thomas (1986) *Planung europäischer Hauptstädte, Zur Entwicklung des Städtebaues im 19. Jh*. Stockholm: Almqvist & Wiksell.

Hall, Thomas (1997a) *Planning Europe's Capital Cities*. London: Spon.

Hall, Thomas (1997b) Brussels, in Hall, Thomas, *Planning Europe's Capital Cities*. London: Spon, pp. 217–244.

Hamnett, Stephen and Freestone, Robert (eds.) (2000) *The Australian Metropolis: A Planning History*. Sydney: Allen & Unwin.

Hardy, Dennis (1983) *Making Sense of the London Docklands: Processes of Change*. London: Middlesex Polytechnic.

Hardy, Dennis (1991) *From Garden Cities to New Towns: Campaigning for Town and Country Planning, 1899–1946*. London: Spon.

Häring, Hugo (1926) Aspekte des Städtebaues. *Sozialistische Monatshefte*, **32**(2), pp. 87–89.

Häring, Hugo and Mendelsohn, Heinrich (1929) Zum Platz der Republik. *Das neue Berlin*, **7**, pp. 145–146.

Häring, Hugo and Wagner, Martin (1929) Der Platz der Republik. *Das Neue Berlin*, **4**, pp. 69–72.

Harrison, Peter (1995) *Walter Burley Griffin: Landscape Architect*. Canberra: National Library of Australia.

Harvey, David (1979) Monument and myth. *Annals, Association of American Geographers*, **69**, pp. 362–381.

Hebbert, Michael (1998) *London: More By Fortune Than Design*. Chichester: John Wiley.

Hegemann, Werner (1930) *Das steinerne Berlin. Die Geschichte der grössten Mietskasernenstadt der Welt*. Berlin: Verlag von Gustav Kiepenheuer.

Hein, Carola (1987) L'implantation des Communautés Européennes à Bruxelles, son historique, ses intervenants. Diploma thesis at the, Institut Supérieur d'Architecture de l'Etat (ISAE), La Cambre.

Hein, Carola (1993) Europa in Brüssel. *Bauwelt*, **84**(40/41), pp. 2176–2184.

Hein, Carola (1995) Hauptstadt Europa. Hochschule für bildende Künste, Hamburg.

Hein, Carola (2001) Choosing a site for the capital of Europe. *GeoJournal*, **51**, pp. 83–97.

Hein, Carola (2002) Brussels encyclopedia of urban cultures in Ember, Melvin and Ember, Carol, R. (eds.) *Cities and Cultures around the World*. Danbury, CT: Grolier, pp. 430–38

Hein, Carola (2004a) *The Capital of Europe. Conflicts of Architecture. Urban Planning, Politics and European Union*. Westport, CT: Greenwood/Praeger.

Hein, Carola (2004b) Bruxelles et les villes sièges de l'Union européenne, in Claisse, Joël and Knopes, Liliane (eds.) *Change, Brussels Capital of Europe*. Brussels: Prisme Éditions.

Hein, Carola (2005a) Trauma and transformation in the Japanese city, in Vale, Lawrence J. and Campanella, Thomas J. (eds.) *The Resilient City: How Modern Cities Recover from Disaster*. New York: Oxford University Press.

Hein, Carola (2005b) La Gare Josaphat sur l'axe Bruxelles-Aéropoort: Perspectives pour un projet européen et multifonctionel/Het Josaphatstation op de as Brussel-Luchthaven: Perspectieven voor en Europees en Multifonctioneel Project, in Laconte, Pierre (ed.) *L'aeroport, le train et la ville: Le cas de Bruxelles est-il unique? De Luchthaven, de Tren et de Stad: is Brussel enig?* Brussels: Fondation pour l'Environnenment Urbain.

Hein, Carola (ed.) (forthcoming, 2006) *Bruxelles – l'Européenne. Capitale de qui? Ville de qui? Cahiers de la Cambre-Architecture*, no 5.

Helmer, Stephen (1985) *Hitler's Berlin: The Speer Plans for Reshaping the Central City*. Ann Arbor: UMI Research Press.

Helsinki City Planning Department (2000) *Urban Guide Helsinki*. Helsinki: City Planning Department.

Helsinki City Planning Office (1997) *Helsinki/City in the Forest. Vision for the Next Generation*. Helsinki: City Planning Office.

Helsinki. City in Forest (1997) Helsingin kaupunkisuunnitteluvirasto.

Herzfeld, Hans (1952) Berlin als Kaiserstadt und Reichshauptstadt 1871 bis 1945, in *Jahrbuch für die Geschichte des Deutschen Ostens*, Vol. 1. Tübingen: Niemeyer, pp. 141–170.

Higher Education Funding Council for England (2002) *Regional Profiles of Higher Education 2002*. Bristol: Higher Education Funding Council for England.

Hilberseimer, Ludwig (1927) *Großstadt Architektur*. Stuttgart: Julius Hoffmann.

Hillis, Kent (1992) A history of commissions: threads of an Ottawa planning history. *Urban History Review*, **21**(1), pp. 46–60.

Hindu, The (2003) The Asian Rome that is Delhi. *The Hindu* On-Line edition, 13 March.

Hines, T.S. (1974) *Burnham of Chicago: Architect and Planner*. New York: Oxford University Press.

Hoffmann, Godehard (2000) *Architektur für die Nation? Der Reichstag und die Staatsbauten des Deutschen Kaiserreiches 1871–1918*. Cologne: Dumont.

Hofmann, Albert (1910) Gross-Berlin, sein Verhältnis zur modernen Grossstadtbewegung und der Wettbewerb zur Erlangung eines Grundplanes für die städtebauliche Entwicklung Berlins und seiner Vororte im zwanzigsten Jahrhundert. *Deutsche Bauzeitung*, **44**, pp. 169–176, 181–188, 197–200, 213–216, 233–236, 261–263, 277, 281–287, 311–312, 325–328.

Holford, William (1957) *Observations on the Future Development of Canberra, ACT*. Canberra: Government Printer.

Holston, James (1989) *The Modernistic City: An Anthropological Critique of Brasília*. Chicago: University of Chicago Press.

Howard, Ebenezer (1898) *To-morrow: A Peaceful Path to Real Reform*. London: Swan Sonnenschein.

Howard, Ebenezer (1902) *Garden Cities of Tomorrow*. London: Swann Sonnenschein.

Huet, Bernard (2001) The legacy, in Takhar, Jaspreet (ed.) *Celebrating Chandigarh: Proceedings of Celebrating Chandigarh: 50 Years of the Idea, 9–11 January 1999*. Chandigarh: Chandigarh Perspectives, pp. 166–170.

Hung, Wu (1991) Tiananmen Square: a political history of monuments. *Representations*, **35**, pp. 84–117.

Hüter, Karl-Heinz (1988) *Architektur in Berlin 1900–1933*. Dresden: Verlag der Kunst.

IAURIF (1991) La Charte de l'Ile-de-France. *Cahiers de l'IAURIF*, Nos 97/98.

IAURIF (1993) *Le Parc des Bureaux en Ile-de-France*. Paris: IAURIF.

Ichikawa, Hiroo (1995) The expanding metropolis and changing urban space. *Creating a City: Considering Tokyo Series*, **5**, pp. 15–85.

Insolera, I. (1962) *Roma moderna*. Torino: Einaudi.

Inter-Environnement Bruxelles Groupement des Comités du Maelbeek (1980) Contre-projet pour la construction du bâtiment du Conseil des Ministres de la CEE aux abords du rond-point Schuman. *AAM*, no. 19.

Irving, Robert G. (1981) *Indian Summer: Lutyens, Baker, and Imperial Delhi*. New Haven, CT: Yale University Press.

Ishida, Yorifusa (1992) *Mikan no Tokyo Keikaku*

(Unfinished Plans for Tokyo). Tokyo: Chikuma Shobo.

Ishida, Yorifusa (2004) *Nihon Kin-Gendai Toshikeikaku no Tenkai 1863–2003* (Historical Development of Modern and Contemporary City Planning of Japan 1868–2003). Tokyo: Jichitai Kenkyusha.

Ishizuka, Hiromichi (1991) *Theories of Japanese Modern Cities: Tokyo, 1868–1923.* Tokyo: Tokyo University Press.

Ishizuka, Hiromichi and Narita, Ryuichi (1986) *The 100 Years of Tokyo Prefecture.* Tokyo: Yamakawa Shuppan.

Ishizuka, Hiromichi (1991) *Theories of Japanese Modern Cities: Tokyo, 1868–1923.* Tokyo: Tokyo University Press.

Jackson, Kenneth (1984) Capital of capitalism: the New York Metropolitan Region, 1890–1940, in Sutcliffe, Anthony (ed.) *Metropolis 1890–1940.* London: Mansell.

Jacobs, Jane (1961) *The Death and Life of Great American Cities.* New York: Random House.

Jacobs, P. (1983) Frederick G. Todd and the creation of Canada's urban landscape. *Association for Preservation Technology (APT) Bulletin,* 15(4), pp. 27–34.

Jacobs, Roel (1994) *Brussels: A City in the Making.* Brugge: Marc van de Wiele.

Jager, Markus (2005) *Der Berliner Lustgarten. Gartenkunst und Stadtgestalt in Preussens Mitte.* Munich: Deutscher Kunst Verlag.

Jauhiainen, Jussi, S. (1995) *Kaupunkisuunnittelu, kaupunkiuudistus ja kolme eurooppalaista esimerkkiä.* Turku: Publicationes Instituti Geographici Universitatis Turkuensis No. 146.

Jinnai, Hidenobu (1995) *Tokyo: A Spatial Anthropology.* California: University of California Press.

Joardar, Souro D. (2002) From Twin Cities to Two-in-one Cities. Paper presented at the IPHS Conference, University of Westminster, London.

Joardar, Souro D. (2003) Urban Space and Ecological Performance of Cities: A Case of Delhi. Paper presented at the Seminar on Ecological Performance of Cities, School of Planning and Architecture, New Delhi.

Joshi, Kiran (1999) *Documenting Chandigarh.* Ahmedabad: Mapin Publishing.

Jung, Bertel (1918) *Pro Helsingfors. Ett förslag till stadsplan för Stor-Helsingfors utarbetat av Eliel Saarinen m.fl.* Helsingfors: Pro Helsingfors säätiö – Stiftelsen Pro Helsingfors.

Kaganovich, L.M. (1931) *Socialist Reconstruction of Moscow and other Cities in the USSR.* Moscow: Cooperative Publishing Society of Foreign Workers in the USSR.

Kähler, Gert (1995) Brüsseler Käse. *Architekt,* **6**, pp. 358–361.

Kahlfeldt, Paul, Kleihues, Josef Paul and Scheer, Thorsten (eds.) (2000) *Stadt der Architektur – Architektur der Stadt. Berlin 1900–2000.* Berlin: Nikolai Verlag.

Kain, Roger (1981) Conservation planning in France: policy and practice in the Marais, Paris, in Kain, Roger (ed.) *Planning for Conservation.* London: Mansell.

Kalia, Ravi (1987) *Chandigarh: In Search of an Identity.* Carbondale and Edwardsville: Southern Illinois University Press.

Kalia, Ravi (2002) *Chandigarh: The Making of an Indian City,* revised ed. New Delhi: Oxford University Press.

Kalman, Harold (1994) *History of Canadian Architecture,* Vol. 2. Don Mills, ON: Oxford University Press.

Käpplinger, Claus (1993) Façadisme et Bruxellisation. *Bauwelt,* **84**(40/41), pp. 2166–2175.

Kervanto Nevanlinna, Anja (2002) *Kadonneen kaupungin jäljillä. Teollisuusyhteiskunnan muutoksia Helsingin historiallisessa ytimessä.* Helsinki: SKS/Suomalaisen Kirjallisuuden Seura toimituksia 836.

Khan-Magomedov, S.O. (1983) *Pioneers of Soviet Architecture.* New York: Rizzoli.

Khilnani, Sunil (1997) *The Idea of India.* New Delhi: Penguin.

King, A.D. (1976) *Colonial Urban Development: Culture, Social Power and Environment.* London: Routledge and Kegan Paul.

Klaus, Susan L. (2002) *A Modern Arcadia : Frederick Law Olmsted Jr. and the Plan for Forest Hills Gardens.* Amherst: University of Massachusetts Press

Kleihues, Josef Paul (ed.) (1986) *International Building Exhibition Berlin (IBA) 1987. Examples of a New Architecture.* London: Academy Editions.

Kleihues, Josef Paul (ed.) (1987) *750 Jahre Architektur und Städtebau in Berlin. Die Internationale Bauausstellung im Kontext der Baugeschichte Berlins.* Stuttgart: Hatje.

Klinge, Matti and Kolbe, Laura (1999) *Helsinki – The Daughter of the Baltic Sea.* Helsinki. Otava Publishing Company Ltd.

Knight, D. (1991) *Choosing Canada's Capital: Conflict Resolution in a Parliamentary System,* 2nd ed. Ottawa: Carleton University Press.

Kohler, Sue, (1996) *The Commission of Fine Arts: A Brief History 1910–1995.* Washington, DC: CFA.

Kohtz, Otto (1920) Das Reichshaus am Königsplatz in Berlin. Ein Vorschlag zur Verringerung der

Wohnungsnot und der Arbeitslosigkeit. *Stadtbaukunst alter und neuer Zeit*, **1**(16), pp. 241–245.

Kolbe, Laura (1988) *Kulosaari – unelma paremmasta tulevaisuudesta*. Helsinki. Otava Publishing Company Ltd.

Kolbe, Laura (2002) *Helsinki kasvaa suurkaupungiksi. Julkisuus, politiikka, hallinto ja kansalaiset 1945–2000. Helsingin historia vuodesta 1945, 3*. Helsinki: Edita Prima AB, pp. 86–88.

Konter, Erich (1995) Verheissungen einer Weltstadtcity. Vorschläge zum Umbau 'Alt-Berlins' in den preisgekrönten Entwürfen des Wettbewerbs Gross-Berlin von 1910, in Fehl, Gerhard and Rodriguez-Lores, Juan (eds.) *Stadt-Umbau. Die planmässige Erneuerung europäischer Grossstädte zwischen Wiener Kongress und Weimarer Republik*. Basel: Birkäuser.

Kopp, A. (1970) *Town and Revolution: Soviet Architecture and City Planning 1917–1935*. New York: Braziller.

Körner, Hans-Michael and Weigand, Katharina (eds.) (1995) *Hauptstadt. Historische Perspektiven eines deutschen Themas*. Munich: Deutscher Taschenbuch Verlag.

Korvenmaa, Pekka (ed.) (1992) *Arkkitehdin työ. Suomen Arkkitehtiliitto 1892–1992*. Hämeenlinna: Rakennustieto Oy.

Kosel, Gerhard (1958) Aufbau des Zentrums der Hauptstadt des demokratischen Deutschlands Berlin. *Deutsche Architektur*, **7**(4), pp. 177–183.

Koshizawa, Akira. (1991) *The Story of City Planning in Tokyo*. Tokyo: Nihon Hyoronsha.

Krier, Léon (1986) The completion of Washington DC: a bicentennial masterplan for the year 2000. *Archives d'Architecture Moderne*, **30**, pp. 7–43.

Krier, Léon, Culot, Maurice and AAM. (1980) *Contreprojets/Contreprogetti/Counterprojects*. Brussels: AAM.

Krinsky, Carol (1978) *Rockefeller Center*. New York: Oxford University Press.

Kropotkin, Peter (1899, 1985) *Fields, Factories and Workshops*. New ed. annotated by Colin Ward. London: Freedom Press.

Kumar, Ashok (2000) The inverted compact city of Delhi, in Jenks, Mike and Burgess, Rod (eds.) *Compact Cities: Sustainable Urban Forms for Developing Countries*. London: Spon Press.

Kuusanmäki, Jussi (1992) *Sosiaalipolitiikkaa ja kaupunkisuunnittelua. Tietoa, taitoa, asiantuntemusta. Helsinki eurooppalaisessa kehityksessä 1875–1917, 2*. Helsinki: Suomen Historiallinen Seura. Historiallisia tutkimuksia 99.

Kynaston, D. (1994) *The City of London: Vol. I A World of its Own 1815–1890*. London: Chatto & Windus.

Kynaston, D. (1995) *The City of London: Vol. II Golden Years 1890–1914*. London: Chatto & Windus.

Lacaze, J.-P. (1994) *Paris: Urbanisme d'État et destin d'une ville*. Paris: Flammarion.

Lacaze, J.-P. (1995) *Introduction à la planification urbaine: imprécis d'urbanisme à la Française*. Paris: Presse de l'ENPC.

Laconte, Pierre (2002) La loi de 1962, quarante ans après/De wet van 1962 feertig jaar later. *A+*, no 176, pp. 18–19.

Ladd, Brian (1997) *The Ghosts of Berlin: Confronting German History in the Urban Landscape*. Chicago: University of Chicago Press.

Lafer, Celso (1970) The Planning Process and the Political System in Brasil: A Study of Kubitschek Target Plan, 1956–1961. PhD Dissertation, Cornell University, Ithaca, NY.

Lagrou, Evert (2000) Brussels: five capitals in search of a place. The citizens, the planners and the functions. *GeoJournal*, **51**(1/2), pp. 99–112.

Lambooy, J.G. (1988) Global cities and the world economic system: rivalry and decision-making, in de Swaan, A. *et al.* (eds.) *Capital Cities as Achievement: Essays*. Amsterdam: Centrum voor Grootstedelijk Onderzook, University of Amsterdam.

Lambotte-Verdicq, Georgette (1978) *Contribution à une anthologie de l'espace bâti bruxellois, de Léopold II à nos jours*. Brussels: Edition Louis Musin.

Lampugnani, Vittorio Magnago (1986) Eine Leere voller Pläne. Die Projekte für das nie verwirklichte Zentrum von Gross-Berlin 1839–1985, in Lampugnani, Vittorio Magnago (ed.) *Architektur als Kultur. Die Ideen und die Formen*. Cologne: Deutsche Verlags-Anstalt.

Lang, Michael H. (1996) Yorkship Garden Village: progressive expression of the Architectural, Planning and Housing Reform Movement, in Sies, Mary and Silver, Christopher (eds.) *Planning the Twentieth Century American City*. Baltimore, MD: Johns Hopkins University Press.

Lang, Michael H. (2001) Town planning and radicalism in the Progressive Era: the legacy of F.L. Ackerman. *Planning Perspectives*, **16**(2), pp. 143–168.

Lang, M.H. and Rapoutov, L. (1996) Capital City as Garden City: The Planning of Post-Revolutionary Moscow. Conference Proceedings, International Planning History Society Conference, 'The Planning of Capital Cities', Thessaloniki, pp. 795–812.

Laporta, Philippe (1986) La CEE à Bruxelles, mariage ou viol. Vers une régionalisation de l'urbanisme, **91**, pp. 19–26.

Larsson, Bo (1994) *Stadens språk. Stadsgestaltning och*

bostadsbyggande i nordiska huvudstader under 1970- och 1980-talen. Lunds: Lunds Universitet.

Larsson, Lars Olof (1978) *Die Neugestaltung der Reichshauptstadt. Albert Speers Generalbebauungsplan für Berlin*. Stuttgart: Hatje.

Laurier, Wilfrid (1989) *Dearest Émilie: The Love Letters of Sir Wilfrid Laurier to Madame Émilie Lavergne*. Toronto: NC Press Limited.

Lavedan, Pierre (1963) Jacques Gréber, 1882–1962. *La Vie Urbain*, January, pp. 1–14.

Laveden, P. (1975) *Historie de l'urbanisme à Paris*. Paris: Hachette.

Law, Christopher M. (1996) *Tourism in Major Cities*. London: International Thompson Business Press.

Lawrence, D.H. (1923, 1995) *Kangaroo*. Pymble: HarperCollins Publishers.

Le Corbusier (1925) *Urbanisme*. Paris: Vicent Fréal.

Le Corbusier (1935) *La Ville Radieuse*. Paris: L'architecture d'aujourd'hui.

Le Corbusier (1946) *Œuvre complète, 1938–1946*. Zurich: Editions d'architecture.

Le Corbusier (1947) *Œuvre complète, 1934–1938*. Zurich: Editions d'architecture.

Le Corbusier (1955) *Ouvre Complete 1946–52*. Zurich: M. Ginsberger.

Le Corbusier (1959) *L'urbanisme des trois établissements humains*. Paris: Minuit.

Lehwess, Walter (1911) Architektonisches von der allgemeinen Städtebau-Ausstellung zu Berlin. *Berliner Architekturwelt*, **13**(4), pp. 123–162.

Levenson, Michael (2002) London 2000: the millennial imagination in a city of monuments, in Gilbert, Pamela K. (ed.) *Imagined Londons*. Albany: State University of New York Press, pp. 219–239.

Levey, Jane (2000) Lost highways. The plan to pave Washington and the people who stopped it. *Washington Post Magazine*, 26 November 26.

Lidgi, S. (2001) *Paris – gouvernance; ou les malices des politiques urbaines (J. Chirac/J. Tiberi)*. Paris: L'Harmattan.

Lilius, Henrik (1984) *The Esplanade During the 19th Century*. Helsinki: Akateeminen Kirjakauppa.

Lindberg, Carolus and Rein, Gabriel (1950) Asemakaavoittelu ja rakennustoiminta. *Helsingin kaupungin historia, III:I*. Helsinki: SKS/Suomalaisen Kirjallisuuden Seuran kirjapaino.

Lissitzky, E. (1986) *Russia: An Architecture for World Revolution*. Cambridge, MA: MIT Press.

Longstreth, Richard (2002) The unusual transformation of downtown Washington in the early twentieth century. *Washington History*, Fall/Winter, pp. 50–75.

Lortie, André (1997) Jacques Gréber (1882–1962) et L' Urbanisme le temps et l'espace de la ville. Unpublished Doctoral Thesis, Université Paris XII.

Lortie, André (1993) *Jacques Gréber urbaniste. Les Gréber: Une dynastie, des artistes*. Beauvais: Musée départemental de l'Oise, catalogue de l'exposition.

Lupano, M. (1991) *Marcello Piacentini*. Roma-Bari: Laterza.

Luzkov, Y.M. *et al.* (1998) *Moscow and the Largest Cities of the World at the Edge of the 21st Century*. Moscow: Committee of Telecommunications and Mass Media of the Moscow Government.

Lynch, Kevin (1960) *The Image of the City*. Cambridge, MA: MIT Press.

Mächler, Martin (1920) Ein Detail aus dem Bebauungsplan Gross-Berlin. *Der Städtebau*, **17**, pp. 54–56.

Magistrat von Berlin (ed.) (1951) *Wettbewerb zur Erlangung von Bebauungsvorschlägen und Entwürfen für die städtebauliche und architektonische Gestaltung der Stalinallee in Berlin*. Berlin: Magistrat.

Magritz, Kurt (1959) Die sozialistische Umgestaltung des Zentrums von Berlin. *Deutsche Architektur*, **8**(1), pp. 1–5.

Malagutti, Cecília Juno (1996) Loteamentos clandestinos no DF: legalização ou exclusão? Master Thesis, Faculdade de Arquitetura e Urbanismo, Universidade de Brasília.

Manacorda, D. and Tamassia, R. (1985) *Il piccone del regime*. Roma: Curcio Editore.

Marchand, Bernard (1993) *Paris, histoire d'une ville, XIXe–XXe siècle*. Paris: Seuil.

Maryland Department of Planning (2001) *Smart Growth in Maryland*. Baltimore MD: Maryland Department of Planning.

Maryland National Capital Park and Planning Commission (2001) *Legacy Open Space Functional Master Plan. Open Space Conservation in the 21st Century*. Silver Spring, MD: MNCPPC.

Mawson, Thomas H. (1911) Town planning in England. *City Club Bulletin*, **4**, pp. 263–269.

Mayer, Albert (1950) The New Capital of the Punjab, Address before Convention Symposium I: Urban and Regional Planning, The American Institute of Architects, Washington DC. In Papers on India, 1934–1975. Unpublished Papers, University of Chicago, 1950, May 11.

Mayor of London (2002) *The Draft London Plan: Draft Spatial Development Strategy for London*. London: Greater London Authority.

Michel, C. (1988) *Les Halles. La renaissance d'un quartier, 1966–1988*. Paris: Masson.

Mikuriya, Takashi (1994) The 50 Years of Tokyo's Administration. Considering Tokyo Series, Vol. 1.
Miller, M. (1989) *Letchworth: The First Garden City*. Chichester: Phillimore.
Miller, M. and Gray, A.S. (1992) *Hampstead Garden Suburb*. Chichester: Phillimore.
Miller Lane, Barbara (1968) *Architecture and Politics in Germany 1918–1945*. Cambridge, MA: Harvard University Press.
Mills, E. (1987) Service sector suburbanization, in Sternlieb, G. (ed.) *America's New Economic Geography: Nation, Region, and Central City*. New Brunswick: Rutgers University, Center for Urban Policy Research.
Mindlin, Henrique Ephim (1961) *Brazilian Architecture*. Roma: Embaixada do Brasil.
Ministère des Travaux Publics (1966) *Études régionales, Notes synthétiques des rapports du Groupe Alpha sur les propositions d'Aménagement et de développement de la région bruxelloise*. Brussels: Ministère des Travaux Publics, Commission Nationale de l'Aménagement du Territoire.
Ministère des Travaux Publics et de la Reconstruction (1956) *Carrefour de l'Occident*. Brussels: Fonds des Routes.
Mission Interministérielle de Coordination des Grandes Opérations d'Architecture et d'Urbanisme (1988) *Architectures capitales: Paris 1979–1989*. Paris: Electa Moniteur.
Miyakawa, Y. (1983) Metamorphosis of the capital and evolution of the urban system in Japan. *Ekistics*, No. 299, pp. 110–122.
Moest, Walter (1947) *Der Zehlendorfer Plan. Ein Vorschlag zum Wiederaufbau Berlins*. Berlin: Verlag des Druckhauses Tempelhof.
Moore, C. (ed.) (1902) *The Improvement of the Park System of the District of Columbia*. 57th Congress, 1st. sess. S. Rept. 166. Washington: US Government Printing Office.
Moore, Charles (1921, 1968) *Daniel H. Burnham: Architect, Planner of Cities*. New York, Da Capo Press.
Moreira, Vânia Maria Losada (1998) *Brasília: a construção da nacionalidade*. Vitória: Editora UFES.
Morris, William (1890, 1970) *News from Nowhere*. London: Routledge & Kegan Paul.
Moscow, Warren (1946) Rockefeller offers UN $8,500,000 site on the East River for skyscraper center. *New York Times*, December 12.
Müller, Peter (1999) *Symbol mit Aussicht. Die Geschichte des Berliner Fernsehturms*. Berlin: Verlag für Bauwesen.

Müller, Peter (2005) *Symbolsuche. Die Ost-Berliner Zentrumsplanung zwischen Repräsentation und Agitation*. Berlin: Mann.
National Capital Authority (2004) *The Griffin Legacy: Canberra the Nation's Capital in the 21st Century*. Canberra: Commonwealth of Australia.
National Capital Commission (1998) *A Capital for Future Generations*. Ottawa: National Capital Commission.
National Capital Development Commission (1970) *Tomorrow's Canberra*. Canberra: Australian National University Press.
National Capital Planning Commission (1961) *A Policies Plan for the Year 2000*. Washington DC: NCPC.
National Capital Planning Commission (1997) *Extending the Legacy: Planning America's Capital for the 21st Century*, Washington DC: NCPC.
National Capital Planning Commission (2001) *Memorials and Museums Master Plan*. Washington DC: NCPC.
National Capital Planning Commission (2002)
National Capital Planning Commission (2004)
National Capital Region Planning Board (NCRPB) (1988) *National Capital Region Plan for Delhi*. New Delhi: Ministry of Urban Affairs and Employment.
National Capital Region Planning Board (NCRPB) (1999) *Base Paper for the Preparation of Regional Plan – 2021, Monograph*. New Delhi: NCRPB.
Nehru, Jawaharlal (1946) *The Discovery of India*. Calcutta: Signet Press.
Nehru, Jawaharlal (1959) Mr. Nehru on Architecture. *Urban and Rural Thought*, **2**(2), pp. 46–49.
Nelson, K. (1986) Labor demand, labor supply and the suburbanization of low-wage office work, in Scott, A.J. and Storper, M. (eds.) *Production, Work, Territory: The Geographical Anatomy of Industrial Capitalism*. Boston: Allen and Unwin.
Nerdinger, Winfried (1998) Ein deutlicher Strich durch die Achse der Herrscher. Diskussionen um Symmetrie, Achse und Monumentalität zwischen Kaiserreich und Bundesrepublik, in Schneider, Romana and Wang, Wilfried (eds.) *Moderne Architektur in Deutschland 1900 bis 2000. Macht und Monument*. Stuttgart: Hatje.
New York City Planning Commission and Richards, Peter (1970) *Plan for the City of New York*, 5 volumes. Cambridge, MA: MIT Press.
New York City Bar Association (2001) Special Committee on the United Nations of the Association of the Bar of the City of New York, 'New York City and the United Nations: Towards a Renewed Relationship', December.

Newhouse, Victoria (1989) *Wallace K. Harrison*. New York: Rizzoli.

Newman, P. and Tual, M. (2002) The Stade de France. The last expression of French centralism? *European Planning Studies*, **10**, pp. 831–843.

Nicaise, Lucien (1985) Vingt et un architectes vont réaliser le projet de l'immeuble C.E.E. à Bruxelles. *Le Soir*, 20 November.

Nicolaus, Herbert and Obeth, Alexander (1997) *Die Stalinallee. Geschichte einer deutschen Strasse*. Berlin: Verlag für Bauwesen.

Niemeyer, Oscar (1960) Fala Niemeyer sobre o plano de urbanização. *Engenharia*, No. 209.

Nikula, Riitta (1931) *Yhtenäinen kaupunkikuva 1900–1930. Suomalaisen kaupunkirakentamisen ihanteista ja päämääristä, esimerkkinä Helsingin Etu-Töölö ja uusi Vallila*. Helsinki. Finska Vetenskaps-societeten.

Nilsson, S. (1973) *The Capitals of India, Pakistan and Bangladesh, Monograph Series 12*. Sweden: Scandinavian Institute of South Asian Studies.

Noin, Daniel (1976) *L'Espace Français*. Paris: Armand Colin.

Noin, Daniel and White, Paul (1997) *Paris*. Chichester: Wiley.

Nossa, Leonêncio (2002) Brasília, do planejamento ao toque de recolher. *O Estado de São Paulo*, 8 December.

Nowack, Hans (1953) Das Werden von Gross-Berlin 1890–1920. PhD Dissertation, Freie Universität, Berlin.

Oishi, Manabu. (2002) *The Birth of the Capital Edo: How the Great Edo Was Built?* Tokyo: Kadokawa Shoten.

Olmsted, Frederick Law Jr. (1911) The City Beautiful. *The Builder*, **101**(July 7), pp. 15–17.

Olsen, Donald (1986) *The City as a Work of Art: London, Paris, Vienna*. New Haven, CT: Yale University Press.

Ottawa Improvement Commission (1913) *Special Report of the Ottawa Improvement Commission, from its inception in 1899 to March 13, 1912*. Ottawa: Ottawa Improvement Commission.

Ottawa-Carleton, Regional Municipality (1976) *Official Plan: Region of Ottawa-Carleton*, Ottawa: Regional Municipality of Ottawa Carleton Planning Department.

Overall, John (1995) *Canberra Yesterday, Today & Tomorrow: A Personal Memoir*. Canberra: The Federal Capital Press of Australia.

Papadopoulos, Alex G. (1996) *Urban Regimes and Strategies: Building Europe's Central Executive District in Brussels*. University of Chicago Geography Research Paper no. 239. Chicago: University of Chicago Press.

Parkins, M.F. (1953) *City Planning in Soviet Russia*. Chicago: University of Chicago Press.

Parliament of Canada (1899) *An Act Respecting the City of Ottawa*, 62–63 Vict. Ch 10. 7 (c), assented 11th August, 1899.

Parliament of Canada (1912) *Report and Correspondence of the Ottawa Improvement Commission relating to the Improvement*. 2 George V. Sessional Paper No. 51A. Ottawa: C.H. Parmelee.

Paul, Suneet (1999) The drubbing of the LeCorb's Chandigarh. *Architecture + Design*, **16**(2), pp. 113–114.

Paviani, Aldo (ed.) (1985) *Brasília em questão: ideologia e realidade*. São Paulo: Projeto.

Paviani, Aldo (ed.) (1987) *Urbanização e metropolização: a gestão dos conflitos em Brasília*. Brasília: EDUnB.

Paviani, Aldo (ed.) (1989) *Brasília: a metrópole em crise*. Brasília: EDUnB.

Paviani, Aldo (ed.) (1991) *A conquista da cidade: movimentos populares em Brasília*. Brasília: EDUnB.

Paviani, Aldo (ed.) (1996) *Brasília: moradia e exclusão*. Brasília: EDUnB.

Paviani, Aldo (ed.) (1999) *Brasília – gestão urbana: conflitos e cidadania*. Brasília: EDUnB.

Pegrum, Roger (1983) *The Bush Capital: How Australia chose Canberra as Its Federal City*. Sydney: Hale & Iremonger.

Pegrum, Roger (1990) Canberra: the bush capital, in Statham, Pamela (ed.) *The Origins of Australia's Capital Cities*. Cambridge: Cambridge University Press.

Perchik, L. (1936) *The Reconstruction of Moscow*. Moscow: Cooperative Publishing Society of Foreign Workers in the USSR.

Perera, Nihal (1998) *Society and Space: Colonialism, Nationalism, and Postcolonial Identity*. Boulder, CO: Westview Press.

Perera, Nihal (2002) Indigenising the colonial city: late 19th-century Colombo and its landscape. *Urban Studies*, **39**(9), pp. 1703–1721.

Perera, Nihal (2004) Contested imaginations: hybridity, liminality, and authorship of the Chandigarh plan. *Planning Perspectives*, **19**(2), pp. 175–203.

Perry, Clarence Arthur (1910) *Wider Use of the School Plant*. New York: Charities Publication Committee.

Perry, Clarence Arthur (1916) *Community Center Activities*. New York: Russell Sage Foundation.

Perry, Clarence Arthur (1929) *Neighborhood and Community Planning*. New York: Regional Plan of New York.

Pescatori, Carolina (2002) *Habitações populares no DF – Atuação do Governo (1956–1985)*. Brasília: PIBIC, FAU/UnB.

Peterson, Jon A. (1985) The nation's first comprehensive city plan. A political analysis of the McMillan Plan for Washington, DC, 1900–1902. *American Planning Association Journal*, **55**(Spring), pp. 134–150.

Peterson, Jon A. (1996) Frederick Law Olmsted Sr. and Frederick Law Olmsted Jr.: the visionary and the professional, in Sies, Mary C. and Silver, C. (eds.) *Planning the Twentieth Century American City*. Baltimore, MD: Johns Hopkins University Press. pp. 37–54.

Piantoni, Gianna (ed.) (1980) *Roma 1911*. Roma: De Luca Editore.

Piazzo, P. (1982) *Roma: La crescita metropolitana abusiva*. Roma: Officina Edizioni.

Piccinato, G. (1987) La nascita dell'edilizia popolare in Italia: un profilo generale. *Storia Urbana*, **39**, pp. 115–133.

Pinon, P. (2002) *Atlas du Paris Haussmannien: la ville en héritage du second Empire à nos jours*. Paris: Parigramme.

Posener, Julius (1979) *Berlin auf dem Wege zu einer neuen Architektur. Das Zeitalter Wilhelms II*. Munich: Prestel.

Prakash, Aditya, and Prakash, Vikramaditya (1999) *Chandigarh: The City Beautiful*. Chandigarh: Abhishek Publications.

Prakash, Ved (1969) *New Towns in India*. Detroit, MI: Duke University Press.

Prakash, Vikramaditya (2002) *Chandigarh's Corbusier: The Struggle for Modernity in Postcolonial India*. Seattle, WA: University of Washington Press.

President's Council on Pennsylvania Avenue (1964) *Pennsylvania Avenue: Report*. Washington DC: US Government Printing Office.

Price, Roger (2002) *The French Second Empire: An Anatomy of Power*. Cambridge: Cambridge University Press.

Proceedings of the Congress of Engineers, Architects, Surveyors and Others Interested in the Building of the Federal Capital of Australia, Held in Melbourne, in May 1901 (1901) Melbourne: J.C. Stephens, Printer.

Punin, N. (1921) *Tatlin: Against Cubism*. St. Petersburg: State Publishing House.

Purdom, C.B. (1945) *How Should We Rebuild London?* London: Dent.

Quilici, V. (ed.) (1996) *E42-EUR. Un centro per la metropoli*. Roma: Olmo Edizioni.

Raatikainen, Voitto (1994) *Meidän kaikkien Stadion*. Helsinki: WSOY Publishing Company Ltd.

Ranieri, Liane (1973) *Léopold II urbaniste*. Bruxelles: Fonds Mercator/Hayez.

Rapoport, Amos (1972) *Human Aspects of Urban Form*. Oxford: Pergamon.

Rapoport, Amos (1993) On the nature of capitals and their physical expression, in Taylor, John, Lengellé, Jean G., and Andrew, Caroline (eds.) *Capital Cities/Les Capitales*. Ottawa: Carleton University Press.

Rapoutov, L. (1998) *Siberian Garden Cities in the Early 20th Century Manuscript*. Moscow: Moscow Architectural Institute.

Rapoutov, L. (1998) *The First Garden City near Moscow Manuscript*. Moscow: Moscow Architectural Institute.

Regional Plan Association of New York (1968) *Second Regional Plan*. New York: RPA.

Reichhardt, Hans J. and Schäche, Wolfgang (1985) *Von Berlin nach Germania. Über die Zerstörungen der Reichshauptstadt durch Albert Speers Neugestaltungsplanungen*. Berlin: Transit.

Reid, Paul (2002) *Canberra following Griffin: A Design History of Australia's National Capital*. Canberra: National Archives of Australia.

Reps, John W. (1967) *Monumental Washington: The Planning and Development of the Capital Center*. Princeton, NJ: Princeton University Press.

Reps, John W. (1997) *Canberra 1912: Plans and Planners of the Australian Capital Competition*. Melbourne: Melbourne University Press.

Ribeiro, Edgar F.N. (1983) The Future of New Delhi: Conservation versus Growth. Paper presented at the Golden Jubilee Celebration of the New Delhi Municipal Corporation.

Ribeiro, Gustavo (1982) Arqueologia de uma cidade: Brasília e suas cidades satélites. *Espaço e Debates*, No. 6.

Robbins, Anthony (1995) World Trade Center, in Jackson, Kenneth T. (ed.) *Encyclopedia of New York City*. New Haven: Yale University Press.

Robert, Jean (1994) Paris and the Ile-de-France: national capital, world city. *Nederlandse Geografische Studies*, **176**, pp. 13–28.

Roncayolo, Marcel (1983) La production de la ville, in Agulhon, Maurice et al. (eds.) *Histoire de la France urbaine, Vol 4: La ville de l'age industriel*. Paris: Seuil.

Rossi, P.O. (2000) *Roma. Guida all'architettura moderna 1909–2000*. Roma-Bari: Laterza.

Rouleau, Bernard (1985) *Villages et faubourgs de l'ancien Paris. Histoire d'un espace urbain*. Paris: Seuil.

Rouleau, Bernard (1988) *Le tracé des rues de Paris*. Paris: Presse du C.N.R.S.

Rowat, D.C. (1966) The proposal for a federal capital territory for Canada's capital, in Macdonald, H.I. (ed.) *The Confederation Challenge*. Toronto: Queen's Printer.

Royal Institute of British Architects (1911) Town Planning Conference, London, 10–15 October 1910: Transactions. London: RIBA.

Sagar, Jagdish (1999) Chandigarh: an overview. *Architecture + Design*, **16**(2).

Sanders, Spencer E. and Rabuck, Arthur J. (1946) *New City Patterns*. New York: Reinhold.

Sanfilippo, Mario (1992) *La costruzione di una capitale. Roma 1870-1911*. Cinisello Balsamo: Silvana Editoriale.

Sarin, Madhu (1982) *Urban Planning in the Third World: The Chandigarh Experience*. London: Mansell Publishing.

Sasaki, Suguru (2001) *The Day That Edo Became Tokyo: The Move of Capital of 1869*. Tokyo: Kodansha.

Sassatelli, Monica (2002) An interview with Jean Baudrillard: Europe, globalization and the destiny of culture. *European Journal of Social Theory*, **5**(4), pp. 521–530.

Sassen, S. (1991) *The Global City*. Princeton, NJ: Princeton University Press.

Sassen, S. (2001) *The Global City*, 2nd edition. Princeton, NJ: Princeton University Press.

Savitch, H.V. (1988) *Post-Industrial Cities: Politics and Planning in New York, Paris, and London*. Princeton, NJ: Princeton University Press.

SCAB (Société Centrale d'Architecture de Belgique) (1979) Quartier Loi-Schuman 'Un Nouveau Berlaymonstre'? *Bulletin hebdomadaire d'information*, no. 6 (numéro spécial de la deuxième série- premier et deuxième numéros spécials de la troisième série), p. 30.

Scarpa, Ludovica (1986) *Martin Wagner und Berlin. Architektur und Städtebau in der Weimarer Republik*. Braunschweig and Wiesbaden: Friedr. Vieweg & Sohn.

Schäche, Wolfgang (1991) *Architektur und Städtebau in Berlin zwischen 1933 und 1945. Planen und Bauen unter der Ägide der Stadtverwaltung*. Berlin: Mann.

Schieder, Theodor and Brunn, Gerhard (eds.) (1983) *Hauptstädte in europäischen Nationalstaaten*. Munich: R. Oldenbourg.

Schinz, Alfred (1964) *Berlin. Stadtschicksal und Städtebau*. Berlin: Westermann.

Schirren, Matthias (2001) *Hugo Häring. Architekt des Neuen Bauens 1882–1958*. Ostfildern-Ruit: Hatje.

Schneer, Jonathan (1999) *London 1900: The Imperial Metropolis*. New Haven: Yale University Press.

Schneider, Romana and Wang, Wilfried (eds.) (1998) *Moderne Architektur in Deutschland 1900 bis 2000. Macht und Monument*. Stuttgart: Hatje.

Schönberger, Angela (1981) *Die Neue Reichskanzlei von Albert Speer. Zum Zusammenhang von nationalsozialistischer Ideologie und Architektur*. Berlin: Mann.

Schoonbrodt, René (1979) ARAU (Atelier de Recherche et d'Action Urbaine): balance and prospects after five years' struggle, in Appleyard, Donald (ed.) *The Conservation of European Cities*. Cambridge, MA: MIT Press, pp. 126–132.

Schoonbrodt, René (1980) Intégrer la CEE à la ville. *Ville et Habitant*, **16**, pp. 1–4.

Schrag, Zachary M. (2001) Mapping Metro, 1955–1968. Urban, suburban and metropolitan alternatives. *Washington History*, Spring/Summer, pp. 4–23.

Schubert, D. and Sutcliffe, A. (1996) The 'Haussmanisation' of London? The planning and construction of Kingsway-Aldwych, 1899–1935. *Planning Perspectives*, **11**, pp. 115–144.

Schulman, Harri (2000) Helsingin aluesuunnittelu ja rakentuminen, in *Helsingin historia vuodesta 1945*. Helsinki: Edita Prima AB, pp. 55–58.

Schultz, Uwe (ed.) (1993) *Die Hauptstädte der Deutschen. Von der Kaiserpfalz in Aachen zum Regierungssitz Berlin*. München: C.H. Beck Verlag.

Schwartz, Joel (1993) *The New York Approach: Robert Moses, Urban Liberals, and Redevelopment of the Inner City*. Columbus, OH: Ohio State University Press.

Schwippert, Hans (1951) Das Bonner Bundeshaus. *Neue Bauwelt*, **17**, pp. 65–72.

Scott, Pamela (1991) This vast empire: the iconography of the Mall, 1791–1848, in Longstreth, Richard (ed.) *The Mall in Washington, 1791–1991*. Washington: National Gallery of Art.

SDAU (1980) *Schéma Directeur d'Aménagement et d'Urbanisme de la Ville de Paris*. Paris Projet 19–20. Paris: Les Éditions de l'Imprimeur/L'Atelier parisien d'urbanisme, pp. 74–92.

Seidensticker, Edward (1983) *Low City, High City: Tokyo from Edo to the Earthquake*. New York: Alfred A. Knopf.

Seidensticker, Edward (1990) *Tokyo Rising: The City Since the Great Earthquake*. New York: Alfred A. Knopf.

Sellier, Henri and Brüggemann, A. (1927) *Le problème du logement, son influence sur les conditions de l'habitation et l'aménagement des grandes villes*. Paris: PUF.

Senate Select Committee Appointed to Inquire into and Report upon the Development of Canberra (1955) *Report of the Senate Select Committee Appointed to Inquire into and Report upon the Development of Canberra*. Parliamentary Paper no. S2. Canberra: Government Printer.

Senatsverwaltung für Bau- und Wohnungswesen Berlin (ed.) (1992) *Hauptstadt Berlin. Zur Geschichte der Regierungsstandorte*. Berlin: Senatsverwaltung für Bau- und Wohnungswesen.

Senatsverwaltung für Bau- und Wohnungswesen Berlin (ed.) (1993) *Parlaments- und Regierungsviertel Berlin. Ergebnisse der vorbereitenden Untersuchungen*. Berlin: Senatsverwaltung für Bau- und Wohnungswesen.

Senatsverwaltung für Bau- und Wohnungswesen, Berlin (ed.) (1996) *Projekte für die Hauptstadt Berlin*. Berlin: Senatsverwaltung für Bau- und Wohnungswesen.

Shvidkovsky, O.A. (ed.) (1970) *Building in the USSR 1917–1932*. New York: Praeger.

Simpson, Michael (1985) *Thomas Adams and the modern planning movement: Britain, Canada, and the United States, 1900–1940*. London: Mansell.

Smith, Henry D. (1979) *Tokyo and London. Japan: A Comparative View*. Princeton, NJ: Princeton University Press.

Sonne, Wolfgang (2000) Ideas for a metropolis. The competition for Greater Berlin 1910, in Kahlfeldt, Paul, Kleihues, Josef Paul and Scheer, Thorsten (eds.) *City of Architecture – Architecture of the City. Berlin 1900-2000*. Berlin: Nicolai pp. 66–77.

Sonne, Wolfgang (2003) *Representing the State: Capital City Planning in the Early Twentieth Century*. Munich: Prestel.

Special Committee on the United Nations of the Association of the Bar of the City of New York (2001) *New York City and the United Nations: Towards a Renewed Relationship*. New York: Association of the Bar of the City of New York.

Speer, Albert (1939) Neuplanung der Reichshauptstadt. *Der Deutsche Baumeister*, **1**(1), pp. 3–4.

Speer, Albert (1970) *Inside the Third Reich. Memoirs*. New York: Simon and Schuster.

Stadtgeschichtliches Museum Leipzig (ed.) (1995) *Das Reichsgericht*. Leipzig: Edition Leipzig.

Starr, S.F. (1976) The revival and schism of urban planning in twentieth century Russia, in Hamm, Michael (ed.) *The City in Russian History*. Lexington, KY: The University Press of Kentucky.

Stein, Clarence S. (1951) *Toward New Towns for America*. Liverpool: Liverpool University Press.

Stenius, Olof (1969) *Helsingfors stadsplanehistoriska atlas*. Helsinki: Pro Helsingfors säätiö – Stiftelsen Pro Helsingfors.

Stephan, Hans (1939) *Die Baukunst im Dritten Reich. Insbesondere die Umgestaltung der Reichshauptstadt*. Berlin: Junker und Dünnhaupt.

Stimmann, Hans (1999) Berlin nach der Wende. Experimente mit der Tradition des europäischen Städtebaus, in Süß, Werner und Rytlewski, Ralf (eds.) *Berlin. Die Hauptstadt. Vergangenheit und Zukunft einer europäischen Metropole*. Berlin: Nicolai.

Stovall, Tyler (1990) *The Rise of the Paris Red Belt*. Berkeley, CA: University of California Press.

Strauven, Francis (1979) Brussels: urban transformations since the eighteenth century, in Appleyard, Donald (ed.) *The Conservation of European Cities*, pp. 104–125. Cambridge, MA: MIT Press.

Striner, Richard (1995) *The Committee of 100 on the Federal City: Its History and Its Service to the Nation's Capital*. Washington DC: The Committee of 100.

Sundman, Mikael (1991) Urban planning in Finland after 1850, in Hall, Thomas (ed.) *Planning and Urban Growth in the Nordic Countries*. London: E & FN Spon.

Süß, Werner (ed.) (1994, 1995, 1996) *Hauptstadt Berlin*, 3 volumes. Berlin: Arno Spitz

Süß, Werner and Rytlewski, Ralf (eds.) (1999) *Berlin. Die Hauptstadt. Vergangenheit und Zukunft einer europäischen Metropole*. Berlin: Nicolai.

Sutcliffe, Anthony (1970) *The Autumn of Central Paris: The Defeat of Town Planning, 1950–1970*. London: Edward Arnold.

Sutcliffe, Anthony (1979) Environmental control and planning in European capitals, 1850–1914: London, Paris and Berlin, in Hammarström, I. and Hall, Thomas (eds.) *Growth and Transformation of the Modern City*. Stockholm: Swedish Council for Building Research.

Sutcliffe, Anthony (1981) *Towards the Planned City: Germany, Britain, the United States and France, 1780–1914*. Oxford: Basil Blackwell.

Sutcliffe, Anthony (ed.) (1984) *Metropolis 1890–1940*. London: Mansell.

Sutcliffe, Anthony (1993) *Paris: An Architectural History*. New Haven, CT: Yale University Press.

Suzuki, Eiki (1992) *The Master Plan for the Modern City of Tokyo: New Tokyo Plan, 1918. Mikan no Tokyo Keikaku*. Tokyo: Chikuma Shobo.

Swyngedouw, Erik (1997) Neither global nor local. 'Glocalization' and the politics of scale, in Cox, Kevin (ed.) *Spaces of Globalization: Reasserting the Power of the Local*. New York: Guilford, pp. 137–66.

Swyngedouw, Erik and Baeten, Guy (2001) Scaling the city: the political economy of 'glocal' development – Brussels' Conundrum. *European Planning Studies*, **9**(7), pp. 827–849.

Szilard, Adalberto and Reis, José de Oliveira (1950) *Urbanismo no Rio de Janeiro*. Rio de Janeiro: O Construtor.

Tafuri, M. (1959) La prima strada di Roma moderna: via Nazionale. *Urbanistica*, **27**, pp. 95–108.

Takhar, Jaspreet (ed.) (2001) *Celebrating Chandigarh: Proceedings of Celebrating Chandigarh: 50 Years of the Idea, 9–11 January 1999*. Chandigarh: Chandigarh Perspectives.

Tamanini, Lourenço Fernando (1994) *Brasília: memória da construção*. Brasília: Editora Royal Court.

Tan, Tai Yong and Kudaisya, Gyanesh (2000) *The Aftermath of Partition in South Asia*. London and New York: Routledge.

Taylor, John H. (1986) *Ottawa: An Illustrated History*. Toronto: J. Lorimer.

Taylor, John H. (1989) City form and capital culture: remaking Ottawa. *Planning Perspectives*, **4**, pp. 79–105.

Taylor, John H. (1996) Whose Plan: Planning in Canada's Capital after 1945. Conference Proceedings, International Planning History Society Conference, 'The Planning of Capital Cities', Thessaloniki, pp. 781–794.

Taylor, John, Lengellé, Jean G. and Andrew, Caroline (eds.) (1993) *Capital Cities: International Perspectives*. Ottawa: Carleton University Press.

Taylor, P.J. (2004) *World City Network: A Global Urban Analysis*. London: Routledge.

Taylor, P.J., Catalano, G. and Walker, D.R.F. (2002) Measurement of the world city network. *Urban Studies*, **39**, pp. 2367–2376

Taylor, Robert R. (1985) *Hohenzollern Berlin. Construction and Reconstruction*. Port Credit, Ontario: Meany.

Thakore, M.P. (1962) Aspects for the Urban Geography of New Delhi. PhD Dissertation, University of London.

Therborn, Göran (2002) Monumental Europe: the national years. On the iconography of European capital cities. *Housing, Theory and Society*, **19**(1), pp. 26–47.

Thompson, Ian (1970) *Modern France: A Social and Economic Geography*. London: Butterworth.

Todd, F.G. (1903) *Preliminary Report to the Ottawa Improvement Commission*. Ottawa: OIC.

Togo, Naotake (1995) *The Creation of the City*. Considering Tokyo Series Volume 5. Tokyo: Toshi Shuppan.

Tokyo Metropolitan University, Center for Urban Studies (1988) *Tokyo: Urban Growth and Planning 1868–1988*. Tokyo: Tokyo Metropolitan University Press.

Tompkins, Sally Kress (1992) *A Quest for Grandeur: Charles Moore and the Federal Triangle*. Washington, DC: Smithsonian Institution Press.

Tsurumi, Shunsuke (1976) *Goto Shinpei*, Vol. 4. Tokyo.

Turnbull, Jeff and Navaretti, Peter Y. (1998) *The Griffins in Australia and India: The Complete Works and Projects of Walter Burley Griffin and Marion Mahony Griffin*. Melbourne: Melbourne University Press.

Ulbricht, Walter (1950) Die Grossbauten im Fünfjahrplan. Rede auf dem III. Parteitag des SED. *Neues Deutschland*, 23 July, quoted in Engel, Helmut and Ribbe, Wolfgang (eds.) (1993) *Hauptstadt Berlin – Wohin mit der Mitte? Historische, städtebauliche und architektonische Wurzeln des Stadtzentrums*. Berlin: Berliner Wissenschafts Verlag.

Unwin, Raymond (1911) Garden cities in England. *City Club Bulletin*, **4**, pp. 133–140.

US (1962) Report to the President by the Ad Hoc Committee on Federal Office Space, May 23, 1962.

Vaes, Benedicte (1980) Extensions C.E.E. (II): sept pouvoirs s'affrontent sur l'aménagement du centre de Bruxelles. *Le Soir*, **27**(8).

Vaitsman, Maurício (1968) *Quanto custou Brasília*. Rio de Janeiro: Posto de Serviço.

Vale, Lawrence (1992) *Architecture, Power and National Identity*. New Haven, CT: Yale University Press.

Vale, Lawrence (1999) Mediated monuments and national identity. *The Journal of Architecture*, **4**, pp. 391–407.

Valeriani, E. (1980) Il concorso nazionale in architettura, in Piantoni, G. (ed.) *Roma 1911*. Roma: De Luca Editore.

Vantroyen, J.-C. (1984) Mêmes ses fonctionnaires refusent que la C.E.E. détruise Bruxelles. *Le Soir*, 6 November.

Varma, P.L. (1950) Letter to Mayer and Whittlesey. In Albert Mayer Papers on India, 1934–1975. Unpublished Papers, University of Chicago.

Varnhagen, Francisco Adolfo de (Visconde de Porto Seguro) (1877, 1978) *A questão da capital: marítima ou no interior?* Brasília: Thesaurus.

Vereinigung Berliner Architekten und Architektenverein zu Berlin (ed.) (1907) *Anregungen zur Erlangung eines Grundplanes für die städtebauliche Entwicklung von Gross-Berlin. Gegeben von der Vereinigung Berliner Architekten und dem Architektenverein zu Berlin*. Berlin: Wasmuth.

Verner, Paul (1960) Grossbaustelle Zentrum Berlin. *Deutsche Architektur*, **9**(3), pp. 119–126.

Vernon, Christopher (1995) Expressing natural conditions with maximum possibility: The American landscape art (1901–c 1912) of Walter Burley Griffin. *Journal of Garden History*, **15**(1), pp. 19–47.

Vernon, Christopher (1998) An 'accidental' Australian: Walter Burley Griffin's Australian-American landscape art, in Turnbull, Jeff and Navaretti, Peter (eds.) *The Griffins in Australia and India: the Complete*

Works and Projects of Walter Burley Griffin and Marion Mahony Griffin. Melbourne: The Miegunyah Press of the Melbourne University Press.

Veuillot, L. (1871) *Paris pendant les deux sièges*, 2 volumes. Paris: V. Palmé.

Vidotto, V. (2001) *Roma contemporanea*. Bari-Roma: Laterza.

Vieira, Denise Sales (2002) *Habitação popular no DF: políticas públicas a partir de 1986*. Brasília: PIBIC, FAU/UnB.

Vinogradov, V.A. (1998) Meaning of the Architecture [sic] Heritage in Luzkov et al. (eds.) *Moscow and the Largest Cities of the World at the Edge of the 21st Century*. Moscow: Committee of Telecommunications and Mass Media of Moscow Government.

Voldman, D. (1997) *La reconstruction des villes françaises de 1940 à 1954: histoire d'une politique*. Paris: L'Harmattan.

Volkov, S. (1995) *St. Petersburg: A Cultural History*. New York: The Free Press.

Wagner, Martin (1929) Behörden als Städtebauer. *Das Neue Berlin*, **11**, pp. 230–232.

Wagner, Volker (2001) *Regierungsbauten in Berlin. Geschichte, Politik, Architektur*. Berlin: bebra Verlag.

Ward, S. V. (2002) *Planning the Twentieth Century City: The Advanced Capitalist World*. Chichester: Wiley.

Waris, Heikki (1973) *Työläisyhteiskunnan syntyminen Helsingin Pitkänsillan pohjoispuolelle*. Helsinki: Oy Weilin & Göös Ab.

Watanabe, Shun-ichi J. (1980) Garden city Japanese style: the case of Den-en Toshi Co. Ltd., 1918–28, in Cherry, Gordon (ed.) *Shaping an Urban World*. London: Mansell, pp. 129-143.

Watanabe, Shun-ichi J. (1984) Metropolitanism as a way of life: the case of Tokyo, 1868–1930 in Sutcliffe, Anthony (ed.) *Metropolis 1890–1940*. London: Mansell, pp. 403–429.

Watanabe, Shun-ichi J. (1992) The Japanese garden city, in Ward, Stephen V. (ed.) *The Garden City: Past, Present and Future*. London: E & FN Spon, pp. 69–87.

Watanabe, Shun-ichi J. (1993) *'Toshi Keikaku' no Tanjo: Kokusai Hikaku kara Mita Nihon Kindai Toshikeikaku* (The Birth of 'City Planning': An International Comparison of Japan's Approach to Modern Urban Planning). Tokyo: Kashiwa Shobo.

Watson, Anne (ed.) (1998) *Beyond Architecture: Marion Mahony and Walter Burley Griffin, America, Australia, India*. Sydney: Powerhouse Publishing.

Wefing, Heinrich (ed.) (1999) *Dem Deutschen Volke. Der Bundestag im Berliner Reichstagsgebäude*. Bonn: Bouvier.

Wefing, Heinrich (2001) *Kulisse der Macht. Das Berliner Kanzleramt*. Stuttgart and Munich: Deutsche Verlags-Anstalt.

Weirick, James (1988) The Griffins and modernism. *Transition*, **24**(Autumn), pp. 5–13.

Welch Guerra, Max (1999) *Hauptstadt Einig Vaterland. Planung und Politik zwischen Bonn und Berlin*. Berlin: Verlag Bauwesen.

Werner, Frank (1976) *Stadtplanung Berlin*. Berlin: Kiepert.

White, Jerry (2001) *London in the Twentieth Century: A City and its People*. London: Viking.

White, Paul (1989) Internal migration in the nineteenth and twentieth centuries, in Ogden, Philip and White, Paul (eds.) *Migrants in Modern France: Population Mobility in the Later 19th and 20th Centuries*. London: Unwin Hyman.

Whyte, William H. (1980) *Social Life of Small Urban Spaces*. Washington DC: The Conservation Foundation.

Wilson, Elizabeth (1992) *The Sphinx in The City*. Los Angeles: University of California Press.

Wilson, William H. (1989) *The City Beautiful Movement*. Baltimore: Johns Hopkins University Press.

Wise, Michael Z. (1998) *Capital Dilemma: Germany's Search for a New Architecture of Democracy*. New York: Princeton Architectural Press.

Wislocki, Peter (1996) Faceless federalism. *World Architecture*, **47**, pp. 84–87.

Wolfe, J. M. (1994) Our common past: an interpretation of Canadian planning history. *Plan Canada*, **34**(July), pp. 12–34.

Wonen-TA/BK (1975) ARAU Brussel = Bruxelles = Brussels. *wonen-TA/BK*, no. 15/16.

Wood, P. (2002) *Consultancy and Innovation: The Business Service Revolution in Europe*. London: Routledge.

Woolf, P.J. (1987) 'Le caprice du Prince' – the problem of the Bastille Opéra (Paris). *Planning Perspectives*, **2**, pp. 53–69.

Wright, Janet (1997) *Crown Assets – The Architecture of the Department of Public Works: 1867–1967*. Toronto: University of Toronto Press.

Yaro, Robert D. and Hiss, Tony (1996) *Region at Risk*. Washington, DC: Island Press.

Young, Carolyn A. (1995) *The Glory of Ottawa: Canada's First Parliament Buildings*. Montréal: McGill-Queen's University Press.

Zwoch, Felix (ed.) (1993) *Hauptstadt Berlin. Parlamentsviertel im Spreebogen. Internationaler Städtebaulicher Ideenwettbewerb 1993*. Basel: Birkhäuser.

Zwoch, Felix (ed.) (1994) *Hauptstadt Berlin. Stadtmitte Spreeinsel. Internationaler Städtebaulicher Ideenwettbewerb 1994*. Basel: Birkhäuser.

Subject Index

airports 12, 50, 53, 90, 121, 123, 166, 220, 241, 256, 259
alignment 134, 184
Athens Charter (see also CIAM) 169, 232
avenues, see streets
axes 16, 21–22, 26–28, 39–40, 47, 51, 53, 97, 116, 127–131, 134–136, 140, 143–146 168, 170, 176, 184–185, 196, 199–202, 205–206, 211, 215, 218, 220, 232, 244, 249–250, 259, 265

Beaux Arts style (see also *Ecole des Beaux Arts*) 17, 20, 26, 36, 160, 258
Bolshevik Revolution 2, 58, 60–62, 67
boulevards, see streets
buildings
  height limits 32, 47–48, 109, 117, 156, 160–161, 196, 211, 243
  preservation 5, 22, 60, 71, 124, 126–127, 174–175, 189–190, 192, 238, 240, 246
  public 5, 9, 33, 131, 159, 162, 186, 201, 203, 218

canals 59, 65, 118, 151–155, 158
capital city planning 5, 40, 47, 115, 201–203, 209–211

capital types 8–14
  de facto 8, 16, 101, 251
  ex-imperial 1, 8, 15, 71, 113
  former 1–2, 8, 10, 15, 71, 203, 253
  global 1, 3, 8–9, 11, 53, 87, 91, 111, 270
  imperial 11, 58, 71, 87, 89, 97, 101
  multi-functional 1–2, 8, 10, 15, 38
  political 1, 3–4, 8, 10–11, 15, 108, 115–116, 127, 187, 253, 270, 272–273
  provincial or state 1, 8–10, 12, 15, 32, 73, 130, 226, 253, 270, 272–273
  super-capital 1, 6–7, 9, 11, 15, 38, 253–257, 263, 265, 267–268
capitol complexes 17, 19, 21, 25–27, 29–31, 51, 117–120, 124, 126–127, 135–137, 140, 142, 144, 205, 231, 233–234, 251
central government, see national government
central business district (CBD) 13, 25, 40, 80, 161, 186, 189, 230, 254
churches 16, 60–61, 63, 71, 75, 97, 99, 131, 136, 167, 212, 258
CIAM (*Congrès internationaux d'architecture moderne*) 6, 169, 205, 226, 230, 260, 265

City Beautiful movement 6, 17, 20–21, 38, 117, 119, 121, 124, 133–135, 152–153, 155
City Efficient movement 121–122, 156
civic centres 26, 36, 136, 153–156, 230, 232
commissions and committees (see also planning commissions; implementation agencies) 44, 79, 83, 112, 151, 152, 153, 156, 167, 205, 247, 261, 267
communist states 2, 16, 45, 58, 61–72, 203, 205–206
competitions (design) 3–4, 22, 26, 32, 44, 73, 76, 80–82, 84, 133–135, 137–139, 141, 145, 167, 197–199, 201, 204–211, 218, 246–247, 250
comprehensive planning 7, 60–61, 68, 79, 121, 125–127, 152, 158, 197, 201, 203, 214–218, 221–222
concert halls 10, 50, 83, 258, 264–267
conferences 64, 117, 205, 246, 248, 271

Depression 17, 84, 115, 140, 156, 216, 255, 257
disasters (see also wars)
  bombing 21, 83, 91, 107–108, 163

earthquakes 3, 21, 106–107
fires 21, 75, 102, 106–108, 146, 148, 154

Ecole des Beaux Arts, Paris 153, 193
edge cities 11, 32, 124
education 8, 10, 38, 40, 47, 50, 68, 74, 96, 102, 108, 122, 153, 159, 198, 203, 215, 254
embankments 48, 59
embassies 125, 146
empires 9, 11, 17, 19, 73, 237, 273
　Austro-Hungarian 1, 8, 237
　British 1, 8–9, 24, 87–91, 97, 99, 132–133, 142, 182–187, 258, 273
　French 9, 38–41
　German 1, 5, 9, 196–198, 201
　Japanese 101–103, 107
　Ottoman, 1, 25, 273
　Portuguese 8–9, 25, 164
　Roman 5, 38, 215, 237
　Russian 9, 11, 16, 50, 58–61, 73–74, 273
　Spanish 8–9
exhibitions 5, 32, 77, 91, 99, 117, 133, 145, 157, 199, 203, 210, 216, 218–219, 223, 271
　World Fairs 43–44, 155, 167, 241, 244, 254, 266–267
expropriation 44, 159, 166, 172, 218, 238

façades 63, 84, 238, 242, 246, 248
federal districts or territories 4–5, 26, 115, 121–122, 126, 133, 144, 150, 154, 159, 166, 168, 170, 173, 175, 190
finger plan 123
focal points 185, 205, 243, 248

garden cities 6, 16, 24, 60, 68, 92, 134, 169, 187, 189, 191, 215, 218, 228–229
gentrification 116, 122, 172
global cities 3, 9, 12, 24, 40, 54, 90, 270–273
globalization 7, 12, 40, 50, 111, 270–273
green spaces (see also parks) 4, 6, 72, 82, 84, 111, 138, 148, 160, 188, 191, 197, 216, 237

greenbelts 10, 68–69, 93, 107, 157–162
grid plans 35, 75, 106, 158, 230, 257, 267
growth pole strategies 45, 52, 54, 250

hinterland 136, 164–166, 170, 176, 273
housing policy 22, 70, 172, 198, 234, 254
housing shortages 6, 67, 80, 107–108, 110, 172–173, 175, 191, 213, 219–220, 234

immigration 60, 254–255, 272
implementation agencies 4, 17, 104–105, 140–141, 145, 151–152, 154, 159, 162
　National Capital Commission (Ottawa) 4, 151, 159–162
　National Capital Planning Commission (US) 3, 19, 122, 125–127
　National Capital Development Commission (Australia) 4, 141–145
　NOVACAP (Brazil) 166–171
industrial districts 45, 49–50, 60, 69, 76, 84, 157, 160, 213, 228, 232, 234
industrialization 9, 22, 59, 70, 73, 95, 188–189, 214, 217, 228, 232
International Monetary Fund 116

landscape design 3, 26–27, 108, 115, 125, 133, 136–138, 140–141, 151, 192, 204, 250
legislature buildings (see also seat of government, capitol complex)
　Assembly (Chandigarh) 27, 233
　Capitol (USA) 19, 33, 117, 127, 251
　Diet (Japan) 104, 107, 112
　European Parliament (Brussels/Strasbourg) 6, 33, 238, 247–249
　General Assembly (UN) 260–261
　Houses of Parliament (UK) 88, 97, 99, 251

　Parliament Buildings (Canada) 20, 150–151, 153–154, 156
　Parliament House (Australia) 21, 131, 136, 139, 143–146, 148
　Parliament House (Finland) 25, 80–83
　Parliament House (India) 186
　Reichstag (Germany) 197, 199, 201, 205, 208–209
libraries 40, 51–52, 59, 118, 142, 145, 158, 160, 261, 265–267
linear cities 12, 143, 169, 209
local government 41, 49, 70, 82, 94–95, 103, 113, 116, 121, 126, 143, 153–154, 156–157, 160–161, 207, 213–216, 221, 238, 245–247
local planning 53, 83, 156, 159, 201

malls 4, 18, 28, 31, 96, 117–119, 127, 135, 168, 202
memorials and monuments 16, 17, 18, 25, 32, 36, 39, 52, 60, 62, 64, 67, 71, 74, 80, 119, 126–127, 134–135, 140, 142, 145, 155, 157, 192, 197–198, 214, 218–219, 233
　war memorials 71, 131, 139–140, 142, 155–158, 194, 201, 242
mixed use planning 29, 257
Modernist design (see also CIAM) 6, 22, 26–27, 36, 76, 122, 124, 140–142, 167–168, 175, 199, 205, 217, 226, 228–230, 260
modernization 3, 14, 22, 39, 46, 48, 73, 101, 103–106, 266–268
municipal council, see local government
municipal planning, see local government planning
museums 17–18, 20, 24, 51–52, 58–59, 71, 75, 83–84, 97, 119, 125, 127, 136, 140, 145, 162, 199, 254, 265, 271

national assembly, see legislatures
national government (including central government; see also presidential involvement, royal involvement) 14, 32, 45, 48–49, 52, 63, 74, 83, 93–94, 105, 112, 133, 189, 199, 203, 206–207, 213–215, 217, 227, 229, 239–240, 245–247, 250, 263

SUBJECT INDEX

nation-states 1, 11, 13, 15, 20, 26, 36, 82–83, 270–273
neighbourhood unit 157, 170, 220, 228, 231
neo-gothic style 20, 150

official plan, see comprehensive planning
official residences (see also palaces) 20, 85, 101, 120, 136, 166-167, 223
Olympic Games 3, 6, 22, 82, 96, 220–221, 223
open space, see parks, green spaces
opera house 24, 51–52, 83, 97, 136, 257, 259, 264–247

palaces 10, 16–18, 20, 22, 24, 26, 30, 40, 52, 59–61, 64–65, 71, 96, 101, 103–105, 107, 197, 199, 205–206, 223, 232
parks 4, 6, 17, 20–21, 68, 71–72, 75–77, 81, 84, 106–107, 117–120, 126, 141–143, 147–148, 151–154, 158–159, 167, 201, 216–217, 221–222, 232–133, 237, 257, 259, 261, 266
parliament buildings, see legislature buildings
planning history 2, 4, 43, 58, 73, 115, 158, 210–211
planning legislation 3, 44, 77, 103, 105–106, 121, 123, 160, 211, 260
planning commissions 3–4 18–19, 50, 105, 118, 120–122, 125–127, 139–145, 147, 151–153, 156–157, 159, 161, 166–167, 170–171, 201, 205, 255–256
plazas, see public spaces
politics and planning 6, 8–9, 19, 22, 30, 87, 99, 116, 159, 202, 209, 215, 226–228, 250, 253, 262
presidents' or prime ministers' involvement in planning
Canada 4, 154–158, 162
Italy 5, 32, 35, 217–219
Russia 2, 58, 62–72, 205–206
UK 89, 94
US 115, 117, 121, 125, 266
private investment 240
property developers 48, 113, 119, 122, 160–162, 173, 214, 240, 244, 247
public health 254
public space 36, 127, 162, 221
gardens 39, 59, 108, 130, 136, 139, 142–144, 186, 213, 215, 218, 232, 234
squares 16, 59, 63–64, 67, 71, 75, 81–82, 84, 96, 110, 118, 125, 146, 152, 155–156, 167–168, 198–201, 203, 205, 207, 264
public transport 65, 68, 96, 105, 155, 161, 168, 218, 222, 232, 246, 250, 256
public-private partnership 13, 126, 221, 267–269

Radiant City 70
railway stations 80, 82–83, 104, 109, 111, 117–118, 123, 136, 153, 158, 201, 213, 221, 230, 232, 238, 247–249, 259
railways 12, 18, 51, 53, 75, 77, 80, 83, 88, 91, 102, 106, 111, 117, 126, 157 8, 165, 183, 201, 249, 254, 262–263
redevelopment 13, 39, 43–44, 111, 122, 145, 189, 191, 242, 257
research institutes 11, 47, 158, 256
royal involvement in planning
Belgium 6, 237–238
Canada 150
Finland 2, 74
France 41, 46–47
Russia 2, 58–61, 72

satellite communities 4, 24, 33, 68, 143, 169, 172–176
seat of government 8, 13, 15, 25, 29, 31, 85, 91, 101, 103, 106, 117, 150–151, 153, 183, 204, 206, 209, 213
senate buildings, see legislature buildings
slums, informal settlements 4, 70, 119, 172–173, 220–221
clearance 32, 122, 157, 175, 256, 264–265
overcrowding 43, 110
urban renewal 113, 117, 120–122, 154, 157, 160, 162, 189, 201, 203–204, 264–266

Smart Growth 126
speculation 111, 125, 135, 213, 215, 218, 220, 223, 244, 246, 249, 259
sprawl, see urban sprawl
squares, see public spaces
stadium 29, 53, 136
stock exchange or *bourse* 88, 111, 136
streets 5, 32, 36, 46, 53, 59, 62, 65, 68, 70, 73, 75, 82, 97, 105–106, 108–109, 117, 122, 126, 140, 143, 151, 155–156, 186–187, 214, 218, 220, 241, 247, 250, 254, 260–263, 267
avenues 6, 26, 32, 59, 80, 124–125, 127, 136, 158, 184, 186, 191, 217, 220, 237, 258–261
boulevards 16–17, 20, 25, 34, 36, 39, 62, 65, 67, 76–77, 157–158, 162, 201, 237, 243
highways (including freeways and expressways) 22, 48, 68, 109, 111, 121–122, 126, 158, 168–170, 189, 203, 220, 241, 254, 257, 262
parkways 4, 140, 151–152, 158, 257
suburbanization 41–42, 47, 50, 77, 83, 104–106, 115, 122–123, 160, 176, 191
suburbs 32, 41–47, 50, 53–54, 67, 75, 77, 80, 83–84, 89, 91, 103, 105–107, 111, 115–116, 121–126, 135, 139, 143–145, 157–161, 169, 172, 176, 188, 191, 197, 237, 241–242, 245, 259
subway systems 52, 65–68, 103, 105–106, 123–124, 169, 221, 244, 257
superblocks 4, 35, 66, 68, 168–170, 230–231, 261–267
super-national groupings 2, 9, 238–239, 260
European Union 6, 10, 32–33, 50, 74, 238–251, 272
United Nations 6, 9, 32–33, 35, 167, 253, 256, 259–261, 265, 267–268
supreme courts 118, 142–143, 147, 159, 197, 204, 214

theatres 10, 59, 61, 63, 75, 83, 136, 158, 259, 264–267
tourism 22, 32–33, 71, 90, 97, 99,

115, 127, 147, 160, 162, 221, 223, 234, 251, 263
traffic 22, 30, 68, 73, 83–84, 90, 120, 125, 155–156, 169–170, 198, 201, 205, 220–221, 241

underground railways, see subway systems
universities 10, 31, 59, 67, 74–75, 136, 218, 254, 264
urban design 1–3, 15–38, 65, 82, 87, 96, 103, 105, 118, 120, 124, 126–127, 141, 147, 155, 162, 168–170, 191, 196, 202–203, 207–209, 211, 247, 253
urban development or growth 2–3, 6, 17, 24, 32, 35, 41, 46, 59, 70, 78, 92, 94–95, 103–106, 108–113, 115, 123, 143, 151, 161, 169, 172–173, 187–191, 197, 210, 215–218, 241–242, 260, 272
urban landscape 40, 115, 199, 250
urban renewal, see slums
urban sprawl 45, 70, 123, 174–175, 186, 189, 191
urbanism 17, 26–27, 167–169, 196, 211, 215–217, 267

vistas 5, 20, 40, 120, 127, 184–186, 189, 191–192
Voisin Plan (Paris) 24, 43, 48

wars (see also disasters) 2, 9, 62, 79, 88, 203, 273

World War I 1, 4–5, 60, 78–79, 139, 154–155, 187, 273
World War II 1, 3–4, 21–24, 32, 45, 64, 67, 71, 91, 107, 116, 127, 156–157, 188, 202–203, 218, 238, 255
waterfronts 19, 24, 84, 90, 122, 125–126, 259
Weimar Republic 5, 196, 198–200, 210
World Cup 53, 223

zoning, 33, 48, 68, 105, 107, 125, 153, 156, 160, 169, 172, 175, 187, 232, 254–255, 258, 260

Index of Towns and Cities

Abuja 13, 17, 28, 273
Agra 183
Amsterdam 8, 10, 12, 59, 271
Aniene 215
Ankara 15, 25
Antwerp 271
Athens 270–271

Baltimore 123, 271
Barcelona 9, 97, 271
Beijing 7, 16, 94, 271–273
Belmopan 13, 15
Bengal 183
Berlin 5, 8–9, 11–12, 14, 16, 22, 35–36, 50, 54, 60, 74, 77, 99, 103, 196–212, 271
    Spreebogen 22, 197, 199, 205, 208–209
Bhubaneswar 228
Bilbao 9
Bonn 8, 11, 13, 204, 209
Brasília 1, 4, 8, 13, 17, 26–27, 31, 34–35, 142, 150, 164–181, 228, 271
    Candangolândia 166, 171, 175
    Ceilândia 172, 175
    Gama, 171–172
    Goiás 165–6, 172
    Guará 172–173
    Núcleo Bandeirante 172–173
    Planaltina 172, 175

    Sobradinho 172, 175
    Taguatinga 172–175
Bruges 237
Brussels 6, 9, 32–33, 35, 50, 237–252, 271
    Marolles 238
Budapest 271–272
Buenos Aires 13, 271

Cairo 271
Calcutta 5, 182–183
Canberra 3–4, 9–10, 13, 17, 20–21, 27–29, 32, 34–35, 130–148, 150, 159, 175, 231, 273
Chandigarh 1, 6, 27, 29, 31, 34–35, 167, 205, 226–236
Chicago 17, 20, 120, 123, 133–136, 138–139, 141, 153, 255, 271
Cologne 237, 271
Constantinople 60
Copenhagen 12, 271

Dar es Salaam 13
Delhi (see also New Delhi) 5, 26, 182–94, 232
Dhaka 30–31
Dodoma 13, 17, 28, 29
Dublin 271–272

Edinburgh 240, 249, 271

Europe 6, 9–12, 22–23, 39, 45, 53–54, 59, 67, 73–74, 77, 89, 96, 103, 105, 117–118, 133, 196, 214, 228–230, 237–252, 259, 267, 272–273

Florence 270
Frankfurt 10–12, 50, 90, 271

Geneva 9, 32, 271
Genoa 213, 271
Greenbelt MD 121, 230

Hague, The 8, 11, 271
Hamburg 10, 271
Haryana 6, 189, 227
Helsinki 2, 25, 35, 73–86, 271
    Pasila 80, 83–84
Hong Kong 271–273

India 5–6, 20, 26, 31, 273
Islamabad 13, 17, 26, 30
Istanbul 25, 271–272

Japan 12–13, 28

Kanpur 232
Kanto 21, 101, 106, 108
Kingston, ON 150
Kyoto 22, 101

Lahore 6, 226, 228–229
Leipzig 197
Liège 238
Lisbon 8, 15, 164, 271–272
London 1, 3, 8–13, 15, 21–22, 24, 34, 50, 54, 62, 68, 87–100, 111, 141, 160, 183–184, 257, 270–271
    Canary Wharf 24, 94, 257
    Greenwich 96–97
Los Angeles 271
    Baldwin Hills 230
Luxembourg 238–240, 245, 248–250, 271

Madrid 8–9, 12, 271, 272
Manila 17, 271
Marseilles 232
Melbourne 8, 10, 130, 133, 139–140, 271, 273
Mexico City 271
Milan 8, 10, 12, 50, 213–214, 265, 271
Minas Gerais 172
Montréal 8, 10, 150, 271
Moscow 2, 8, 16, 19, 34–35, 58–72, 201, 205, 271
    Kremlin 16, 19, 58, 60–66, 68, 71
    Red Square 61, 64, 71
Mumbai (Bombay) 183, 192, 232, 271, 273
Munich 8–10, 271
Nagoya 12–3, 108
New Delhi (see also Delhi) 5, 17, 19–21, 26–27, 34–5, 182–195, 202, 228, 231, 271, 273
    Connaught Place 190
New York 1, 6, 9, 32, 34, 64, 68, 90, 94, 97–111, 119, 167, 253–269, 270–272
    Battery Park City 263–264
    Lincoln Center 253, 256–257, 264–268
    Manhattan 32, 34–35, 97, 254–259, 261–264, 267
    Rockefeller Center 6, 253, 256–258, 261–262, 265, 267–268

World Trade Center 253, 256, 261–264, 267
Nouakchott 13, 15

Osaka 12, 108, 271–272
Oslo 271
Ottawa 4, 8, 10, 20, 32, 35, 150–163, 175
    Hull 20, 151–153, 156, 158–162
    Outaouais 161

Pakistan 6, 13, 26, 30, 226
Paris (see also Versailles) 2, 8–13, 15, 17, 24, 32, 34, 38–57, 62, 76, 84, 94, 97, 99, 103, 153, 155, 169, 198, 232, 257, 270–271
    La Défense 24, 32, 40, 45, 47, 51–53, 84, 169, 257
Philadelphia 8, 259, 271–272
Portugal 9, 164
Prague 271
Prussia 103, 198
Punjab 6, 183, 226–227, 229

Québec, Ville 31, 151

Radburn, NJ 170, 230
Rajputana 183
Reston 28, 123
Reval, (Tallinn) 74, 273
Rio de Janeiro 4, 8, 13, 26, 164, 166–169, 271
Rome 1, 5–6, 9, 17, 32, 34–35, 60, 99, 187, 213–25, 239, 266, 271
    EUR 5–6, 32, 218–21
Russia 9, 16, 58–59, 62, 64, 71–74, 79, 259

San Francisco 120, 259, 271
São João del Rei 164
Sendai 13
Seoul 271, 273
Shahjahanabad 182, 184–185
Shanghai 271–272
Singapore 271
Smolensk 60

Soviet Union 1–2, 9, 16, 19, 22, 58, 62–3, 65, 67–68, 82, 91, 228, 273
St. Petersburg 2, 8, 58–52, 65, 67–68, 71–72, 74, 271
Stockholm 8, 74, 77, 271
Strasbourg 9, 238, 240, 249–250
Stuttgart 8, 271
Sydney 8, 10, 131–133, 137, 138, 140, 271, 273

Tokyo 1, 3, 8, 11–13, 21–22, 34, 90, 101–114, 270–271
    Edo 3, 101, 103–104, 106, 112
    Ginza, 102, 105
    Hibiya 104
    Shinjuku 109
Toronto 8, 10, 150, 271
Turin (Torino) 8, 223, 271
Tver, 60

USA 9–10, 17–31, 89, 115–117, 133, 228–230, 253, 255, 259, 265, 272–273

Vatican 5, 221–222
Venice 10
Versailles 17, 118
Vienna 8–9, 11, 74, 99, 237, 271

Warsaw 22–23, 271–272
Washington 3, 8, 10, 13, 17–19, 21, 27–29, 32, 34, 36, 115–129, 133, 135, 150–152, 156, 159, 187, 202, 271
    District of Columbia 19, 115, 117, 121–122
    Georgetown 116, 126
    Maryland 115, 118, 121–124
    Pennsylvania Avenue 124–125
    Pentagon 121, 127
    Virginia 32, 115, 123–124

Yokohama 101, 106, 111
Yugoslavia 1, 273

Index of Persons

Aalto, Alvar 2, 82–84
Abercrombie, Patrick 3, 92
Adams, Thomas 156
Adenauer, Konrad 204
Ahern, Thomas 154
Albuquerque, Cavalcante de 166, 176
Aldrich, Winthrop 259
Alexander I, Tsar of Russia 2, 74
Ambroise-Rendu, M. 47
Anspach, Jules 237

Backheuser, Everardo 165
Baker, Herbert 17–18, 185–187
Barlow, Montague 92
Bartholomew, Harland 121–122
Bech, Joseph 238
Bennett, Edward 4, 20, 120, 133, 153–154, 156
Besme, Victor 237–238
Böckmann, Wilhem 102
Borden, Robert 152–154
Brinckmann, Albert Erich 198
Bunshaft, Gordon 262, 266
Burnham, Daniel 17, 19–20, 117–118, 120, 122–123, 133–135, 153

Caro, Antony 99
Catherine the Great, Empress of Russia 59
Cauchon, Noulan 152, 156

Charles, Duke of Lorraine 238
Chirac, Jacques 41, 48, 52
Coelho, Polli 165
Conklin and Rossant 28
Coulter, Robert 131–132, 137–138, 142, 148
Costa, Lúcio 4, 26, 166–173
Crewe, Earl of 183
Cruls, Luis 165, 176
Curzon, Lord 183

Dewey, Thomas 262
Diefenbaker, John 160
Drew, Jane 226, 232
Dutra, Eurico 165

Eberstadt, Rudolf 197
Ehrenström, Johan Albert 74
Eisenhower, Dwight 266
Ende, Hermann 102
Engel, Carl Ludwig 74, 77, 85
Evans, A. 130–131, 142

Finley, William 122–123
Firme, Raul de Penna 166
Fletcher, A.L. 226
Forshaw, J.H. 92
Foster, Norman 99, 208
Frampton, Kenneth 232
Fry, Maxwell 226
Fukuda, Shigeyoshi 105–106

Gabriel, Jacques-Ange 119
George, Pierre 47
George V, King of UK 183
Giolitti, Giovanni 215
Giscard d' Estaing, Valéry 48–49, 51
Giurgola, Aldo 21, 143
Goebbels, Joseph 201
Goodman, Charles 122
Gorbachev, Mikhail 69
Goto, Shinpei 106
Gottmann, Jean 14
Gravier, Jean-François 41, 46
Gréber, Jacques 4, 20, 155–158, 160–161
Grey, Earl 152
Griffin, Marion Mahony 133, 135, 137–140
Griffin, Walter Burley 3–4, 20–21, 130–131, 133–148
Guilherme, Eliseu 165
Gutheim, Frederick 124

Hall, Peter 1, 7, 38, 127
Hardinge, Charles 5, 183–187
Häring, Hugo 199
Harrison, Peter 140–141
Harrison, Wallace 257, 259, 261–267
Hastings, Warren 183
Haussmann, Georges-Eugène 2, 34, 39–44, 46–48, 54, 97, 103, 223
Hebbert, Michael 94

Hénard, Eugène 42
Hitler, Adolf 5, 22, 36, 200–203, 210
Hobrecht, 196
Hoffmann, Ludwig 197
Holden, Charles 92
Holford, William 4, 92, 141–142
Holt, Herbert 153
Horta, Victor, 238
Howard, Ebenezer 92, 169

Ivan III, Tsar of Russia 60

Jacobs, Jane 262
Jansen, Hermann 25
Jeanneret, Pierre 226
Joshi, Kiran 227
Jung, Bertel 2, 77–80
Justement, Louis 122

Kaganovitch, L.M. 63, 67–69
Kahn, Louis 30–31, 209
Kalia, Ravi 227
Kallio, Oiva 80
Kennedy, Jackie 125
Kennedy, John 124–125
Khrushchev, Nikita 64, 70, 72
Kitchen, John 156
Koenigsberger, Otto 229
Kohtz, Otto 199
Kosel, Gerhard 206-207
Kropotkin, Peter 92
Kubitschek, Juscelino 4, 26, 166–167, 170–171

L'Enfant, Pierre Charles 17, 19, 115, 117–120, 122, 126–127, 135
Lacombe, Roberto 166
Laurier, Wilfrid 151–154, 156
Le Corbusier, 6, 24, 27, 29, 43, 48, 62, 70, 113, 167, 169, 199, 205, 226–233, 235, 260–261, 267
Lenin, Vladimir 2, 58, 62–64, 67
Leopold II, King of Belgium 6, 237–238
Leontii, Benois 60
Ligen, Pierre-Yves 48–9
Livingstone, Ken 94–5
Lutyens, Edwin 5, 17–120, 184–189, 191–192

MacArthur, Douglas 107–108
Mächler, Martin 199, 201
Mackenzie King, William Lyon 4, 154–158

Marx, Karl 58, 61
Matos, Renato 171, 175
Mawson, Thomas 134, 152–153
Mayer, Albert 6, 226, 229–233
McKim, Charles 117
McMillan, James 3, 17, 18, 115, 117–122, 125–127
Meiji, Emperor of Japan 101–103, 113
Menzies, Robert 4, 140–142
Meredith, Colborne 152–153
Mills, Robert 119
Minobe, Governor of Tokyo 110
Mitterrand, François 49, 51
Möhring, Bruno 197–198
Moore, Charles 119–120
Morris, William 92, 95
Moses, Robert 256–257, 259, 262, 264, 266
Moynihan, Patrick 124
Mussolini, 5, 32, 35, 217, 219

Napoleon III, Emperor of France 46–47
Nathan, Ernesto 215
Nehru, Jawaharlal 6, 226, 228, 229, 235
Niemeyer, Oscar 4, 166–167, 261
Northcote, Stafford 183
Nowicki, Mathew 6, 226, 229–233

O'Malley, King 133, 138
Olmsted, Frederick Law Sr. 117, 127
Olmsted, Frederick Law, Jr. 19, 121, 151

Pei, I.M. 122, 125
Peter the Great, Tsar of Russia 2, 58, 59
Piacentini, Marcello 5, 218
Pitt, William 164
Pompidou, Georges 48, 51–52
Purdom, C.B. 93
Pushkin, Alexander 58–59
Reis, José de Oliveira 166
Rockefeller, John D. 257, 260–261, 264, 266
Rockefeller, Nelson 31, 256–257, 262–263, 265–267
Roosevelt, Theodore 117

Saarinen, Eero 123 266
Saarinen, Eliel 2, 78–80

Sanjust, Edmondo 214–215
Sarin, Madhu 227
Scharoun, Hans 203, 205
Schneer, Jonathan 89
Schultes, Axel 208–209
Scrivener, Charles 133
Sella, Quintino 213
Shepard, Alexander 117
Silva, Andrade e 164–165
Sirén, J.J. 81–82
Sixtus V, Pope 17
Speer, Albert 5, 22, 36, 201–203, 210
Spengelin, Friedrich 204–205
Stalin, Josef 2, 58, 63–72, 205–206
Sulman, John 139
Sutcliffe, Anthony 39, 42, 97

Tallberg, Julius, 79
Tange, Kenzo 22, 28, 109
Thapar, T.N. 226
Thatcher, Margaret 89, 94
Tobin, Austin 256
Todd, Frederick 4, 151–152
Töpfer, Klaus, 209
Trudeau, Pierre Elliot 162

Unwin, Raymond 134, 152
Varma, P.L. 226
Victoria, Queen 87, 150
Vivian, Henry 152
Viviani, Allesandro 213–214

Wagner, Martin 199
Wagner, Otto, 77
Ward, Stephen 115
Washington, George 115, 121
Washington, Walter 116
Waters, Thomas, 102
Weese, Harry 123
Wells, H.G. 89
White, Jerry 88, 90
Wilde, Oscar 168
Willhelm III, German Kaiser 197–199
Woodward Smith, Cloethiel 122
Wright, Frank Lloyd 62, 133

Yamasaki, Minoru 262
Yeltsin, Boris 69, 71

Zeckendorf, William 122, 259

A library at your fingertips!

eBooks are electronic versions of printed books. You can store them on your PC/laptop or browse them online.

They have advantages for anyone needing rapid access to a wide variety of published, copyright information.

eBooks can help your research by enabling you to bookmark chapters, annotate text and use instant searches to find specific words or phrases. Several eBook files would fit on even a small laptop or PDA.

NEW: Save money by eSubscribing: cheap, online access to any eBook for as long as you need it.

Annual subscription packages

We now offer special low-cost bulk subscriptions to packages of eBooks in certain subject areas. These are available to libraries or to individuals.

For more information please contact webmaster.ebooks@tandf.co.uk

We're continually developing the eBook concept, so keep up to date by visiting the website.

www.eBookstore.tandf.co.uk

U.W.E.L. LEARNING RESOURCES

12/04

Harrison Learning Centre
Wolverhampton Campus
University of Wolverhampton
St Peter's Square
Wolverhampton WV1 1RH
Telephone: 0845 408 1631

1 4 MAY 2008

Telephone Renewals: 01902 321333
Please RETURN this item on or before the last date shown above.
Fines will be charged if items are returned late.
See tariff of fines displayed at the Counter. (L2)